A
HISTORY
OF
METALLURGY

2nd Edition

A
HISTORY
OF
METALLURGY
Second Edition

R. F. Tylecote

MANEY

FOR THE INSTITUTE OF MATERIALS

Book B0789
First published in paperback in 2002 by
Maney Publishing
1 Carlton House Terrace
London SW1Y 5DB
for the Institute of Materials

First published in 1976
Reprinted in 1979
2nd edn published 1992
© The Institute of Materials 1992
All rights reserved

ISBN 1-902653-79-3

Printed and bound in the UK by
Antony Rowe Ltd

Contents

Preface to the Second Edition

The first edition was published in 1976 and an enormous increase in the general interest in the subject of archeometallurgy has taken place since then. Much of this relates to the early phases and has been discussed in Proceedings of International Conferences. For this reason therefore the main changes between the two editions can be seen in such proceedings which have been intensively quoted in the following prehistoric sections.

The post Roman chapters needed some revision to reflect excavational work in Europe and additions have been made to the chapter on the Industrial Revolution. The last three chapters however, have been left much as they were in the original edition. The completion of K.C.Barraclough's new work entitled 'Steelmaking 1850–1900' has covered an important revision that might have gone into chapter 10. To those who want to know more about the new methods of steel making this must be highly recommended. The remaining chapters of the role of the metallurgists and their problems stand unchanged. This also will be a subject inviting consideration by others from time to time.

Finally, what I said in 1976 about the transition to non-metallic materials is even more true than I thought. This Institute will soon be one of materials.

R. F. Tylecote

Foreword

As Professor Tylecote has pointed out in his Preface to this edition, general interest in archeometallurgy has grown enormously since the first edition of 'A History of Metallurgy' was published by The Metals Society in 1976. This has also been reflected in the success of The Institute of Metals' Historical Metallurgy Series of which Professor Tylecote was General Editor.

It is with great satisfaction, therefore, as well as regret at the loss of a valued friend and accomplished scholar, that The Institute of Materials publishes this second edition. It is hoped that this will serve in recognition of an unrivalled personal contribution to the development of the subject, based on wide-ranging knowledge and extensive practical experience. Ronnie Tylecote is greatly missed by his many colleagues and friends in the Institute and beyond.

On a practical note, readers of this Volume may be interested to know that in addition to 'Steelmaking: 1850-1900', to which Professor Tylecote refers in his Preface, a further related source of information on the general topic is provided by 'The Industrial Revolution in Metals' recently published by the Institute, which Professor Tylecote contributed to and edited with Mrs Joan Day.

Keith Wakelam
The Institute of Materials
London

Acknowledgements

A book of this kind is never entirely the product of its author, but is dependent on the work of many scholars all over the world who have contributed directly or indirectly. I wish to thank all those who have helped in one way or another, in particular:

Dr Beno Rothenberg, Professor Alexandru Lupu, Dr V. Karageorghis, J. R. Marechal, Ing. R. Thomsen, Bernard Fagg, H. H. Coghlan, Professor Peter Shinnie, T. A. Wertime, Dr R. Moorey, Dr Hugh McKerrell, Dr Radomir Pleiner, Professor W. U. Guyan, Dr Inga Serning, and Professor N. Barnard. The later chapters owe much to the work of the late G. R. Morton.

I would especially like to thank my wife, Elizabeth Tylecote, for her linguistic assistance and for assembling much of the material. Most of the book has been typed by Mrs Edna Oxley, to whom I am exceedingly grateful.

The reproduction of many of the illustrations would not have been possible without the permission of the owners of the copyright. My thanks are due to all those who have helped in this way, and I am glad to make the following acknowledgments; *Metals and Materials* for Fig. 2, which is from my paper 'Early Metallurgy in the Near East', p.291, Fig. 4, July 1970; Dr Beno Rothenberg for Figs. 3 and 7; Michael Thompson for Fig. 8, which appeared in his translation of A. L. Mongait's 'Archaeology in the USSR', published by Penguin Books Ltd, Harmondsworth, in 1961. Dr V. Karageorghis for Figs. 9 and 18, which are from my paper on Cypriot copper smelting in the Report of the Department of Antiquities, Cyprus, 1971; The American School of Classical Studies at Athens for Fig. 10, which appeared in J. L. Caskey's paper on Lerna in *Hesperia*, vol.XXIV, 1955, plate 14; Dr Beno Rothenberg and Professor A. Lupu for Fig. 16; The Trustees of the British Museum for Figs. 19 and 27; Dr C. Storti for part of Fig. 24 which appeared in a paper by H. H. Coghlan and J. Raftery in *Sibrium*, vol. VI, 1961; *The American Journal of Archaeology* for Fig. 25, which appeared in Herbert Maryon's paper in AJA, 1961, 65, 173, Fig. 17; *Inventaria Archaeologica* for Fig. 26; Professor M. J. O'Kelly for Fig. 31; George Jobey and *Archaeologia Aeliana* for Fig. 32; The Iron and Steel Institute for Fig. 33; The Science Museum, London, for Fig. 35 (Crown copyright); The Pitt Rivers Museum, Oxford, for Figs. 36 and 37 which appeared in their Occasional Paper No. 8 on 'Early iron in the Old World', by H. H. Coghlan; Bernard Fagg for permission to use Fig. 38, which first appeared in the *Bull. HMG*, 1968, 2, (2), 81, Fig. 1; The Iron and Steel Institute for Fig. 40; Robert Thomsen, Varde, for Fig. 42; Edward Arnold (Publishers) Ltd, for Fig. 43; The Museum of Antiquities, University of Newcastle upon Tyne,

and South Shields Corporation, for Fig. 46; George Boon for Fig. 48, which first appeared in his paper in *Apulum*, 1971, 9, 475; Professor H. O'Neill and Edward Arnold Ltd for Fig. 51; Martin Biddle, Director of the Winchester Research Unit, for Figs. 53 and 54; Professor C. S. Smith and the Chicago University Press for Fig. 55, from Smith and Hawthorne's version of 'On Divers Arts' by Theophilus; *The Ulster Journal of Archaeology* for Figs. 61 and 62 which are taken from K. Marshall's paper in *UJA*, 1950, 13, 66, Figs. 1-5 and 11; Dr A. Raistrick and The Newcomen Society for Fig. 64; D. W. Crossley for Fig. 71; The Science Museum, London, for Fig. 72 (Crown copyright); M. Davies-Shiel and the Historical Metallurgy Group for Fig. 74; The Institution of Metallurgists for Fig. 75; Professor C. S. Smith for Fig. 77; The Public Record Office for Figs. 81 and 82 (Ref. SP12/122/63, Crown copyright); The Newcomen Society for Fig. 84; The Iron and Steel Institute for Fig. 85, which appeared in G. R. Morton and W. A. Smith's paper in *J. Iron Steel Inst.*, 1966, **204**, 666, Fig. 7; The Iron and Steel Institute for Fig. 90, which appeared in the paper by G. R. Morton and N. Mutton in *J. Iron Steel Inst.*, 1967, 205, 724, Fig. 1; The Newcomen Society for Fig. 93; Dr J. D. Gilchrist for his drawing of the Cowper stove in Fig. 99; J. R. M. Lyne and The Historical Metallurgy Group for Fig. 104; The Iron and Steel Institute for providing the copy of Fig. 106 from Réaumur; The Cumberland County Record Office for Figs. 108, 110, and 112 which have been taken from the Curwen collection, Reference D/Cu. 5/96; The Boulton and Watt Collection, Birmingham Reference Library, for Fig. 109, which came from Portfolio 239; Professor Aubrey Burstall and P. Elliott for Fig. 113, reprinted by permission of Faber and Faber Ltd, from 'A History of Mechanical Engineering'; J. K. Harrison and the North East Industrial Archaeology Society for Fig. 114; Professor Sten Lindroth for Fig. 117; D. W. Hopkins for Fig. 121; E. J. Cocks for Fig. 125; Edward Arnold (Publishers) Ltd, for Fig. 134; The Iron and Steel Institute for Fig. 138; Firth Brown Ltd, Sheffield, for Fig. 139; Dr E. G. West for Fig. 140; Ronald Benson and Edward Arnold (Publishers) Ltd, for Fig. 143; Dr R. N. Parkins for Fig. 145; Professor J. F. Nye and the Royal Society for Fig. 146; The Addison-Wesley Publishing Co. Inc. and L. H. Van Vlack for Fig. 148.

The sources of the other illustrations are as follows: Fig. 12 is plate XVIII from the 'Life of Rekhmara', 1900, London, by P. E. Newberry; Fig. 34 is from 'Technolgie' by H. Blumner, 152, vol. 4, Fig. 8; Fig. 49 is from plate LVIII in W. Gowland's paper on the 'Early metallurgy of silver and lead', pt. I (Lead), in *Archaeologia*, 1901, 57, (2), 359; Fig. 53 is from the Bayeux Tapestry; Fig. 60 is from

J. H. Lefroy: *Arch. J.,* 1868, 2S, 261; lithograph by the Royal Artillery Institute, London; Fig. 83 is from A. Fell: 'The early iron industry of Furness and District', 1908, plate opposite p. 241; Fig. 91 is from Percy: 'Iron and steel', Figs. 127 and 129; Fig. 97 is from John Gibbons: 'Practical remarks . . .', 1839, Corbyn's Hall, Staffordshire; Fig. 98 is from T. Turner: 'The metallurgy of iron and steel', 1895, London, Fig. 28, p. 117; Fig. 99a *is* from the same source, Figs. 3 and 35, p. 122; Fig. 101a is from J. H. Hassenfratz: 'La Sidérotechnique', vol. 2, 165,1812, Paris; Fig. 101b is from Jars: vol. 1, plate I, Fig. 1; Fig. 102a is from Dufrenoy: vol. 1 , plate VI, Figs. 6 and 7; Fig. 103 is from Percy: 'Iron and Steel', Fig. 61 (redrawn); Fig. 105 is from Jars: vol. 1, plate Vl, Fig. 6; Fig. 106 is from R.A.F. de Réaumur: 'L'Art de Convertir le Fer Forge en Acier et L'Art D'Adoucir le Fer Fondu', plate 13,1722, Paris; Fig. 107 is from C. Tomlinson: 'Cyclopaedia of Useful Arts', 345, 1852, vol. 1, pt. II; Fig. 111 is from Dufrenoy: vol. 2, plate III, Fig. 4; Fig. 115, source unknown; Fig. 116 is from Diderot: 'Recueil de Planches sur les Sciences, Les Arts Liberaux et les Arts Méchaniques, avec leur explication', 1767, Paris, 559, 1969, Readex Microprint Corporation; Fig. 118 is from L. Ercker: 'Beschreibung aller furnemisten mineralischen Ertzt und Bergwerck sarten', 1629, Franckfurt a.M.; Fig. 119 is from Schluter: vol. II, plate XLII; Fig. 120 is from Schluter: vol: II, plate LI; Fig. 122 is from Diderot, 563; Fig. 124 is from Dufrénoy: vol. 2, plate XVIII, Figs. 1-5; Fig. 126 is from Diderot: 564; Fig. 127 is from Diderot: 297; Fig. 128 is from Dufrénoy: vol. 2, plate XIV, Figs. 3 and 4; Fig. 129 is from Dufrénoy: vol. 2, plate XIV, Figs. 5 and 6; Fig. 130 is from Percy: 'Lead', Figs. 56 and 57; Fig. 131 is from Dufrénoy: vol.2, plate XIV, Fig. 1; Fig. 132 is from Dufrénoy: vol. 2, plate XVI, Figs. 5 and 6; Fig. 133 is from Percy: 'Lead', Fig. 11, p. 126; Fig. 135 is from Diderot: 567; Fig. 136 is from Percy: 'Refractories', 1875, plate 2; Fig. 141 is from W. Gowland: 'The metallurgy of the non-ferrous metals', 1918, London, Fig. 14; Fig. 142 is from *Phil. Trans. Roy. Soc.,* 1829,119,1, plate I; Fig. 147, D. 1. Mendeleeff: *Journal of the Russian Chemical Society,* 1869, 1, 60.

Introduction

This book is an introduction to the history of metallurgy from the earliest times to the most recent. A study of this magnitude, embracing many of the countries in the world, cannot be treated adequately in one volume and it is intended in the course of time to cover this fertile field in a multi-volume work. However, it is felt that there is a demand for a single volume work from students of metallurgy and archaeology, not to mention the far greater numbers interested in the study of the history of technology in general.

The present author has previously discussed in detail the earlier periods of the subject up to AD 1600, with special reference to the British Isles. In the present volume we are concerned with such important subjects as the rise of metallurgy in the Near East and the Industrial Revolution in Western Europe.

The treatment of the subject has followed the pattern of the earlier work, with the maximum importance being attached to the material evidence and relatively little to the documentary evidence. This is not to decry the value of the latter, but merely to accept the fact that documtary evidence omits the details, such as the design of the furnaces and the composition of the metals produced, that seem so important to a practicing metallurgist. In the medieval and post-medieval periods, especially, we do get detailed descriptions of the techniques used, and these have been used to the full.

Today, metallurgy would be described as the science of metals, but up to the 18th century it was only concerned with the practice of metallurgy, which consists of the traditional methods of smelting, melting, and working of metals. The spread of this knowledge was not an even one throughout the world but depended on the ability of civilizations to invent new and exploit known techniques.

The fact that we find a well marked metallic sequence occurring over widely spaced areas of the globe may suggest that we are dealing with a diffusion process. For example, the sequence pure copper-arsenical copper-tin bronze-iron, occurs in different places and in some cases at times separated by over 2000 years. Not always, however, do we get the full sequence, for in parts of Africa we go from Neolithic technology with the making of flint tools and pottery to an Early Iron Age civilization. Of course there are good reasons for this; Neolithic civilizations did not always reach the technological standards upon which they could build a Copper Age nor, alternatively, did they have the trade contacts to benefit from the current diffusion trend.

However, geology suggests another possible reason for the metallic sequence. The main primary copper ore-bodies of the world have weathered into a succession of layers in which, going from top to bottom, there is an oxidized zone containing native (metallic) copper and oxidized minerals such as malachite, and lower down, an enriched sulphide zone in which such impurities as arsenic are present in strong concentrations. This may account for the ubiquity of the strongly arsenical coppers in the second phase of our metallic sequence. Later—probably in the Late Bronze Age—the smelters had to be content with the lower grade sulphide ores at the base of the deposit, as indeed we have to be today.

For a long time the theory of the diffusion of ideas and techniques has been generally accepted by the archaeologists. Today, however, as our detailed knowledge increases, the general diffusional thesis is being questioned. But it would appear that as far as metallurgy is concerned there is still good reason to accept the general validity of diffusion, but at the same time remember that as native copper was available to anyone who could use it, many people could have made the first few steps in metallurgical development independently. It is probable that these groups were overtaken by the mainstream of technological advance before the local people had time to develop their own metallurgy.

The use of radio-carbon dating techniques has caused some revision of the previous dating of archaeological sites and, consequently, the beginning of industrial processes. Radio-carbon dates differ appreciably from some of the conventional historical dates in parts of the world, and the radio-carbon technique itself is in the process of continuous refinement. The ^{14}C dates now tend to be earlier than was originally calculated on the basis of the earlier assumptions about the constancy of the cosmic ray activity and the halflife of the ^{14}C isotope.

The technique of copper smelting developed in Anatolia or Iran probably as early as the 6th millennium BC, and had reached the British Isles and China by the beginning of the 2nd millennium. It is possible that the appearance of smelted copper in South and Central America at the beginning of the first millennium AD is a case of independent development. In Anatolia the Iron Age started between about 1500 and 1000 BC and reached China, Britain, and Nigeria by about 400 BC. North and South America and Australasia obtained their knowledge of ironworking with European colonization, beginning about the 15th century AD (*see* Appendix 4).

This book has been arranged on the basis of archaeological periods up to the coming of the Iron Age, and for this reason the reader should not be surprised to find a reference to the native copper-using American Indians of the 14th century AD included in the same early chapter as the beginning of the Near-Eastern copper-using civilizations of the 6th millennium. For the same reason, the Early Iron Age cultures of Nigeria and Japan come into the same chapter on the Early Iron Age as European ironmaking, but after the Roman period, when we are dealing mainly with Europe, it is necessary to divide the subject into historical periods.

To try and make the special metallurgical terms understandable to the archaeologist and the historian I have included a technical glossary (*see* Appendix 1), but the more normal technical terms will be found in most technical dictionaries.

Chapter 1
Metals and ores in the Neolithic period

When inquiring into the beginnings of metallurgy, it is necessary to look into certain techniques in everyday use in lithic societies, since the use of metals arose out of the experiences of lithic peoples with metallic materials. The use of red oxide of iron in ritual and funerary practice is well known over a wide area from the earliest times. Neolithic people also decorated walls with it, and at Eridu and Susa pieces of hematite were used for burnished pottery by about 4000 BC. The greens and blues of copper minerals would certainly appeal, and their use as cosmetics in Egypt and Mesopotamia is well attested. In Crete, small pieces of azurite were discovered in a habitation layer dated to *c.* 6000 BC.[1]

There is little doubt that when the neolithic pottery stage was reached, green minerals were used for decorating pottery; their instability would soon be discovered for, unlike the red oxides of iron, copper-base minerals turn black when heated under oxidizing conditions.

The effect of reducing conditions in producing globules of metallic lead during the firing of lead glazes is well known, and one wonders whether smelting could have been discovered in this way. There is no evidence for this, but the fact that enclosed kilns, in which sufficiently reducing conditions would be possible, were not known until the Copper Age might suggest that such an accident was responsible for the origin of copper smelting.

Glazes were known before glass, and glazed steatite (soapstone) is known from one of the earliest periods in Egypt (the Badarian period in *c.* 5000 BC), but there is no evidence for the use of copper in such glazes.[2] However, copper has been detected in the glazed quartz frit known as 'faience', but not before the 18th Dynasty (1600–1300 BC). A clay tablet containing two recipes for green glazes has been found in the Tigris region, dated to about 1600 BC.[3] Both these recipes contained substantial amounts of copper.

Copper

Now let us turn to metals themselves. The existence of native copper and meteoric iron is well known, but the frequency of finding native copper is probably greater than is generally realized. All large and small deposits of copper ores seem to produce their quota of native copper. Small objects of copper, such as beads, pins, and awls, turn up sporadically in very early contexts around the ninth–seventh millennia BC, as at Ali Kosh in Western Iran,[4] and Cayönü Tepesi near Ergani in Anatolia.[5]

It is not possible to differentiate between melted native copper and copper smelted from pure ores, but as long as the native copper is not melted a distinction is possible. Native copper was once thought to be much purer than copper produced from the purer ores. Recent work has shown that this distinction is extremely dubious. Many analyses have been published,[7] and Table 1 shows some recent analyses of native metal of comparatively low purity.

The main problem with native copper is its extreme heterogeneity. Some pieces consist of very large grains but others are of small angular crystals which have large cavities or deposits of gangue (unwanted minerals such as calcite) in their interstices. For this reason elements such as calcium, aluminium, magnesium, and silicon have been omitted from the analyses shown in Table 1 because it is almost certain that they are present as particulate impurities. Micro-examination also shows the presence of irregular areas which etch differently from the mass, and it is believed that these correspond to minor variations in composition, perhaps of arsenic and silver. Contents of silver of 0.052, 0.042, 0.116, and 0.024%, respectively, have been reported all on the same lump.

While silver seems to attain values as high as 0.6%, gold is usually undetected or absent. Arsenic, nickel, lead, antimony, and iron may be present in considerable amounts, although antimony is rare. The first four comprise the main diagnostic elements for the origin of early metallic artefacts; it seems that the composition of native copper largely reflects the deposit in which it occurs.

A native copper nodule taken from the smelter at Ergani Maden in 1968 weighed 1.4 kg and consisted of about 12 'zones'. Each zone consisted of an aggregate of crystals, some showing signs of twinning. Electron probe analysis showed that the copper contained considerable sulphur and some Fe, As and Ni (*see* Table 1). The gangue phase between the zones was highly titaniferous together with iron oxides and silica, but Sb was absent. This contrasts with the Iranian copper from Talmessi and Anarak, which contains up to 15% As but is low in all but Ag (0.1%).

It is becoming clear that the impurities in native copper can be highly segregated and the actual analysis obtained depends a lot on the method of analysis used.

Of course, most native coppers are of a much higher purity than those shown in Table 1 and the ores are generally of a lower degree of purity. Therefore, if a very

Table 1 Composition of some less pure native coppers, %

Element	Talmessi (Iran)[4]	Rhodesia[7]	Aran Moor (Donegal)[7]	Anarak (Iran)[10]	Takhtul Chalgan (USSR)[8]	Ankara (Turkey)[18]	Ergani (Turkey)[41]
Ag	0·023 0·014	0·004	0·005	~0·1	0·6	—	ND
Au	—	—	—	tr.	nil	—	ND
As	0·08	ND	0·002	0·1–1·0	nil	—	0·15
Sb	—	ND	ND	ND	0·4	—	ND
Pb	<0·0001	0·0005	0·0005	—	0·3	—	ND
Ni	ND	ND	0·0003	ND	0·5	—	0·84
Co	—	0·1	ND	—	—	—	—
Hg	0·001–0·01	—	—	—	—	—	—
Fe	0·005	0·1	~0·2	present	0·02	0·17	0·4
Bi	0·00005	0·0006	0·0003	ND	0·003	—	ND
Sn	—	—	—	—	—	tr.	ND
Zn	ND	—	—	—	0·1	—	ND
S	—	—	—	—	—	—	0·13

ND not detected; – not sought; tr. trace
Limits of detection vary but with good techniques they are usually <0·0001

pure artefact is found in an early context there is a high probability that it has been made from native copper.

If an artefact has been made from hammered native copper, provided it has not been heated above a certain temperature, certain characteristics of native copper will prevail. During annealing, the grain-boundary impurities tend to fix the grain boundaries and the soluble segregates do not diffuse away until temperatures of about 600°C are exceeded. No doubt the silver content is mainly responsible for raising the recrystallization temperature of native copper, which is invariably found in a worked state—the hardness varying from 63 to 102 HV (Vickers scale).

Experiments have shown that native copper cannot be extensively worked without intermediate annealing.[6] Much depends on the size of the original grains and the possibility of a crack spreading along a grain boundary. It has been shown that, as long as incipient cracks are immediately removed during forging, the thickness of pieces of Michigan and Iranian copper can be reduced by 96% and thus it is possible to produce beads or even a small axehead.[4]

A small copper bead from Ali Kosh in Iran, dated to the seventh millennium BC and now completely oxidized, was examined and found to have been made of rolled-up copper sheet about 0.4 mm thick.[4] Its polygonal shape was the same as that reproduced by experiment and there is little doubt that it was made from native copper. A pin from the north mound at Sialk was also examined; this was in much better condition and was dated to the middle of fifth millennium BC. The microstructure showed that it was heavily worked native copper identical to that of a newly worked piece of Iranian copper. The chemical composition was very similar, and its hardness was 109 HV.

A more typical object undoubtedly made of native copper is the awl from Tell Magzallia, also in Iraq.[14] This is a high-purity copper with 0.2% Ag, some of which is segregated. It is heavily corroded and the elements in the residual metallic core are: Sn 0.09, Pb 0.01, Zn 0.08, Fe 0.02, Ni 0.001 and Bi 0.0001%. The microhardness is 106 HV which, together with the structure, shows the effect of severe cold work. Dated to the seventh millennium BC, this is the oldest piece of worked native copper yet known.

When native copper is melted the gangue components separate and float out. The soluble segregates tend to dissolve and the metal becomes homogeneous, with a typical cast cored structure and lower hardness. In one case, the hardness was reduced from 84 to 37 HV.

Most of our knowledge of native copper artefacts comes from the New World. Well-crafted objects were being made in large quantities from unmelted native copper in the Lake Superior region of North America between 3000 BC and AD 1400.[11,12] Spearheads belonging to the Old Copper Culture (3000–1500 BC) have been examined.[4] In some cases these are tanged (pointed to take the shaft), and in others the tang has been carefully worked into a half socket. All show extensive work hardening, and the average hardness of the individual artefacts varies from 59 to 108 HV. As the observed structures were those of annealed material with and without final cold work, it was concluded that the normal procedure was to hammer and anneal the copper until the final shape was very nearly obtained, and then finish the object by localized or overall cold working. In a few cases they had been left in a fully annealed condition. This shows that a lithic civilization was able to go one step further than merely hammering native copper, and make use of the knowledge that heating in a fire caused it to soften so that further work could be done. Although the annealing temperature was often as high as 800°C no melting had been done: to reach the melting point (1 084°C) a forced draught would have been necessary.

Iron

Meteoric iron was also available to lithic people. This usually contains about 10% Ni and is therefore much harder and more difficult to work. The nickel content varies from 4 to 26%, and can be easily detected as there seem to be no ores capable of giving this level of nickel in a homogeneous form by direct smelting.

It has been calculated that there are at least 250 t of meteoric material extant, and that 99.4% of this is malleable.[13] However, the problems are much the same as with native copper. When cold hammered, the material tends to crack along well-defined crystal planes. With care and annealing, or by hot forging, small artefacts can be made and lithic peoples in the Cape York area of West Greenland have been making tools from meteoric material containing about 8%Ni. While prolonged heat treatment at elevated temperatures will destroy the structure of meteoric iron in time, the material will be always recognizable by its composition.

Another source of iron is telluric iron. Small grains occur in some basalts and are large enough to forge into flakes which can be inserted into an organic hilt to make a knife with a serrated edge. Such grains rarely contain more than 3% nickel.

Greenland has provided examples of both meteoric iron and telluric iron. The principal source of the former is the so-called Cape York meteorite, which broke up in the atmosphere and was scattered over a wide area. This was a typical medium octahedite with a nickel content of 8%. Some pieces of this meteorite, and a number of Eskimo tools made from it, exist in Scandinavian collections, and some have been examined by Buchwald and Mosdal. Most have been cold worked but none have been hot worked. The reason for the latter is probably the shortage of fuel. Hardnesses range from 200 HV for an 'unworked meteorite' to 330 HV for the cold-forged blades. Annealing reduces the hardness to 155 HV. Clearly the original material had been strained before it touched earth. There was no difficulty reducing the thickness to 92% of the original with a 2.2 kg steel hammer.

Telluric iron occurs on Disko Island off the Greenland coast in two forms. The first, a malleable form, contains on average about 0.3–2.5% Ni and 0.1 to 1.0% C. The second, a non-malleable white cast iron, contains 1–2% Ni and 1–4% C. The maximum nickel content is too low for meteoric iron, and the carbon too high. The first occurs as small grains 1–5 mm diam. The second type fragments on hammering, but it would seem to occur in larger pieces in the basalt boulders.

A recent find, showing not only the use of meteoric iron but its identifiable character despite almost complete mineralization after 3000 years, is that fond in a Chinese Shang Dynasty bronze axe.[16] A small piece of meteoric iron had been used to improve the cutting edge, and the characteristic alpha-gamma widmanstätten structure was still identifiable when using the electron probe, as the high Ni/low Ni lamellae were still present in the rust. Clearly, very little diffusion of Ni had taken place, but some overall loss must have occurred, as one would not expect a piece of meteoric iron to have such a low Ni content as 1.76%. Penetration of bronze in the cracks in the iron show that the axe was made by casting the bronze on to the iron blade inserted in the mould.

A number of iron objects have been found in pre-Iron Age contexts, and many of those analysed have been found to contain nickel. These are listed in Table 2, but some have been found to be devoid of nickel and have therefore created a problem. Either they have got into earlier levels by accident and have been dated wrongly or, if found in Copper or Bronze Age levels, have possibly resulted from smelting under such conditions that iron has been reduced from ferruginous fluxes or slags during copper smelting. This possibility will be discussed in connection with the beginning of the Iron Age.

One of these pieces has been found in the pyramid at Gizeh and has been dated to the third millennium BC.[21] It is certainly not meteoric and consists of reduced pieces of iron oxide hammered together, without the usual telltale slag that is normally present in wrought iron. It would seem that, as early as this, some smiths were capable of making and working non-meteoric iron.

Meteoric iron was probably used unknowingly well into the Early Iron Age and perhaps even later.[22] A socketed iron axehead, dated to the Hallstatt period (800–300 BC), was found[23] to have a mean nickel content of 4%. The nickel content was due to a lamination which contained 8–10%Ni in the centre of the axehead, which is well within the meteoric range. The smiths of this period often welded up pieces of iron of different origin and it would seem that here one of their pieces had been of meteoric origin. However, it may be possible for thin layers of high nickel content to be obtained in wrought iron by surface enrichment of nickel during the oxida-

Table 2 Composition of iron artefacts believed to be of meteoric origin

Artifact	Provenance	Date used	Composition,%				Ref.
			Fe	Ni	Co	Cu	
Dagger	Ur	3000 BC	89.1	10.0	-	-	18
Beads	Gerzeh	3500 BC	92.5	7.5	-	-	18
Knife	Eskimo	Recent	91.47	7.78	0.53	0.016	17
Knife	Deir el Bahari	2000 BC	-	10.0	-	-	2
Knife	Eskimo	AD 1818	88.0	11.83	tr	tr	19
Axehead	Ras Shamra	1450 - 1350 BC	84.9	3.25	0.41	nil	20
Dagger, }	{Tutankhamun,	1340 BC	-	pres.	-	-	20
Headrest }	{Thebes	1340 BC	-	pres.	-	-	20
Plaque	Alaca Hüyük	2400 - 2200 BC	-	3.44 (NiO)	-	-	20
Macehead	Troy	2400 - 2200 BC	-	3.91 (NiO)	-	-	20

Table 3 Composition of natural gold

Provenance	Shape of occurence	Composition %			Ref
		Ag	Cu	Rest	
Urals	Stream	0.16	0.35	Fe; 0.05	27
Urals	Crystal	4.34	≯0.33		27
N. Tagil	Crystal	8.35	≯0.29		27
Archangel	Not known	9.45	0.35	SiO_2; 0.08	27
Donetz	Not known	14.71	≯0.08		27
Urals	Crystal	20.34	≯0.66		27
Urals	Crystal	28.30	≯0.84		27
Altai	Vein	38.38	0.31	Fe; 0.033	27
Urals	Nugget	5.78	≯0.21		27
Urals	Nugget	3.96	≯0.12		27
Germany	Sheet	6.60	nil	Pt; 0.007	27
Germany		8.42	0.02	Fe; 0.16	27
Italy, Po	Sand	4.69	-		27
Italy, Po	Sand	6.40	-		27
Italy, Po	Nugget	6.37	-		27
Italy, Genoa		10.30	1.4		27
Finland	Stream	1.79	≯0.21		27
Finland	Stream	9.61	≯0.89		27
Finland	Stream	21.90	≯1.0		27
Ireland	Nugget	5.10	≯2.9*	2.9	28
Ireland	Stream	6.17	≯0.73	Fe; 0.78	27
Ireland	Stream	8.85	nil	SiO_2; 0.14	27
Ireland	Stream	8.10	tr.	Fe; 2.1	27
England, Devon	Dendrites	1.89	nil	nil	27
England, Devon	Dendrites	7.47	nil	nil	27
England, Devon	Dendrites	8.41	nil	nil	27
England, Cornwall	Stream	9.05	nil*	$SiO_2 + Fe_2O_3$; 0.83	27
Wales, Clogau	Vein	9.26	tr.	0.58	27
Wales, Clogau	Vein	9.24	<0.02*	0.91	27
Wales, Dollgellau	-	13.99	tr.	1.12	27
Scotland, Wanlockhead	-	12.39	<0.66*	Fe; 0.35	29
Scotland, Sutherland	'Fine'	18.47	nil*	SiO_2; 0.26	27
Scotland, Sutherland	'Fine'	19.86	nil*	Fe; 0.12	27
Scotland, Sutherland	-	20.78	nil*	nil	27
Romania, Siebenburgen	Vein	26.36	≯0.2		27
Romania, Siebenburgen	Vein	14.68	0.04	Fe; 0.13	27
Romania, Siebenburgen	Vein	20.90	≯1.0		27
Romania, Siebenburgen	Vein	29.40	≯1.0		27
Romania, Siebenburgen	Sheet	27.60	nil		27
Romania, Siebenburgen	Sheet	33.20	nil	SiO2; 0.42	27
Romania, Siebenburgen	Sheet	38.74	≯0.77		27
India, Kolar	Ingot	6.38	0.38		30
India, Kolar	Ingot	7.30	0.93		30
Australia, Coolgardie	Nugget	1.05	-		31
Brazil	Nugget	20 (app)	-		32
West Africa	Grains	2.20	-		42
Greece, Thassos	Grains	16.50	0.1		33
Longcleugh, Lanark	Grains	c. 5.0	<0.01	Hg; 0.03	34
Sperrin, N. Ireland	Grains	9.80	<0.1	Sn; <0.01	35

* By difference;   tr. trace

tion of the more oxidizable iron when it is preheated for forging. High nickel laminations can be introduced by piling, i.e. by the bending over, hammering, and welding of the same piece many times.

Gold

The other native metals are gold and platinum. Gold almost entirely occurs in native metallic form. However, unless it is found in the form of nuggets it is difficult to use in a primitive context, as the fine dust-like particles collected by washing in stream beds are not easily seen, and also not easily consolidated by melting. As nuggets are still to be found on the surface today (in 1983 one weighing 63 kg was found in Brazil)[32] they must have been even more common in early times. The composition of nugget material is essentially the same as deep-mined material, so there is no way of distinguishing early nugget material from deep-mined gold. Neither, considering the analyses shown in Table 3, is there a safe way of provenancing gold finds by composition. All natural gold is impure, usually assaying about 10% of silver and up to 1% of copper. One examination[24] of a series of European gold artefacts showed that there were at least two groups. One was similar to the gold alleged to have come from the Wicklow mountains and another, of which the majority was composed, came from some other source, possibly not Irish. But a further series, of Hungarian–Romanian origin, was found to be similar to the first group.[25] There is little doubt that there are considerable variations in the impurity content of native gold but no evidence that this has any archaeological significance.

In order to offset the whitening effect of the silver in natural gold, early people very soon learnt to add copper. The natural copper content, according to analyses by Eluère, can be less than 0.1%.[26] But some natural golds, as we see from Table 3, can have higher values than this —up to 1%. When the copper content of an artefact exceeds this, it is very possible that is has been intentionally added.

It is an interesting fact that there is no gold artefact which is dated earlier than the end of the fifth millennium BC, yet gold nuggets must have been noticed and, because of their malleability, used by lithic peoples. One can only conjecture that, for early people, gold was too valuable to be buried in graves or was soon pilfered and put back into circulation. This suggests that much third –fourth millennium gold was re-used material.

Platinum

Platinum occurs, like gold, in the form of water-borne grains in alluvial gravels. These contain from 50 to 80%Pt, the remainder consisting of metals of the platinum group together with small amounts of base metals. Most of the known occurrences have been in South America, particularly Colombia and Ecuador.[36]

Another source is the Ural mountains of the USSR. Small nuggets weighing up to 36 g have been found and these have formed the basis of recent Russian coinage.[37] There is no evidence for the intentional use of platinum in this region in antiquity, although it does occur as a common inclusion in gold artefacts and must have caused a great deal of trouble to early gold workers.[38,39] Unfortunately the composition of the inclusions in no way indicates the provenance of the gold.

Objects containing large amounts of platinum have been found in South America, and even small pieces have been found in Egypt.[2] It is not thought, however, that the metal was recognized as a separate entity by early people. An alloy containing gold, platinum, and silver, in the proportions 70:18:11, was found amongst early material from parts of Ecuador that had been in Inca possession, and the platinum content of objects made by pre-Colombian Indians[40] varied from 26 to 72%. It would seem that these alloys were made by melting a naturally occurring mixture of precious metals panned out of alluvial deposits. Sometimes, the platinum grains were extracted to avoid getting a silvery gold. In some cases gold artefacts were 'plated' with platinum, possibly by hammer welding thin platinum sheet on to the gold.

References

1   J. D. EVANS: *Atti CISPP*, 222, 1965, 2
2   A. LUCAS: 'Ancient Egyptian materials and industries', 4 ed., (revised J. R. Harris), 1962, London, Edward Arnold
3   C. J. GADD and R. CAMPBELL THOMPSON: *Iraq*, 1936, 3, 87
4   C. S. SMITH: Actes XI Congr. Int. d'Hist. Sciences. Warsaw–Krakow, 237, vol.VI, 1965,
5   M. J. MELLINK: *AJA*, 1965, **69**, 138
6   H. H. COGHLAN: 'Notes on the prehistoric metallurgy of copper and bronze in the Old World', 1975 (2nd Ed.), Pitt Rivers Museum Occ. Paper No.4, Oxford
7   H. H. COGHLAN: *PPS*, 1962, **28**, 58
8   I. R. SELIMKHANOV: *ibid.*, 1964, **30**, 66
9   E. VOCE: *Man*, 1948, **48**, 19
10   R. F. TYLECOTE: *Metals and Materials*, 1970, **4**, (7), 285
11   D. L. SCHROEDER and K. C. RUHL: *American Antiquity*, 1968, **33**, 162
12   K. WINTERTON: 'Dating of some museum objects by metallurgical means', Jul. 1957, Ontario Research Foundation, Department of Engineering and Metallurgy
13   G. F. ZIMMER: *J. Iron Steel Inst.*, 1916, **94**, (11), 306
14   N. V. RHYNDINA and L.K. YAKHONTOVA: 'The earliest copper artefact from Mesopotamia', Soviet Arch, 1985,(2), 155–165
15   V. F. BUCHWALD and G. MOSDAL: 'Meteoric iron, telluric iron and wrought iron in Greenland', **Man and Society**, 1985, **9**, 3–49
16   LI CHUNG: 'Studies on the meteoric iron blade of a Shang dynasty bronze Yueh-axe', Peking Institute of Iron and Steel Technology, 1976, 259–289
17   T. A. RICKARD: *JIM*, 1930, **43**, 297
18   C. H. DESCH: 1st Report, Sumerian Committee, British Association, 1928, 437
19   H. H. COGHLAN: 'Notes on prehistoric and early iron in the Old World', 1977, (2 ed.), Pitt Rivers Museum Occ. Paper No.8, Oxford
20   J. K. BJORKMAN: 'Meteors and meteorites in the ancient Near East', Centre for Meteorite Studies, Arizona State Univ. 1973, June 30
21   EL S. EL GAYAR and M.P. JONES: 'Metallurgical investigation of an iron plate found in 1837 in the Great Pyramid at Gizeh', *JHMS*, 1989, **23**, (2), 75–83

22   J. ZIMNY: Z. *Otchlany Wiekow*, 1966, **32**, 29

23   J. PIASKOWSKI: *J. Iron and Steel Inst.*, 1960, **194**, 336

24   A. HARTMANN: 'Some results of spectrochemical analysis of Irish gold', Ogam Tradition Celtique, 1965, (98), Suppl. to Celticum, 12

25   A. HARTMANN: *MAGW*, 1964–5, **93–5**, 104

26   CHRISTIANE ELUERE: *Les Ors Prehistoriques*, Paris, 1982

27   H. OTTO: 'Die chemische Untersuchungen des Goldringes von Gahlstorf un seine beziehungen zu anderen funden', *Jahresschrift des Folkmuseums*, Bremen, 1939, 48–62

28   G. A. J. COLE: 'Memoir and maps of localities of minerals of economic importance and metalliferous mines in Ireland', *HMSO*, Dublin, 1922

29   A. J. S. BROOK: 'Technical description of regalia of Scotland', *PSAS* 1889–90, **12**, 89–92

30   B. L. MUNJANATH (Ed.), 'The Wealth of India', *CSIR*, New Delhi, 1957

31   A. LIVERSIDGE, 'The crystalline structure of gold and platinum nuggets and gold nuggets', *J Chem. Soc*, 1897, **71**, 1125

32 *The Guardian*, 22, Sept., 1983

33   G. A. WAGNER, E. PERNICKA and W. GENTNER: 'Naturwissenschaften', 1979, **66**, 613

34   JOAN J. TAYLOR: 'Bronze Age Goldwork of the British Isles', Cambridge, 1980

35   S. BRIGGS, J. BRENNAN and G. FREEBURN: 'Irish prehistoric goldworking', *JHMS*, 1973, **7** (2), 18–26

36   D. McDONALD and L. B. HUNT: 'A history of platinum and its allied metals', London, 1982

37   H-G. BACHMAN and H. RENNER: 'Nineteenth century platinum coins: an early use of powder metallurgy', *Plat. Met. Rev.* 1984, **28** (3), 126–31

38   J. M. OGDEN: 'Platinum group inclusions in ancient gold artefacts', *JHMS*, 1977, **11** (2), 53–71

39   N. D. MEEKS and M.S. TITE: 'The analysis of platinum-group inclusions in gold antiquities', *J. Arch. Sci.*, 1980, **7**, 267–275

40   P. BERGSOE: *Nature*, 1936, **137**, 29

41   R. F. TYLECOTE: 'The evolution of the metallurgy of copper and copper-based alloys', in: Journées de Paleometallurgie, Universite de Technologie de Compiègne, 22–23 Feb. 1983, Compiègne, 1983, 193-221

42   T. K. ROSE, 'The metallurgy of gold', London, 1915

Chapter 2
The technique and development of early copper smelting

It has been customary for archaeologists to call the first metal age the Bronze Age because it was originally thought that all early copper-base artefacts were bronzes, i.e. copper–tin alloys. However, it is now known that it took a long time for a true Bronze Age to get under way, and analyses show that before bronze was ever used there was a long period when smelted coppers of relatively high purity, or coppers containing substantial amounts of arsenic or antimony, were used. In many areas there was an overlap between such metals and true bronzes, e.g. in the British 'Wessex' period. This chapter will deal, therefore, with non-tin-bearing coppers.

It is always difficult to say precisely when a culture begins. In archaeological works one meets with the terms Chalcolithic and Eneolithic, which have the same meaning and imply a transition period between a Neolithic technology and a Copper Age. Should the first appearance of a piece of hammered copper mark the beginning of the Copper Age? As we see, this has already been answered in the negative, and the smelting of copper ore is now our only criterion. But how much use must be made of a material for it to have an age called after it? There was a long period while copper was proving itself against the merits of stone and flint, while more and more deposits of copper ore were being found and while the technique of extractive metallurgy spread in an essentially lithic society. Metallurgically, the distinction between the Chalcolithic and the Copper Age is only one of scale, and both are considered in this chapter.

Initial Stages

The fact that no permanent pottery kilns are known from the Neolithic period, and that the start of extractive metallurgy can be equated with the use of such kilns, supports copper extraction was started either by accident, or by intention, i.e. in a pottery kiln. But if this was the case and the result was successful it would soon have been realized that a kiln was not a satisfactory medium in which to continue the process.

A possible process of transition between the melting of native copper and the smelting of pure oxide ores could be the melting of heavily-weathered pieces of native copper from the surface of a deposit. Such occurrences can be found in the Sahara in Niger, where the native copper is heavily oxidized and, when melted in a crucible under reducing conditions, would give more copper than expected. This would be noted and gradually it would be found that it was possible to obtain metal from the oxide alone.[1]

It has been found that copper can be obtained from pure oxide minerals in a crucible by direct reduction.[2] Excess charcoal would be used and the product comminuted. The excess charcoal would be washed away and the copper prills recovered. There are few early sites, such as Tal i Iblis,[3] which show signs of crucible smelting with the presence of large numbers of used crucibles which are difficult to explain in any other way. So we cannot rule this out as an early production process.

In order to be certain about the initiation of extractive techniques we have to concentrate on the slags, in view of the impossibility of distinguishing between melted native copper and copper smelted from pure oxide ores. Slaggy material was found in association with the copper at Çatal Hüyük in Anatolia.[4] This site dates from 7000–6000 BC and is probably the earliest known site to produce a vitrified copper product. It is claimed[5] that this is a smelting slag and not merely a crucible melting slag which could have resulted from melting native copper. Crucible slags result from the reaction between the alkali in the fuel ash and silicates from the fabric of the crucible, and usually carry large amounts of entrapped copper. But if smelting was done in a crucible without fluxes, then slags, such as those given in Table 4, would be produced. They would be viscous and separation of copper would be poor, thus giving a large amount of residual copper, as we see in the slag from Feinan.[6]

A more efficient way of getting slag/metal separation was to add a flux, either iron oxide or managanese oxide, and thus turn the high silica content into a Mn or Fe silicate, which melts in the temperature range 1100–1200 BC. This principal was practised at Feinan in the Chalcolithic period.

When native or man-made copper is melted it tends to absorb oxygen from the atmosphere, and most early copper objects contain cuprous oxide as a result. Apart from the beads from Çatal Hüyük, which may well be hammered native copper, the earliest dated smelted copper artefacts are probably those from Tepe Yahya in

Table 4 Early copper slags believed to be from smelting

%	Afunfun Niger 800 BC	Feinan Chalcolithic 35-3200 BC	Shiqmim Chalcolithic 35-3200 BC
SiO_2	49.0	26.2	16.85
TiO_2	0.6	0.12	-
Al_2O_3	7.7	1.56	2.07
FeO	1.2	5.11	17.84
MnO	0.3	0.14	0.02
MgO	8.9	0.69	1.09
CaO	22.7	1.11	4.84
BaO	-	0.02	-
Cu	5.7	53.50	26.97
K_2O	1.2	0.21	-
Na_2O	1.1	-	-
P_2O_5	-	0.16	-
Ni	-	0.07	-
Zn	-	0.12	0.02
Pb	-	0.05	0.03
S	-	0.17	-
Sn	-	-	0.02
Total	98.4	89.23	-

- = not sought

Iran (c. 3800 BC) (see Table 5).[7] Thus, there seems to have been a long interval between these two sites' periods of activity, although one suspects that this will be filled when objects from earlier sites have been chemically and metallographically analysed. It is clear that the

Chalcolithic period lasted for a long time, with metal being used mainly for small objects like beads, awls, pins and, later, knives and daggers. The metallurgical evidence we have for any period before about 3500 BC is comparatively slight, and most of our material dates from about 3500 BC.

Table 6 shows some of the early sites on which copper artefacts have been found, and gives the conventional archaeological date together with the calibrated ^{14}C date where available. The dating of the later sites in the Near East is fairly reliable and ^{14}C dates are not expected to differ much from the generally accepted archaeological dates based upon historical events. Many of the early sites can only be dated by ^{14}C techniques but, in some cases, ^{14}C dating has considerably altered previously held dates, particularly in South East Europe.

Much of the early metalwork could be made from native copper and many of the smaller artefacts are merely hammered native copper. Some of the larger items, such as the hammer-axes, are cast native copper, and it is only when there are substantial traces of As or Ni that we can be sure that we are concerned with copper smelting, and therefore in a true Copper Age.

It was only after 3500 BC that the Sumerian city states of the lower Euphrates and Tigris were established. They made use of the copper from the highlands to the north and east, since the alluvial delta region was devoid of minerals and fuel. The needs of these cities could have encouraged the highlanders to seek minerals to satisfy this demand and at the same time, increase the local use

Table 5 Some examples of early smelted copper artefacts

Object	Provenance	Date, BC	Composition, % As	Sb	Pb	Ni	Weight, kg	Reference
Flat axe	Egypt	3500	0·49	ND	0·17	1·28	1·56	25, 26
Reamer	Syria (Amuq)	3400	1·35	ND	0·003	0·93		30
Reamer	Syria (Amuq)	3400	0·04	ND	0·01	0·16		30
Spatula	Iran (Yahya)	3800	1·7	<0·1	0·05	<0·01	0·065	95
Chisel	Iran (Yahya)	3800	3·7	<0·1	0·05	<0·01	0·095	95
Awl	Iran (Yahya)	3800	0·3	<0·1	0·05	<0·01	0·029	95
Hammer-axe	Prague		0·77	0·77	0	0·022		55
Hammer-axe	ČSSR (Tibava)	3000	1·15	0·0	0	0·01		"
Hammer-axe	ČSSR (Tibava)	3000	3·10	–	–	–		"
Axe-adze	Romania (Tirgu-Ocna)		0·8	0·0	0	0·01		"
Staffhead	Israel (Beersheba)	3300	12·0	0·72	tr.	0·05		34
Ornament	Israel (Mishmar)	3300	3·5	0·18	0·034	0·17		"
Ornament	Israel (Mishmar)	3300	11·9	0·61	0·039	1·22		"
Tool	Israel (Mishmar)	3300	1·92	tr	–	–		"
Tool	Israel (Mishmar)	3300	–	–	–	1·90		"
Axe	Israel, Kfar Monash	3300	1·15	–	0·06	0·71		98
Axe	Kfar Monash	3300	4·07	–	–	1·25		"
Axe	Kfar Monash	3300	–	–	–	1·01		"
Chisel	Kfar Monash	3300	3·55	–	–	–		"
Saw	Kfar Monash	3300	–	–	–	0·49		"
Spearhead	Kfar Monash	3300	2·20	–	–	0·70		"
Piece	Alishar Hüyük	2800	2·43					97
Vessel	Ur	2900	0·65			tr.		97
Pin	Tel Asmar	2500	2·08			0·9		96

ND = not determined; tr. = trace

Table 6 Spread of copper metallurgy in the Near East and SE Europe

Place		Type of product	Date, BC	
			Conventional*	Calibrated [14]C
Asia Minor	Çatal Hüyük	Beads } native?		7000–6500
	Suberde	Wire }		7000–6500
	Ali Kosh	Beads; hammered native		7000–5800
	Sialk I	Pin; cast copper	4500	
	Anau I	Awls, needles, knives	4500	5500
	Anau II	Spearheads, axes	4000	5300–4300
	T. Gawra	Blade, awl, ring		3500
	T. Giyan	Ni-containing	4500–3000	3700
	T. Yahya	As-containing	3800	4000
	T. Yanik	Impure Cu	3500	
	Troy I	Tools etc. (As)	3000	
Egypt	Badarian	Awls and pins	5000–4000	
	Predynastic	Axe (Ni, As)	4000	
Syria	Brak	Pins and sheet	4500	
	Amuq F.	Tools (+Ni)	3500–3100	3400
Palestine	Mishmar	Tools and ornaments (As)	3200	
	Kfar Monash	Tools (As)	3300	
	Beersheba	Tools etc. (As)	3500	
Hungary	Tiszapolgár	Hammer-axes (No As)	3000	4500
Bulgaria	Karbuna	Beads and axes	3000	
Slovakia	Tibava	Hammer-axes (As)	3000	
Hungary	Baden	Awls and axes	2000	2500–3000

* In some cases this column includes ^{14}C dates with old (5570 year) half life

of copper. The spread of technique must have been fairly rapid since we find a large number of places some distance away from Mesopotamia showing evidence of metal usage.

In the Near East this period coincides with a substantial increase in trade, and the existence of artefacts in a particular place is no evidence of local production. We must look for workshops, furnaces, slags, and moulds. Many of the larger excavated sites have yielded workshop areas, moulds, and furnaces. Sites far from mineral deposits will have received their metal in ingot form and cast it locally.

We are beginning to see a number of levels of smelting technique. These could have been used at the same time and represent differing stages of cultural development within different groups or tribes. It is easy to recognise the later techniques that produced large amounts of black, glassy, fayalitic slags high in iron. But there are areas free of these, where copper ores were processed leaving little, if any, residue. It is now accepted that the smelting technique used in these cases was that of a crucible in which high-grade oxide ores free of iron were reduced with charcoal fuel.[8] The next stage would be one in which less pure ores were used with high silica or iron oxides, using silica or iron as a flux and producing large quantities of slag as well as prills of copper metal. We will refer to these three stages as:

   (a) Crucible smelting
   (b) Smelting in non-slag tapping furnaces using fluxes or self-fluxing ores
   (c) Smelting in slag-tapping furnaces making high-iron or manganese slags.

Today one would define a crucible as a vessel heated from the outside, with the heat conducted through the wall. But in early times crucibles were often heated from above, and the heat radiated downwards onto the charge. Indeed this system continued in Japan into the 18th century[9] and one might say that the 19th century refining furnaces used in Europe embodied the same principle.

The distinction between smelting in a crucible and smelting in a non-tapping furnace is not always clear—by inserting a tuyere from the top well down into the crucible it becomes a furnace. Furthermore it is possible to cause smelting reactions in a crucible while at the same time heating it from the outside.[10]

It is not unreasonable to consider the first type of smelting furnace as a hole in the ground. Although its Chalcolithic date is not accepted by all scholars[11,12], a typical example of this type of furnace is that found at Timna, site 39 (*see* Fig. 1).

Many of the products of the Chalcolithic smelting era, as well as some of those produced by melting native copper, contain high levels of arsenic. In fact the preponderance of artefacts containing arsenic is so great that one must face one or other of two possibilities; (a) the addition of minerals of high arsenic content to molten copper under reducing conditions; or (b) the selection of arsenical copper minerals. If one examines the analyses one notices that very few of the artefacts exceed about 7% As. (The exceptions here are some beads from the Caucasus[13] which contain 14–24% As.)

Unfortunately many analyses have had to be confined to the surface layers only, and these can be misleading, as some elements such as arsenic can be segregated into the surface layers.

High arsenic alloys could be made by co-smelting a copper oxide with an arsenical sulphide such as orpiment (AsS).[14]

1 Chalcolithic-type copper smelting hearth at Timna, Israel

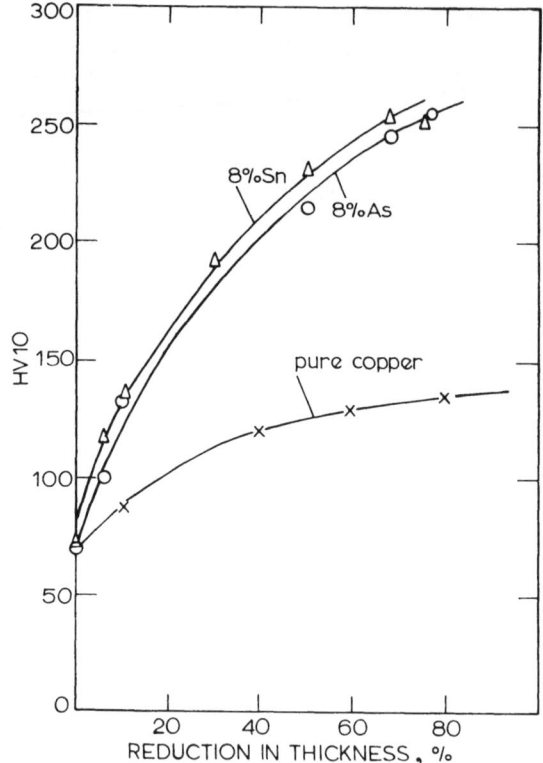

2 Effect of tin and arsenic addition to copper on the hardness after cold working (after Marechal [98])

Metallic arsenic is a relatively volatile substance with a boiling point of 613°C, and arsenic trioxide (As_2O_3), boiling at 457°C, is even more so. However, if arsenical oxidized copper minerals are smelted—and smelting of such ores can only occur under reducing conditions—one would expect the product to retain some of the arsenic. In fact, calculation shows that most of the arsenic present below 7% will be retained. Even if the arsenical copper, after smelting, is held in a deep crucible under reducing conditions little arsenic is lost, and this only slowly from the surface. In fact, arsenic is lost at appreciable rates only when arsenical copper is hot forged. It has been shown[15] that the recovery of arsenic during the smelting of a Helgoland arsenical oxide ore, containing 1–2%As, is about 100%.

Sulphide ores, on the other hand, must be roasted at some stage during the extraction process, and one would expect a considerable loss of arsenic during this stage. Therefore, one would not expect the recovery of arsenic to be as high as in the case of an oxide ore.

Why are arsenical coppers so desirable, or were the more easily mined copper ores arsenical? The mechanical properties of arsenical copper in the cast condition are not much better than those of pure copper. There is, however, a great difference in the wrought condition owing to the more rapid work hardening of Cu–As alloys. This effect is also shown by the tin bronzes[16] (see Fig.2). Most pure and arsenical copper artefacts show considerable amounts of working—mainly hot but some cold—and this can be seen in the elongation of the slag and oxide particles present in the metal. It is probable that the preference for working rather than casting developed out of the working of native copper. It was a good technique for the material available and clearly persisted until the development of tin bronzes.

The second question is best answered by describing the nature of copper deposits. Most, perhaps all, copper ores have started as sulphides. In a typical sulphide deposit, such as that at Ergani Maden in Turkey, the surface minerals consist of gossan or iron oxides which represent the oxidized ferrous component of the sulphide deposit (see Fig.3). In these near-surface layers may be found precious metals, native copper and some oxidized copper minerals, but much of the copper will have been washed down to enrich the zone below, i.e. the secondary enrichment zone. This is the zone which provides copper in the highest concentrations and, as the arsenical and antimonial minerals are relatively soluble, it usually contains these also in a high concentration. This zone will often contain minerals of the fahlerz type, i.e. copper–arsenic–antimony sulphides such as $(CuFe)_{12}(AsSb)_4S_{13}$, or solid solutions of tetrahedrite, $(CuFe)_{12}Sb_4S_{13}$, and tennantite, $(CuFe)_{12}As_4S_{13}$. The lowest zone represents the original deposit and contains copper sulphides in low concentration, usually about 1–4%. This is the grade of ore most worked in recent times.

So one can see that, while the surface deposit would be the first to be used and would give native copper or

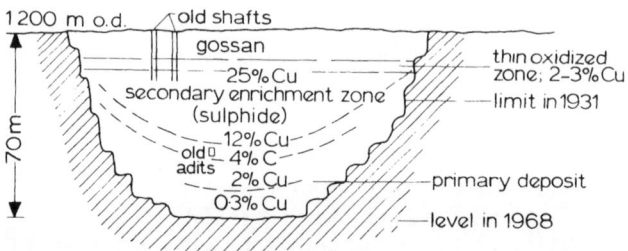

3 Section through a typical open pit copper mine showing the primary sulphide ore at depth and enrichment below the gossan (Ergani Moden, Turkey)

easily smelted oxide ore yielding a pure metal, the next zone would be preferred because it had a higher concentration and was found to give harder copper. This indicates that the technique of extracting metal from sulphides was mastered at an early date. This type of ore requires a long roast at a temperature not exceeding 800°C to convert both iron and copper sulphides to oxides, and the slagging of the iron oxide with silica (sand) to form the ferrous silicate (fayalite) under smelting conditions. This is the reverse of the previous process of adding iron oxides to flux silica.

It would seem, therefore, that the main reason for the metallic sequence Cu, Cu–As–Sn, Cu–As, etc. lay in the nature of the deposit. The native copper in the gossan would be used first, the oxidized ores next, and the sulphidic secondary enrichment zones last. Once the primary sulphide deposit was reached the metal produced would be weaker than arsenical copper and alloying, with arsenic or tin, would be essential for the tools and weapons of the period.

An amphora containing orpiment was found in the LBA Kas wreck found off the coast of southern Turkey.[17] Another arsenical substance is 'speiss', which is mainly an arsenide of Ni, Fe or Cu. This can be produced in the smelting of *fahlerz*-type ores, and could have been saved for reuse in a diluted form as an additive to pure copper[18] (a 'hardener'). Such material has been found at Guschau in Germany where pieces of cast copper-base alloy bar of roughly hemispherical section and about 5 cm long contained 17%As and 13%Sb, together with 16–19%Ni. It is probable that nickel tends to stabilize the antimony and arsenic as compounds and thus reduce the losses during smelting This effect is very marked in modern copper blast furnaces where 'speiss' is produced.

Not all copper deposits are primary, like that at Ergani Maden. The Israeli deposits near Timna are secondary and consist of disseminated oxide minerals in a white sandstone. Some of this has been concentrated to form nodules of very high copper content, not unlike the ironstone nodules of the carboniferous coal measures of Europe. Since the sandstone is very soft, the isolated nodules can be easily released by hammering away at the soft rock face, giving rise to enormous dumps of sand.

The antimony-rich fahlerz ores were clearly not as common as the arsenical ores, but there are several instances of high antimony content in Early Bronze Age (EBA) artefacts. For example, we have antimonial copper ingots and pieces of antimony from Velem St. Vid in Hungary,[20] but there is no evidence for the mines having been worked before the Middle Bronze Age (MBA) (1800 BC). Ore from these mines contains 16.6% Sb and 17.4%Cu.

There are also pieces of copper-base material, possibly ingot, from Parre, which are now in the museum at Bergamo, Italy.[21] This material contained 14.5%Sb, 0.42%Sn, and 6.2%Ag. Since it was found together with plano-convex ingots of tin bronze it must be dated to the EBA or later. There are many other cases of artefacts containing relatively high antimony. Since antimony hardens copper to the same extent as arsenic it would be just as desirable, and because it is not so volatile a greater proportion would be retained on working.

The Spread of Metallurgy

In the central area surrounding Anatolia, the diffusion of ideas amongst the developing cultures of that region was fairly rapid. In Iran, a number of sites give early dates for copper artefacts. Amongst these is Sialk, where the first phase (Sialk I, *c.* 4500 BC) has yielded several pure wrought copper artefacts.[22] The first was merely a pinhead which consisted of a cast and wrought copper containing cuprous oxide; the second was an arrowhead which had been annealed or hot worked. At Tepe Yahya, south of Kerman, a level dated to 3800–3500 BC was found to contain large arsenical–copper implements such as chisels and awls.[7] There is no doubt that these represent smelted copper.

At Tal i Iblis in Iran, under a furnace giving a ^{14}C date of 3792±60 BC was a dump 60 cm thick and 100 m long consisting of fragments of small copper-stained crucibles and dross. These were dated to 4091±74 BC at the bottom and 4083±72 at the top. C.S. Smith[3] comments:

'The curious shape of the crucible fragments, the copious residues of charcoal, the sparsity of the clay and the lack of any furnace debris indicate that the smelting practice was quite different from later standard methods.'

This would seem to be one of the few examples of crucible smelting known, and its early date fits the technological sequence. The crucibles were shallow and boat-shaped and it appears that they were filled with high grade copper oxide ore and charcoal, placed under charcoal and heated to the melting point of copper with the aid of bellows.

In a fourth millennium level at Tepe Giyan near Nehavend, two copper artefacts containing 1.0 and 1.35%Ni, respectively, were found as well as two tin bronzes.[23] At Yanik Tepe in North West Iran, a flat tanged dagger blade was found containing 98%Cu and a 'fortuitous trace of tin'. This was dated to *c.* 3500 BC.[24] The absence of smelting furnaces and slag suggests that some of the copper implements had been made from native copper melted in a crucible .

Naturally, the more advanced civilizations, such as Egypt, would be amongst the first exploiters of the new metallurgical techniques. The earliest Egyptian copper finds are dated to the period 5000–4000 BC, but these consist entirely of awls and pins and could have been made from native copper. By the middle predynastic times (*c.* 4000 BC) axes were being made of impure smelted copper. An early predynastic axe contained 1.28%Ni+Co, 0.49%As, and 0.17%Pb, and these elements have not been found in native copper in such proportions.[25] By the First Dynasty (*c.* 3000 BC) arsenical coppers with a higher arsenic content were making their appearance. Bronzes began to appear during the Fourth Dynasty (2600 BC).

A metallographic examination of a First Dynasty flat axe containing 1.5%As showed it to have been cast worked, and annealed at 700°C, or hot worked and lightly cold worked to give a hardness of 80–90 HB in the centre, and 92–112 HB on the edge.[26] The earlier predynastic flat axe weighing 1.56 kg, with a high nickel and lower arsenic content, had much the same structure with a hardness of 63–73 HB in the centre, and 85 HB at the edge.

Amongst the earliest copper finds in Mesopotamia are some from Tepe Gawra, north east of Mosul.[27] These comprise a blade, an awl, a ring from level XVII and a button from level XII, both dated to the Ubaid period (4000–3500 BC) but not analysed. Al Ubaid is a mound not far from Ur, and the metals analysed from the early levels were found to be slightly impure coppers free of arsenic.[28] The Ubaid culture at Ur (3500–3200 BC) has yielded axeheads of 8.1 and 11.1%Sn, undoubtedly the earliest tin bronzes yet known.[29]

In Northern Syria,[30] copper tools were found at Amuq in a level dated to c. 3500 BC. These were essentially arsenical coppers with a nickel content of 0.4–2.05%. One object contained 1.52%Sn and another was a straight tin bronze without nickel or arsenic. A crucible fragment from a level dated to 3100–2800 BC was found to contain a deposit which gave 0.5%As, 1%Ni, and 5%Sn. At Brak, another Syrian site, copper pins and sheet were found, of similar date and composition, but without tin.[31] A basalt mould for casting chisels was found in a level dated to 2300–2100 BC.

At Abu Matar in the Beersheba valley of Palestine have been found a number of Chalcolithic sites dated to the Beersheba–Ghassulian Culture (c. 3500 BC). Copper oxide mineral, which had been roasted and partially reduced was found together with crucibles and parts of a furnace.[32] But the usual fayalite type of smelting slag was not present and it would seem that some sort of melting operation had been carried on here, in crucibles in a cylindrical furnace.[33]

Some of the earliest Chalcolithic objects from this region are cultic and date from about 3500 BC. A hoard, found in a cave at Natal Mishmar Engedi,[34] on the west shore of the Dead Sea, comprised 240 maceheads, 80 staffheads and 20 chisels or narrow axes. It was found that the cultic objects such as the maceheads had high As levels, while the tools had much lower As levels to make them less brittle and more serviceable. The iron content of these was high, which suggests that smelting was carried out with high iron slags, unlike the slags from Abu Matar. It is clear that there was a marked development during the later Chalcolithic towards proper fluxing with iron oxides producing fayalite slags, and a parallel improvement in casting technology towards hollow, thin-walled, cored objects. Some of the maceheads could have been silvery due to their high As content. Another hoard, from Kfar Monash, consisted mainly of tools and weapons contemporary with the First Egyptian Dynasty (3200–2750 BC). Many of these show As in the range 1.15–4.07%, a useful range for tools.[35]

The furnace found at Timna Site 39, near Eilat, seems to fit the Chalcolithic period with its high iron fayalite slags containing copper prills[37] (see Table 7). But the ores

in the Wadi Arabah, i.e. those from Timna and Fenan, do not carry high levels of As, and the objects found associated with the 11th Century BC Egyptian mining temple at Timna demonstrate this fact,[38] although they do have the high iron contents expected from fayalite slags. Therefore it is almost certain that the high As levels found in some of the objects of the Mishmar and Monash hoards were made elsewhere or by the addition of high As minerals brought from other regions.

The early development in Egypt did not at once spread further west, and we must return to Anatolia and the Troad and follow the spread of metallurgy along the Black Sea coast to the Danube, and across the Aegean to Northern Greece. In Troy itself, metal appears in the remains of the first city (Troy I, 3000 BC) in the form of arsenical copper, but the badly corroded condition gives only a qualitative significance to the analyses. By Troy II (2500–2200 BC), it is quite clear that we are dealing with wrought Cu–1%As alloys with about 0.015%Ni, probably from an Anatolian source.[39] The finding of open and closed moulds of steatite and clay, together with crucibles, confirms the existence of local manufacture but not local smelting.

In South East Europe we have evidence for two early copper mining sites, that at Rudna Glava in Yugoslavia and Ai Bunar in Bulgaria. The two are of roughly the same date — the period 4500–4000 BC, the Late Vinca culture.[40] The mine at Rudna Glava was worked by firesetting and then breaking the face with the stone tools. The ore was oxidic and would have needed iron ore for fluxing. This was proved experimentally, but we still have no evidence for smelting on the site.[41]

Ai Bunar belongs to the Karanova IV period giving it a date of 4700 BC. Mining of oxide ores was done with antler and metal tools. Again there was no sign of smelting on the mining site; but in this case slag was found in nearby settlement sites.[42]

The early dates of these two sites clearly show that the exploitation of copper ores began in Europe much earlier than has been previously thought. Whether this is evidence of independent progress from native metal is still in doubt.

The dating of large copper artefacts, such as shaft-hole axes, is far earlier than at Troy which, as we see, is

Table 7 Composition of copper smelting slags from Site 39, Timna, Israel (after Lupu[37])

Composi-tion, %	Number 09	72a	58C	64
FeO	22·38	34·91	30·37	43·12
MnO	–	0·80	29·33	0·16
SiO$_2$	44·78	36·77	30·11	16·26
CaO	11·42	12·1	12·17	20·80
MgO	2·65	1·42	0·98	0·23
Al$_2$O$_3$	3·93	0·98	–	0·37
Cu	11·1	2·64	0·73	8·36
H$_2$O	ND	4·04	ND	ND
Na$_2$O	ND	1·36	ND	ND

ND = not determined
— = not detected

relatively late. But the artefacts themselves are of pure copper[43] and, although large and cast with cored shaft holes, could have been made from native copper. So it is possible that we have in this region an early native copper tradition which originated independently but did not develop into a true copper age using smelting techniques. We know that pure copper artefacts were being buried in Hungarian graves of the Tiszapolgar culture before 3000 BC.[44] Unfortunately, there is no evidence of smelting in the area and it is possible that this was native copper, worked locally.

Bulgaria, however, has yielded stone moulds for flat axes and copper tools dated to 3000 BC.[45] From Moldavia comes the Karbuna hoard of 852 objects, 444 of which were of copper, of the same date.[46] Most of these were beads and there were only two axes, so it would appear that metallurgy, based on a cast native copper tradition, was widespread in South-East Europe by this time.

About this period, diffusion was taking place, across the Aegean from Anatolia, and we find moulds and crucibles, dated between 3000 and 2600 BC, on the island of Lesbos.[47] The eleven artefacts analysed were of impure copper and four of them had tin contents of between 0.16 and 1.65%. Most of the finds in the Greek Peninsula and in Crete are not earlier than 2500 BC (Early Minoan and Early Helladic II — both Early Bronze Age cultures) and this seems to be also true of those in Cyprus.

The presence of small amounts of tin in some Chalcolithic artefacts raises the question of the supply of tin-containing minerals in the Near East. In 1987 it was announced that stannite-containing minerals had been found in the Bolkerdag region of the Taurus mountains in southern Turkey.[48] The mean tin content was 0.2% which is not low for a tin ore. Later, in 1989, another find was made in the same mountain range at Kestel, this time of the more common mineral cassiterite.[49]

Furthermore, the mine yielded archaeological material which was dated to the fourth–third millennium BC (late Chalcolithic—EBA). In the alluvial deposits below the mine the cassiterite concentration was 0.25%. If in sufficient quantity, this level would certainly be workable today.

There were probably many small sources of tin like these in the world, most of which were worked out and are now unknown. This matter will be dealt with more fully in the next chapter.

The Caucasus seems to have been a great barrier to the spread of ideas from Iran. The inhabitants of the Trans-caucasian sites in the Lower Volga and the Ukraine did not use copper until about 2200 BC and most of the artefacts were highly arsenical.[50] Caucasian sites, dated to about 2000 BC, have yielded spearheads, knives, pins, and adzes with arsenic contents between 0.5 and 10.0%. Only five out of 54 objects from these sites contained tin, and the analyses gave 1.82, 2.00, 10.08, 11.54, and 12.6%Sn, respectively.[50] One of the daggers had silver rivets.

Returning to South East Europe, we have seen that Late Neolithic cultures, with some use of copper for ornaments, were giving way to Chalcolithic cultures, where copper was being used for tools. Even so, in

Eastern Slovakia at about 3000 BC, the economic importance of copper was negligible. At Tibava, however, every grave in a cemetery contained a copper adze-axe, some 18 cm long, which was made of ore from Central Slovakian deposits.[51] These graves belong to the Tiszapolgar culture, which we have already met in Hungary, and it seems that the metallurgical traditions of this culture were beginning to spread westwards.

While many of the axes were high-purity copper and could therefore be made from native copper, some of these were arsenical. These must have been made from smelted copper.

The Tiszapolgár culture was superseded by the Baden–Bodrogkeresztur culture with quadrangular awls and flat axes. At Vuçedol on the Drave, a melting furnace and a flat-axe mould were found.[52] The beginning of the Baden culture can be dated to just before 2000 BC.

In Italy, the earliest copper-using cultures were the Remedello, from near Bologna, and the Rinaldone from Tuscany, which appeared between 3000 and 2500 BC. Analyses of artefacts [53,54] from these cultures show sufficient arsenic and antimony to prove that they used smelted copper (*see* Table 8).

In Iberia, the El Argar culture of the south east was a transitional Copper–Early Bronze Age culture, and many of the coppers have been analysed for arsenic (*see* Table 9). Arsenic was found in both the mineral (1.86%As_2O_5) and the slag (0.25%As_2O_5) and there is no doubt that smelting was done at this time.[57] The earlier sites of the Los Millares culture and that at Vila Nova de S. Pedro, have also produced arsenical copper.[58] These could be dated to about 3000–2500 BC and may be independent of the mainstream of east–west diffusion.

Table 8 Copper objects from the Rinaldone and Remedello cultures of Italy (after Cambi, [53,54] and Otto and Witter[94])

Object	Composition, %				
	As	Sb	Ni	Ag	Pb
Rinaldone					
1 Axe	0·0	0·0	tr.	0·23	0·0
2 Dagger	0·19	0·42	tr.	0·08	0·33
3 Needle	0	0·21	tr.	0·28	tr.
4 Axe	tr.	0·30	tr.	0·21	0·0
5 Dagger	0·11	0·29	0·07	0·08	0·0
6 Dagger	0·17	0·40	tr.	0·29	0·24
7 Dagger	1·5	0·0	0·052	0·008	0·0
8 Axe	0·05	0·0	0·022	0·010	0·0
11 Dagger	0·18	0·16	0·05	0·17	0·29
12 Piece	0·30	0·43	0·05	0·25	0·0
Remedello					
Dagger	7·80	0·0	tr.	0·20	0·0
Flat axe	tr.	0·0	tr.	0·10	tr.
Flat axe	tr.	0·3	0·02	0·10	0·3
Flat axe	tr.	tr.	tr.	0·05	0·8
Axe	0·0	0·6	tr.	0·05	0·09
Axe	0·0	0·5	tr.	0·05	0·07
Flat axe	0·15	0·0	tr.	0·16	0·0
Flat axe	0·40	0·0	tr.	0·13	0·0
Flat axe	0·17	tr.	tr.	0·05	1·2
Dagger	8·0	tr.	tr.	0·20	0·05
Dagger	4·6	0·6	tr.	0·12	0·06

Table 9 Early Iberian copper: El Argar and earlier pre-Beaker cultures

Provenance-object	Composition, %					Reference
	As	Sb	Ni	Ag	Pb	
Los Millares						
Flat axe	1·1	tr.	0·03	0·14	0·0	58
Vila Nova de S. Pedro						
Dagger	2·1	tr.	–	0·08	–	56
Estramaduro						
Dagger	0·75	tr.	0·0	0·02	0·0	55
Point	2·6	0·0	0·0	0·02	0·0	
Point	3·7	tr.	tr.	0·01	0·0	
Point	2·8	0·02	0·0	0·01	0·0	
Point	tr.	0·0	0·0	<0·01	0·0	
Dagger	6·4	0·02	0·0	0·01	0·0	
Pin	3·5	0·0	0·0	0·01	0·0	
Chisel	2·4	0·0	<0·01	0·01	0·0	
Chisel	0·9	0·01	0·0	0·0	0·0	
El Argar						
Flat axe	0·64	0·0	0·014	0·54	0·023	57
Halberd	3·7	0·0	<0·01	<0·01	0·018	
Dagger	1·65	0·0	tr.	~0·01	0·0	
Chisel	1·6	0·0	0·0	<0·01	0·0	
Flat axe	1·8	0·0	0·0	0·014	0·0	
Halberd	2·8	0·0	<0·01	0·39	2·3	
Dagger	2·2	0·0	0·0	<0·01	0·0	
Halberd	3·5	tr.	0·0	<0·01	0·0	

It is in Iberia that we seem to have more evidence for crucible smelting. At El Ventorro, near Madrid,[59] the remains of three pots were found which contained copper dross up to 4 mm thick and which were made of a fabric differing from the domestic pottery. Although decorated, it seems certain that these were used as crucibles for smelting. These Beaker Culture crucibles were about 10 cm dia. and 5 cm deep, and are quite unlike those from Tal i Iblis. Another site is Almizeraque, a Chalcolithic site in Almeria. Rich oxide ores were smelted and the products contained 2–10 % As and low iron. The normal high iron slags were non-existent and there were no tuyeres. The temperatures inside the crucibles were higher than those outside, but this is a normal feature of early crucibles used for melting. It is suggested that polymetallic minerals with and without As were reacted in the crucibles.[60] Pazoukhin[14] has shown how this may be done, but is is doubtful whether it could be done without a forced draught.

The spread across Europe to the north must have been fast, since copper artefacts were found in Beaker graves in Holland which have ^{14}C dates of 1940–625 BC.[61] Admittedly, it is not certain whether the influences came from west or east. As far as the British Isles are concerned, they appear to have an Iberian–Atlantic stream and a Continental stream meeting at about 1900 BC. There is some evidence that the Atlantic route from Iberia via Brittany to Ireland established a metal industry in Ireland before the Continental influence had finally taken root in England.

Metallurgy came rather late to Scandinavia. The pieces of copper sheet found in Jutland (Konens Hfj), and dated to 2900 BC, must be an import or made from native copper.[62] Metals were not produced in this region until about 1500 BC.

Melting of native copper was taking place at Pegrevna in Soviet Karelia in the third millennium BC.[63] There is some doubt about the actual sources of the native copper in this region as Fennoscandia was covered by an ice sheet which removed the oxidized layers of the copper deposits which are normally the seat of native coppers. A ring of very high purity copper was found in the Finnish site of Suovaara which was dated by its associated pottery to 3350–2800 BC.[64] The fact that this site is adjacent to the largest copper mine in Finland at Outokumpu suggests a local source (native copper can be formed by secondary reactions).

Russia and the Far East

Copper-base alloys appeared in China between 3000 and 2500 BC, probably nearer the latter date. The diffusion route was either a Northern or Central Asiatic one, starting from the Anatolian–Iranian area and either traversing the Caucasus, or going through Iran, up the Amu-Darya, over the Tien-Shan to Kashgar. Some consider the more northerly route the more likely, and there is some Russian evidence to substantiate this theory.[65] Two sites favour the central route; Tepe Yahya, already mentioned, and Anau (now Anau-Namazga) just north of the Iran–USSR border in Turkmenistan. Anau I gives a Copper Age date of between 5500 BC and 4500 BC.[66,67] The finds consisted of awls, needles, and knives and could have been made from native copper from Iranian sources. Anau II (about 5300–4000 BC) produced flat copper spearheads, a mould, and axes similar to those from Sialk III of a later date.

The northern route may have started from the Kuban culture of the Northern Caucasus, which flourished at the end of the third millennium. From these sites some copper ingots, slag, and moulds with arsenical copper implements have been dated to about 2000 BC. In the Lower Volga, the Timber grave culture was flourishing at this time and in the Kalinovka barrow north of Volgagrad (2000–1800 BC) we find a founder's grave containing stone moulds, tuyeres, and crucibles (Fig.8).[68] The asymmetrical shaft-hole axe is a type that has turned up in East Germany, where it was found to be made of a Cu–1%As alloy. Russian sites are remarkable for their assemblages of metalworking material, much of it belonging to the Bronze Age. In North West Siberia, west of the Urals, a Turbino culture site dated to 1500 BC produced a tuyere, copper drops, moulds, and crucibles.

In the area of the upper Yenesei and the Altai mountains a Chalcolithic culture (Afanasevskaja) existed at about 2000 BC. These people were of European rather than Mongoloid type, and it would appear that there had been a gradual penetration along the southern steppe zone from very early times.

Pre-Shang China (before 1600 BC) shows evidence for the use of Cu, Cu–Sn and Cu–Zn alloys, possibly going back to 3000 BC.[69] Brass is a bit of a surprise in the period

2400–2000 BC in Shangdong, but experiments using the available ores show that it is possible.[70]

India

The two great prehistoric civilizations of the Indian Subcontinent, both in Northern India in the Indus valley, are at Harappa and Mohenjo-Daro. Because of their comparatively early dating (3000–1500 BC) it is probable that the knowlege of metalworking came through Southern Iran or along the Persian Gulf, rather than over the difficult route through Northern Baluchistan or Afghanistan.[71] Some of the Baluchi sites give early dates[72] but most of the metallurgical evidence is from later times.[73,74] In spite of the early appearance of copper at Tepe Yahya, the Copper Age starts about 3000 BC with the Amri-Nal culture of South Baluchistan and Sind, and many of the coppers contain nickel rather than arsenic. Two hoards, containing flat axes, chisels, saws, and a knife, were found in the Nal cemetery. A fragment of a flat axe found nearby contained 4.80%Ni and 2.14%Pb. All were dated to the first half of the third millennium BC.

The Mohenjo-Daro-Harappa civilization was rather conservative in its forms, and flat axes were used in preference to shaft-hole types, showing that connections with Mesopotamia or Iran, if present in the early phases, were not maintained.[75] In Harappa itself, three alloys were in use during the period 2500–2000 BC, i.e. (Cu–Ni alloys with 1.27%Ni, Cu–As alloys with 4.42%As, and bronzes with 11%Sn and low arsenic. Of 64 of the specimens examined from Mohenjo-Daro, 20 contained nickel in the range 0.3–1.49%, and only nine were bronzes: coppers and bronzes were present at all levels.[76] Lothal is the only Harappan site which has produced crucibles and moulds.[77]

The borders of Afghanistan, Iran and Pakistan meet in the delta of the river Helmand. In this area are numerous copper smelting sites with slag heaps. On the Iranian side we have Shahr-i-Sokhta, and on the border of Afghanistan and Pakistan we have the Shela Rud area. Here the heaps vary from those representing a few seasons' activity to areas covering several square kilometres. [14]C dates in these areas vary from 3950 BC to AD 800. The most likely ore source would be the Chagai mountains in Pakistan: the slags are very high in copper and the mines are associated with various types of stone tool.[78]

In the Ganges basin, metal hoards were still substantially of copper in the period 1500–300 BC. The number of implements found was about 500–600 from 20 sites, and were mainly flat axes and harpoons of primitive type.[79] In North East Thailand, copper-base objects, including socketed axes, crucibles, and moulds, have been found[80] in a context which is roughly in agreement with the Harappa dating (3000–2300 BC).

South and Central America

In South America the first metal to appear in archaeological sites was gold. Fragments of hammered sheet were found in Early Formative sites (500–1 BC) at Supe and Viru in Peru: more elaborately welded objects were found at sites to the north of the Lamboyeque valley.[81] On the Bolivian (west) side of Lake Titicaca, at Tiahuanaco, excavations in late levels (pre-Columbian AD) have yielded gold and silver ornaments. In this period, Cu–Au alloys have been found in Peru at Salinar and on the central coast, and pure copper at Cavernas and Chiripa. The Cu–Au alloys must be intentional and they have a wide range of composition. They could have been made by melting gold and native copper; by dissolving out the copper a gold-rich surface would be obtained.[82]

The Mochica culture on the north coast (Classic period, *c.* AD 500) used as weapons unhafted flat copper axes with a hole in the butt, by which they were attached to the belt with a thong.[81] Copper-tipped lances and copper-shod digging stocks also appear. At Tiahuanaco, the enormous building stones are held in place with 13 cm long copper cramps, similar to the iron cramps used in classical Greece and Iran more than a thousand years earlier.[83,84]

The copper smelting site at Batan Grande in Peru has recently been excavated by Shimada and his colleagues[85] and the metallurgical aspects reported by Merkel.[86] A row of smelting furnaces was found, which seem to have been blown by blowpipes rather than bellows. It is generally accepted that the human being is capable of blowing intermittently at a rate of 40 l/min which should be enough for two blowers to maintain a well-insulated furnace at a temperature of 1 200°C. Three blowers could probably have maintained a continuous supply of air.

Bellows seem to be rare in this part of the world and blowpipes have had a traditional role in metallurgical operations[87] (*see* Fig. 4). The ores smelted seem to have had a high As content and this may have been obtained by blending. The product was a mixture of slag and prills of copper with 1–20% As. These were separated from the slag by crushing and remelted to give small plano-convex ingots. The dating is to the Sican period— AD 900 to the Conquest.

Since Bolivia and Argentina both have tin deposits, bronze appeared comparatively early in the metal

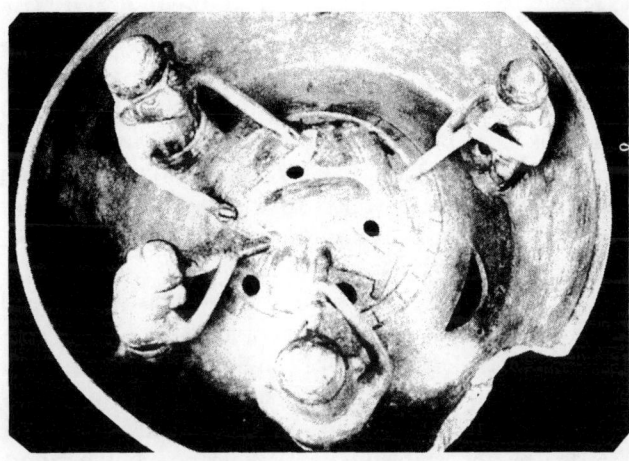

4 Decorative bowl from Batan Grande in Peru, depicting the use of blowpipes in the copper smelting process

Table 10 Composition of South American antimonial and arsenical coppers

(a) Mainly from the West Coast	Composition, %			Reference
	Cu	Sb	As	
1 Socketed spearhead (Titicaca)	97·4	–	2·14	88
2 Axe (Lima Museum)	95·2	–	4·43	83
3 Hoe (Pacasmayo)	98·4	–	1·55	"
4 Hoe (Trujillo)	95·95	–	4·03	"
5 Blade of hoe	95·6	0·08	4·27	"
6 Sheet (Andes)	90·75	–	5·30	89
7 Hoe (Peru Coast)	98·2	0·7	–	83

(b) From Argentina (from Caley[90])

Object	As	Sn	Sb	Pb	Zn
Armband	3·81	1·57	2·94	nil	0·30
Headband	3·40	2·05	0·42	nil	1·22
Axehead	3·37	nil	tr.	1·30	0·78
Axehead	2·65	nil	nil	0·12	0·20
Awl	2·12	nil	nil	1·02	0·37
Axehead	6·88	0·91	0·32	nil	nil
Axehead	2·09	nil	nil	nil	nil
Tweezers	1·75	nil	nil	nil	nil
Wedge	1·30	1·02	0·38	nil	nil
Breastplate	1·17	1·56	0·47	nil	nil

sequence in South America. This aspect is dealt with in the next chapter. Analysis of early artefacts has shown many to be of pure copper, some of arsenical copper, and some antimonial. Even in Inca times (after AD 1200) copper and arsenical coppers continued to be used after the introduction of high-tin bronzes, closely paralleling the situation in certain parts of the Old World.

The structure and mechanical properties are what we expect from such materials. Items 5 and 7 in Table 10 were examined metallurgically:[83] the hardness of the 4.27%As copper hoe had been increased by cold working to a maximum of 128 HB near the edge. Elsewhere, the hardness ranged from 72–107: it could be reduced to 62 HB by annealing. The antimonial hoe (no.7) had a maximum hardness of 106 and a minimum of 73 HB. Annealing reduced the hardness to 50 HB. It is worth noting that the maximum hardness of bronze implements tested and coming from the same area was 150 HB for a 13.4%Sn bronze. A rattle from Supe, on the Peruvian coast, was found to be 100%Cu and to have been made from two hemispheres of sheet, joined by hammer welding near the melting point of the metal in a charcoal fire to provide a protective atmosphere.

Metallurgy probably spread to Mexico about AD 1500 via the copper-using Chimú people on the North Peruvian coast.[91] By the time of the Spanish conquest, the Aztec civilization had not reached a Bronze Age, probably because tin is not available in Mexico and regular trade was not established with their Inca contemporaries to the south.

The Aztec civilization began in Mexico about AD 1000. It is said that copper helmets were worn in one of the previous cultures dated to AD 100–500, but this has not been confirmed archaeologically. Cold-hammered copper was used for axes, needles, and ornaments from Middle Culture times (AD 300), but it is possible that these were made from native copper. By Aztec times, the crescent-shaped tumi knives were in use here also and, like gold dust, are believed to have been used as media of exchange. Three small bells from Yucatan[92] were found to be pure copper while one from Mexico contained 19 3%Pb.

While bronze was unknown, a considerable amount of gold and silver was available and the alloying of copper and gold was practised as in Peru. Gold was collected in nugget form and panned as dust, and calabashes of gold dust are shown in Aztec art as a means of exchange.

References

1   R. F. TYLECOTE: 'Early copper slags and copper-base metal from the Agadez region of Niger', JHMS, 1982, 16, (2), 58–64

2   R. F. TYLECOTE: 'Can copper be smelted in a crucible' JHMS, 1974, 8 (1), 54

3   J. R. CALDWELL: 'Tal i Iblis', 1967, Iran, 5, 144–146

4   J. MELLEART: 'Çatal Hüyük', 1967, London, Thames and Hudson

5   H. NEUNIGER, R. PITTIONI and W. SIEGEL: 'Fruhkeramikzeitliche Kupfergewinnung in Anatolia', Arch. Aust., 1964, 35, 98–110

6   A. HAUPTMAN: 'The earliest periods of copper metallurgy in Feinan, Jordan', Der Anschnitt, 1989, Beiheft 7, Old World Archaeometallurgy, p119–136

7   C. C. LAMBERG-KARLOVSKY: 'Excavations at Tepe Yahya, Iran, 1967–69', Progress rep. I, Bull.27, Amer. School of Prehist. Res. Harvard, 1970

8   P. T. CRADDOCK: 'Evidence for Bronze Age metallurgy in Britain', Curr. Arch., 1986, 9 (4), 106–109

9   MASUDA TSUNA and KUDO ZURUKU: (Ed. C. S. Smith), Burndy Lib Conn. 1983

10   U. ZWICKER and F. GOUDARZLOO: 'Investigation on the distribution of metallic elements in copper, etc.', Archaeophysika, 1978, 10, 360–375

11   J. D. MUHLY: 'Timna and King Solomon', Bibliotheca Orientalis, 1984, 3/4, 275–292

12   B. ROTHENBERG: Bull. Mus. Haaretz, 1966, (8), 86

13   I. R. SELIMKHANOV: Soviet Arch., 1962, (1), 67

14   V. A. PAZOUKHIN: 'Uber den Ursprung des alten Arsenkupfers', Metallurgija i gornoedelo, I, Moskva, 1964

15   W. LORENZEN: 'Helgoland und das fruheste kupfer des Nordens', Ottendorf, Niederelbe, 1965, Abs. Bull. HMG, 1967, 1, (7), 13

16   J. R. MARECHAL: Métaux-Corrosion-Ind., 1958, (397), 377

17   G. BASS: 'Splendors of the Bronze Age', Nat. Geogr. Mag., 1987, Dec. 172, (6), 731

18   U. ZWICKER: 'Investigations on the extractive metallurgy of Cu–Sb–As ores', In: Aspects of early metallurgy, (Ed. A. Oddy), BM Occ. Pap. 17, 1977

19   J. R. MARECHAL: Prehistoric Metallurgy, Lammersdorf, 1962, 22

20   O. DAVIES: Man, 1935, 35, (91), 86

21   S. STORTI: *Sibrium*, 1960, **5**, 208

22   R. GHIRSHMANN: 'Fouilles de Sialk',1938, Vol.2, 205, Geuthner, Paris,

23   G. CONTENAU and R. GHIRSHMANN: 'Fouilles de Tepe Giyan', 1935, 135, Geuthner, Paris,

24   C. A. BURNEY: *Iraq*, 1961, **23**, 138

25   H. C. H. CARPENTER: *Nature*, 1932, **130**, 625

26   H. C. H. CARPENTER: *ibid.*, 1931, **127**, 589

27   A. J. TOBLER: 'Excavations at Tepe Gawra II',1950, Philadelphia, University of Pennsylvania, University Mus. Monograph

28   C. H. DESCH: *Brit. Assn.*, 1936, 1–3

29   C. F. ELAM: *J. Inst. Metals*, 1932, **48**, 97

30   R. J. BRAIDWOOD et al: *J. Chem. Educ.*, 1951, **28**, 88

31   M. E. L. MALLOWAN: *Iraq*, 1947, **9**, 1

32   J. PERROT: *ibid.*, 1955, **5**, 17–40; 73–84; 167–189

33   R. F. TYLECOTE et al.: *JHMS*, 1974, **8**, (1), 32

34   C. A. KEY: *Science*, 1964, Dec. 18, **146**, 1578–1580

35   C. A. KEY: *IEJ*, 1963, **13**, 289–290

36   S. SHALEV and J. P. NORTHOVER: 'Chalcolithic metal and metalworking from Shiqmim', In: *Shiqmim*, (Ed. T. E. Levy) BAR Int. Ser. No. 356, 1987, 357–371, 689

37   A LUPU, *Bull. HMG*, 1970, **4**, 21

38   B. ROTHENBERG: 'The Egyptian Mining Temple at Timna', *Inst. Archaeomet. Studies*, London, 1989

39   *Bull. HMG*, 1966, 1 (7)

40   B. JOVANOVIC: 'The technology of primary copper mining in SE Europe', *PPS*, 1979, **45**, 103–110

41   B. JOVANOVIC: 'Rudna Glava', Bor-Beograd, 1982

42   E. N. CERNYCH: 'Gornoe delo i metallurgiya v. drevneyshey Bulgari',Sofia, 1978

43   E. VOCE and V. G. CHILDE: *Man*, 1951, **51**, (234), 139

44   IDA BOGNAR-KUTSIAN: 'The Copper Age cemetery of Tiszapolgar-Basatanya', 1963, Budapest, *Akad. Sci. Hung.*

45   G. I. GEORGIEV and N. J. MERPERT: *Antiquity*, 1966,**40**, 33

46   T. P. SERGEEV: *Soviet Archaeology*, 1962, **1**, 135

47   W. LAMB: 'Excavations at Thermi, Lesbos', 1936, Cambridge, BSA 1928–30, **30**,1

48   K. A. YENER and H. OZBAL: 'Tin in the Turkish Taurus mountains; the Bolkerdag mining district', *Antiquity*, 1987, **61** (232), 220–226

49   K. A. YENER et al: 'Kestel; an Early Bronze Age source of tin ore in the Taurus Mountains, Turkey', *Science*, 1989, April 14, **244**, No. 4901

50   I. R. SELIMKHANOV: *PPS*, 1962, **28**, 68

51   E. and J. NEUSTUPNY: 'Czechoslovakia', 62,1961, London, Thames and Hudson

52   R. R. SCHMIDT: 'Die Burg Vuçedol', 1945, Zagreb

53   L. CAMBI: *Studi Etruschi*, 1959, **27**, 200

54   L. CAMBI: *ibid.*, 1959, **27**, 415

55   S. JUNGHANS et al.: 'Kupfer und Bronze in der fruhen Metallzeit Europas', (SAM 2), 1968, Berlin, Verlag Gebr. Mann

56   S. JUNGHANS et al.: 'Metallanalysen Kupferzeitlicher und fruhbronzezeitlicher Bodenfunde aus Europa', (SAM 1),1960, Berlin, Verlag Gebr. Mann

57   H. and L. SIRET: 'Les Premiers Ages du Metal dans le Sud-est de l'Espagne, 1887, Anvers

58   A. DE PACO: *Zephyrus*, 1955, **6**, 27

59   R. HARRISON et al.: 'Beaker metallurgy in Spain', *Antiquity*, 1975, **49** (196), 273–278

60   S. ROVIRA: In paper presented to the conference on 'The Discovery of Metals at St Germain en Laye', Jan. 1989

61   J. J. BUTLER and J. D. VAN DER WAALS: *Paleohistoria*, 1966, **12**, 51

62   B. STURUP: *Antiquity*, 1967, **41**, 315

63   A. P. JOURAVLEV: 'The earliest workshop for the manufacture of copper in Karelia', *Soviet Arch.*, 1974, (3), 242 –246

64   J. P. TAAVITSAINEN: 'A copper ring from Suovaara in N. Karelia', *Fennoscandia antiqua*. I, 1982, 41–49

65   E. H. MINNS: *Proc. Brit. Acad*, 1942, **28**, 47

66   V. MASSON: *Atti CISPP*, 1962, **2**, 205

67   J. MELLAART: *CAH*, 1967, No. 59

68   M. GIMBUTAS: 'Bronze Age cultures in Central and Eastern Europe', 1965, London, Mouton

69   R. F. TYLECOTE: 'Ancient metallurgy in China',*Metall. Mat. Tech.*, 1983, Sept., 435–439

70   SUN SHUYUN and HAN RUBIN: 'A preliminary study of early Chinese copper and bronze artefacts', Kaogu Yuebao, 1981, (3), 287–302

71   S. RATNAGAR: 'Encounters: The Westerly Trade of the Harappa Civilisation', OUP, Delhi, 1981

72   B. DE CARDI: *Antiquity*, 1959, **33**, 15

73   W. A. FAIRSERVIS: 'Archaeological studies in the Siestan basin of SW Afghanistan and East Iran', Anthrop. Papers of Am. Mus. Nat. Hist., 1961, Vol.48

74   J. M. CASAL: 'Fouilles de Mundagik', 1961, 2 Vols, Paris, Klincksieck

75   S. PIGGOTT: 'Prehistoric India',1950, London, Penguin

76   C. H. DESCH: *Brit. Assn*, 1929, 264;1931, 269

77   R. F. TYLECOTE: 'Early metallurgy in India', *Metallurgist and Mat. Tech.*, 1984, July, 343–350

78   W. TROUSDALE: Personal communication

79   S. PIGGOTT: *Antiquity*, 1944, **18**, 173

80   W. G. SOLHEIM: *Current Anthrop.*, 1968, **9**, 59

81   G. H. S. BUSHNELL: 'Peru', 1956, London, Thames and Hudson

82   W. C. ROOT: *J. Chem. Educ.*, 1951, **28**, 76

83   E. v. NORDENSKIOLD: 'The Copper and Bronze Ages in South America', 1921, Goteborg

84   H. LECHTMAN: 'Traditions and styles in Central Andean metalworking. In The Beginnings of the use of metals and alloys', (Ed. R. Maddin), 2nd Int. Conf. Zhengzhou, China, 1986, London, 1988, 344–378

85   I. SHIMADA, S. EPSTEIN and A. K. CRAIG: 'The metallurgical process in ancient N. Peru', *Archaeology*, 1983, Sept/Oct., 38–45

86   J. MERKEL and I. SHIMADA: 'Arsenical copper smelting at Batan Grande, Peru', *IAMS Bull.*, 12, 1988, 4–7

87   C. B. DONNAN: 'A pre-Columbian smelter from N. Peru', *Archaeology*, 1973, Oct. 26 (4), 289–297

88   M. LOEB and S. R. MOREY: *J. Amer. Chem. Soc.*,1910,**32**, 652

89   W. WITTER: *Metall und Erz.*, 1936, **33**, (5),118

90   E. R. CALEY: 'Analysis of ancient metals',1964, Oxford, Pergamon

91   G. C. VAILLANT: 'The Aztecs of Mexico', 149,1950, London, Penguin

92   A. J. FISKE: *J. Amer. Chem. Soc.*, 1911, **33**, 1115

93   T. E. LEVY and S. SHALEV: 'Prehistoric metalworking in the southern Levant', *World Arch.*, 1989, **20** (3), 352–372

94   H. OTTO and W. WITTER: 'Handbuch de altesten vorgeschichtlichen metallurgie in Mitteleuropa', 44, 1952, Leipzig, Johann Ambrosius Barth

95   R. F. TYLECOTE and H. McKERRELL: *Bull. HMG*,1971, **5**, 37

96   C. H. DESCH: *Brit. Assn.*, 1932, 437

97   C. H. DESCH: *Brit. Assn.*, 1935, 340

98   R. HESTIN and H. TADMOR: *Israel Excav. J.*, 1963, **13**, 265

Chapter 3
The Early Bronze Age

The fact that tin conferred on cast copper objects considerable additional strength in the as-cast state without the necessity of cold working was, without doubt, a great discovery. But the idea probably developed very slowly, and in the Near East we have a period when arsenic and small amounts of tin were used together.

In some countries it is possible to subdivide the Bronze Age into Early, Middle, and Late on the basis of the metal typology, but this is an exception rather than the rule. In the British Isles, the metals used in the early period were arsenical coppers and straight tin bronzes without arsenic or lead. In the Middle Bronze Age, the alloy used was a straight tin bronze often exceeding 10%Sn. In the Late Bronze Age the alloy normally contained 10%Sn, but lead was introduced into castings. The introduction of lead was by no means universal, and in Britain seems to have been limited to the south-east. In other countries lead is often found in bronzes of any period. It is therefore better to divide the period into two: the earlier, experimental age, and the later, full Bronze Age.

Extractive Techniques

While the deposits of arsenical and, to a lesser extent, antimonial copper ores are comparatively common, as shown by their use in early copper objects, tin is not often found together with copper. In fact, tin deposits are something of a rarity. The tin deposits of the world are associated with certain types of granite and occur in only a few well known localities such as Malaya, China, Bolivia, Cornwall, Saxony-Bohemia, and Nigeria. In all these places tin is found mainly as the mineral cassiterite (SnO_2) which is white in the pure state, but is more often contaminated with greater or lesser amounts of iron which render it brown or black. In certain deposits small amounts of the sulphide, stannite, are found but this is comparatively rare. Today, the greater proportion of all tin comes, as undoubtedly in early times, from alluvial or mined deposits of cassiterite. This oxide is relatively stable and has a high specific gravity (7.0) so that it has collected like gold in the beds of streams or in gravels and sands. For this reason it has probably been recovered for as long as gold, although originally it would have been discarded as worthless.

It is noteworthy that at least two of the well-known deposits of tinstone, i.e. in Cornwall and Saxony, contain copper ores as well, and it would be possible for copper ores from these regions to be accidentally con-

taminated with tin. A prehistoric copper slag from Ranis, Saxony, contained 0.62% of metallic copper in the form of globules. This was perfectly normal for a copper smelting slag.[1] While the overall tin content of the slag was only 0.05%, it was found that the tin content of the copper globules was 1.2%. This showed that most of the tin in the slag was present in the copper globules, and that the smelted copper would have contained 1.2%Sn. Copper ores from Cornwall, smelted in South Wales in the 19th century AD, were also found to give copper containing 0.7%Sn.[2] One of the ores smelted contained 0.94%Sn and 12.3%Cu, giving a ratio of copper to tin of 93:7, and was therefore capable of producing a 7%Sn bronze.

Many early copper-base alloys from widely-separated areas of Eurasia contain small amounts of tin, often together with arsenic (see Table 11). There is little doubt that most of the tin content is the result of smelting copper ore contaminated with tin minerals, although later some of this contamination could have been caused by the addition of tin bronze as scrap.[3]

While it is possible that deposits containing small amounts of tin are responsible for the low-tin copper-base alloys shown in Table 11, it is unlikely that the real tin bronzes originated from this type of deposit. It is far more likely that the tin needed came from well known deposits in Italy, Bohemia, Saxony, Malaya or even Nigeria. It must be remembered that high-tin bronzes did not enter the archaeological arena until quite late (after 3000 BC at Sumer[4]), and that many of the earliest metals hailed as bronzes in the 19th century have now been found to be arsenical coppers. For this reason it is highly probable that the earliest date for the use of the standard 7–10%Sn bronze in the Near East coincides with the ability of these civilizations to trade over considerable distances.

Until we come to know more about trade to and from the civilizations of Asia Minor in the second millennium, this sort of possibility cannot be discounted. Even Malaya might be included in the range of possibilities. Trade was certainly coming through the straits of Hormuz and up the Persian Gulf by 2500 BC to supply the needs of the cities of Mesopotamia, and tin bronzes with 10% of tin did not appear much before this.[5]

Unfortunately, we have few instances in the Near East of tin as a metal in its own right or as its oxide cassiterite. There is the tin bracelet from Thermi, Lesbos,[6,7] dated to about 3000 BC.

Table 11 Examples of low-tin bronzes with and without arsenic

Area	Site	Object	Date, BC	Composition, % Sn	As	Reference
Iran	Tepe Hissar I	Pin	3900–2900	1·74	–	49
Iran	Geoy Tepe	Dagger	2000	~0·5	~0·3	52
Iran	Tepe Yahya	Dagger	3000	3·0	1·1	110
Iraq	Ur	Blade	2800–2500	2·40	–	4
Iraq	Ur	Pin	2800–2500	1·0	–	4
Iraq	Kish	Fragment	2800	2·52	–	106
India	Mohenjo-Daro	–	2100–1700	1·2	–	82
Aegean	Lesbos	–	2800	1·65	–	7
Iraq	Tel Asmar	Dagger	2500	2·63	0·15	107
Turkey	Troy II	–	2500–2000	2·18	0·97	111
Turkey	Troy III	–	2500–2000	2·90	1·50	111
Egypt	Tutankhamun's tomb	–	2000–1800	1·80	tr	3
Spain	Ifre	Pendant		1·07	–	40
Argentina	–	Armband	c. AD 800	1·57	3·81	95
Peru	–	Knife	c. AD 800	2·49	–	97
England	Idmiston	Dagger	1700	1·54	2·9	17
Brittany	Ploudaniel	Flat axe	1800	0·09	4·07	81

– = not sought

It is possible to make tin bronze in two ways. The first is by mixing cassiterite with copper and heating to the melting point of bronze under charcoal, when the cassiterite is reduced to tin and absorbed by the copper. The sulphide, stannite, can also be used in this way.[8] The second method is to add metallic tin in the right proportions to molten copper. This supposes that tin has been smelted, and evidence for this is sparse. In the UK, a Cornish Bronze Age site at Caerloggas has produced tin slag in a ritual context and there is no doubt that tin was smelted in the UK in the Middle Bronze Age (1600 BC).[9] On the other hand, in Plymouth harbour in the adjacent county of Devon, two small ingots of copper containing small amounts of tin have been found, in one of which there are residual cassiterite crystals suggestive of the use of the first process.[10]

We also have another piece of evidence for trade in tin or cassiterite from the wreck found off the coast of Cap Gelidonia in southern Turkey, dated to the Late Bronze Age (1200 BC). About 16 kg of white material containing 14%SnO_2 and 71% of $CaCO_3$ were recovered from the sea bed. It is believed that this represents corroded material from a tin ingot having a cross-section of about 6 cm square.[11,12] Also included in the cargo were copper and bronze ingots and scrap metal. The ship had been travelling in a westerly direction and it has been suggested that it was a Syrian ship taking copper from Cyprus to the Mycenaean civilizations in Crete or Greece. There is no doubt that the tin was not originally obtained in Cyprus and must have been traded from more distant places and picked up at one of the ports of call.

The second wreck found off this coast, that from Ulu Burun near Kas, has also produced plano-convex (bun-shaped) and oxhide ingots of tin.[13]

Recently, small tin deposits have been located in Sardinia,[14] Egypt, and in the Taurus mountains of Turkey. These, like most tin deposits, are of comparatively low grade. In one, the main tin mineral is stannite

(SnCuFeS) associated with Pb and Zn, giving tin concentrations up to 3400 ppm with a mean of about 0.2%. Slag collected from the area showed more than 500 ppm Sn. Copper sources containing 3.93% Cu were located within 700 m of the tin.[15] Even so, concentrating the stannite to give a bronze containing 8% Sn would be quite difficult. Such processes were, of course, well known in early times and we now know that it is easier to make tin bronzes by the addition of pure stannite rather than the more common cassiterite.[8] The second deposit in the same area contained the more normal mineral, cassiterite.[16]

The low-tin alloys can be found in nearly all early civilizations of the world, as shown in Table 11. In the case of those found in England, Spain, Brittany and Peru there is not much doubt as to their origin, but in the Near East their source is more in doubt and their existence does tend to confirm the view that there have been sources of copper ores yielding metal with 1–3%Sn in Turkey, Iran or Iraq.

Where the ores already contain arsenic and tin contamination has taken place, we find the so-called arsenical bronzes with 1–2%Sn and 1–4%As. The effect of tin and arsenic on the mechanical properties is more or less additive, and it is found that cold-worked alloys of this type are a good deal stronger than pure or slightly impure coppers. The addition of 1%Sn in solid solution would confer about the same increase in hardness upon workings as 1%As. A 2.4%Sn alloy blade from the Royal Graves at Ur (2800 BC) was found to consist of wrought and annealed material. This provided evidence of the continuation of the arsenical copper tradition of forging rather than casting.[4]

The next stage was the addition of tin, either as oxide or metal, to arsenical copper. Eight British Bush Barrow (Wessex) daggers had an average content of 2%As and 7.5%Sn,[17] which would have given a strength approximating to that of a 9.5%Sn bronze. In order to retain the arsenic, care would have to be taken to avoid oxidation and it is probable that the ingot material had a slightly higher arsenic content. Usually, much of the arsenic and some tin is in the slag inclusions and is, therefore, not available for strengthening.

In studying development from the Early Bronze Age through the Middle Bronze Age one sees a steady reduction in the arsenic content of copper-base artefacts in those parts of the world that originally used deposits of arsenical ores. This may have been due to a change in smelting technique, whereby the metal is held longer under oxidizing conditions, or to changes in the nature of the ore body with depth.

During the early phases of the Early Bronze Age (if this is the correct term for an era using alloys containing only 1–4%Sn) arsenical coppers often recovered their previous position. This suggests that the tin trade was subject to frequent interruptions and could not be relied upon. It is only with the steady trade connections of the Mesopotamian cities that we begin to find consistency in the tin content, and the establishment of the standard 10%Sn bronze.

When it was possible to make tin additions of this order it was no longer necessary to smelt in such a way as to retain the arsenic in the ore. No doubt, in many cases, this stage coincided with decreasing amounts of arsenic in the ore as the primary sulphide deposit was reached at lower levels. In these cases, the arsenic alloys would have been made by the addition of arsenical minerals.

The expansion of the area in which bronzes were replacing arsenical coppers occurred rapidly, probably because tin deposits were more common outside the Near East than within. Suppliers of tin ores might have wondered what these were being used for, and the techniques for the addition of tin to copper would soon have spread.

Extractive Techniques

The evidence for such techniques lies in the furnaces and slag heaps. Unfortunately, it is often extremely difficult to date the slag heaps and sometimes even the furnaces. As we have already said, copper smelting slag has been of fairly uniform composition throughout metallurgical history, i.e. fayalite $(2FeO.SiO_2)$, with greater or lesser amounts of alumina and lime. The copper content is usually low, i.e. 2–4%, and often much less. One would be lucky to find a stratified heap containing pottery, but most metalworkers seem to have been poor people and only in civilizations where pottery is extremely plentiful are we likely to find sherds in the heap. Since most slag has entrapped pieces of charcoal, a ^{14}C date will often be obtainable.

While it has not been easy to find evidence for very early smelting in some countries, the mines and their waste heaps have often given very early ^{14}C dates, as we see in Wales and Ireland[18] (see Table 12).

With the furnaces themselves we may be more lucky, since some of them are in settlements or attached to workshops within palaces. This, however, only happens

in places where there were good communications and organization since it is generally easier to smelt in the mountains rather than carry the fuel and the ore to the settlements.

Melting Furnaces and Crucibles

For the early periods we have no satisfactory evidence of copper ingot shape. In the Late Bronze Age, oxhide and plano-convex shaped ingots were used in the Aegean, but only small plano-convex ingots were used elsewhere. In the Roman period we find plano-convex ingots weighing as much as 20 kg.

If arsenical coppers were the direct product of smelting and not of alloying one should expect to find ingots of such material. Although plano-convex ingots of arsenical copper are known they are a comparative rarity and are not necessarily restricted to early periods. It is thought by archaeologists that the ingot torcs or 'oesenhalsringe', described as 'cast neckrings with re-coiled ends',[19] represent the earliest ingot material known. These bars or rings, shown in Fig.5, are admittedly for the most part arsenical[20,21] but it does not seem reasonable to expect an early smelter to have formed his product into a long narrow bar (28 cm long) in the bottom of his smelting furnace, nor to have tapped it into such a long narrow mould. Although these are often referred to as 'cast', the one from Leiten in Austria (now

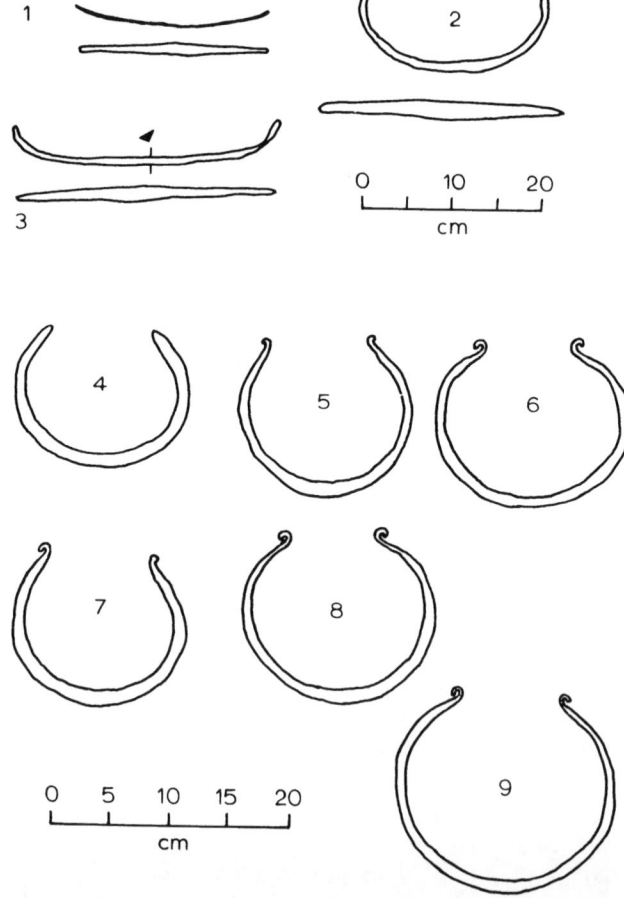

5 Neckrings and semi-finished bars of arsenical copper (after Otto and Witter[1])

at Linz) has been wrought,[1] and one would expect them to have been worked from more compact pieces of ingot. The shape has to be compared with the iron 'currency bars' of later date which were made as intermediate products suitable for swords and general smithing. However, since copper is not so easily welded by smithing, the long narrow bars and neckrings would not have been of much use for making axes or halberds but only for the pins and the wire coils, used as armrings, which the hoards often contain. We seem to have no evidence for an ingot form that could have been used as a starting point both for the neckrings and for the production of more massive objects. Perhaps the Chalcolithic and Copper Age smelting furnaces never produced massive 'bun ingots' but merely scraps of metal amongst the slag, which was broken up to release these pieces of metal. These might then have been melted in crucibles and cast into a short bar ingot which could then be forged into a long bar or neckring. These bars could be easily broken up by the melter, put into a crucible, and cast into more massive objects such as axeheads and halberds. Some of the early stone moulds show bar-like cavities which would have served this purpose.

Crucible Furnaces

For the purpose of melting scraps of metal in crucibles, a ring of stones, a pile of hot charcoal, and a clay tuyere connected to bellows is all that would have been required. Nothing would have remained of this arrangement except perhaps for the clay tuyere. It is possible to obtain sufficiently high temperatures for melting bronze (950°C) with a cylindrical furnace with a grate rather like a brazier. The air would have entered underneath the furnace, assisted by the chimney effect of the cylindrical portion (*see* Fig.6). Usually there would have been a side hole through which the crucible could be placed in the hottest part of the fire a short distance above the grate. The grate would of course be of clay, pierced with holes rather like a pottery kiln. The earliest known crucible furnace is perhaps that from Abu Matar, the Late Chalcolithic site near Beersheba (3300–3000 BC), where the remains were found of circular furnaces 30–40 cm diameter with 3 cm thick vertical walls at least 12–15 cm high.[22,24] The inside surfaces had become well vitrified owing to the presence of wood ash and metal. The crucibles were oval with rounded bottoms 11 × 8 cm outside and 7 cm deep internally. They were made of grey clay which had been mixed with chopped straw. No moulds or tuyeres were found, but a copper flat axe, awls, and maceheads were recovered from this and nearby sites. Apart from this site we have no further evidence of crucible furnaces until we come to the finds in the Timna area of Israel of the Late Bronze Age–Early Iron Age transition period.[25,26] The crucible furnaces at Timna consisted of square stone boxes into which four stone slabs had been set into the ground on edge with a kerb of small stones on top (*see* Fig.7). This kerb may have been a later heightening of the furnace due to build-up of material around it. The box was full of wood ash and, unless assisted by a blast from a tuyere, little

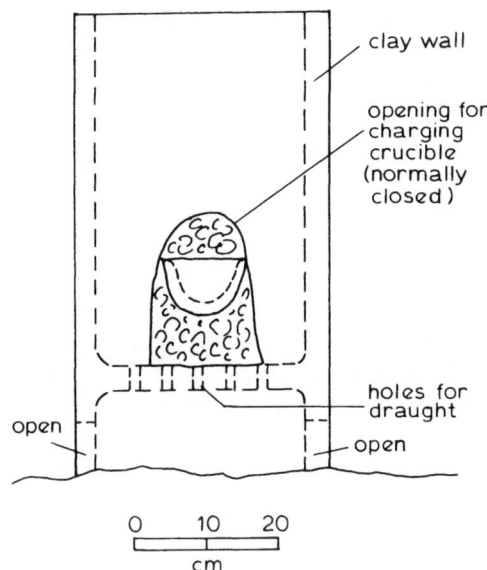

6 *Reconstruction of chalcolithic crucible furnace from remains found at Abu Matar, Israel (after Perrot [22])*

draught would have reached the crucible in the furnace as found. One must assume that the fire was cleaned out frequently when in use and that there was some opening for air into the bottom of the furnace, as in the Roman period forge furnaces. Similar furnaces have been found in Soviet Karelia dating to the Chalcolothic–EBA period.[27]

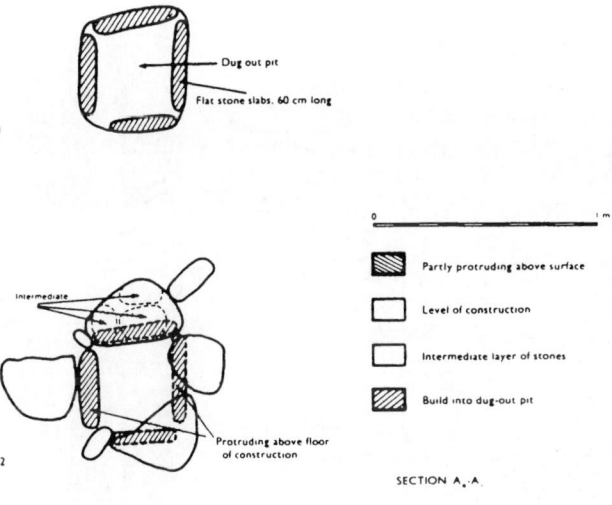

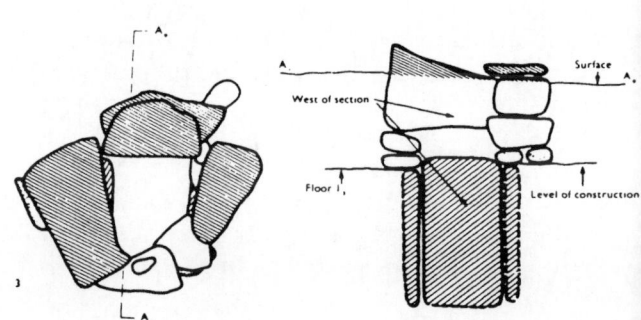

7 *Square stone crucible furnace from Timna (after Rothenberg and Lupu[113])*

Smithing and Founding

The tuyere, being made of hard-burnt clay, is another artefact that has had a good survival rate. The simplest of these is a conical piece of clay with a cylindrical hole. More complicated ones have a hole which is large at one end to take the bellows nozzle, and smaller at the end which enters the furnace: this end is often vitrified and slagged.

Copper and Bronze Age tuyeres seem to be no different from those of any other period, and perhaps the best examples are those found in the Early Timber Grave Culture burials at Kalinovka[28] which are dated to about 2000–1800 BC. These are crucible hearth tuyeres and show the features just described in a very clear manner. It is interesting that they belonged to a bronze founder and were buried with him in his grave, showing that the smith had an important social position. Figure 8 shows the whole assemblage of moulds, tuyeres, and crucibles found in barrow no.8 at Kalinovka, north of Volgagrad. This is one of the most interesting metallurgical deposits yet found and it will be used to illustrate many points in the following pages. First, we shall consider the tuyeres. The end which takes the bellows nozzle is about 20 mm dia. but the hearth end seems to be extremely narrow (5.0 mm) compared with the 20 mm dia. tuyeres from smelting sites such as those at Timna[26] and in Sinai.[29] Most of the tuyeres from the barrow have some decoration, which one can also see on tuyeres from Timna and other sites, and this feature may aid bonding into the furnace linings. Here, it is probably incidental and just something that craftsmen cannot avoid. Similar narrow-ended tuyeres were found in another barrow, but these were not decorated.

Some tuyeres are 'D'-shaped in section and have a flat side. This is presumably so that they could be laid on the ground, as shown in Fig.9. The crucibles used in such a smithing hearth would have been heated mostly by radiant heat from the fire above, and for this reason would have been large and shallow.[30] But they would have needed to be poured very quickly when removed from the fire as loss of heat would have been rapid.

Such furnaces would have to have been very well insulated so that heat was not conducted through into the ground. The Japanese 18–19th century hole-in-the-ground type show how important this was.[31]

Many sites have produced elbow-shaped tuyeres. These can be used in many ways but one such makes possible the blowing of a melting furnace by a smith, working bag bellows with both hands. This technique was convincingly demonstrated by Ph. Andrieux at Beaune in the Spring of 1988, when a crucible of the type shown in Fig. 13H was poured into a mould using a charred stick inserted in the socket.

Crucibles may be made from clay or stone. The earliest is probably from Abu Matar, and the remains indicate a hemispherical oval crucible about 11 × 8 cm across and 7 cm deep, with a wall thickness of 1 cm and a definite lip for pouring.[22]

In the second millennium at Meser in Palestine,[32] and at Thermi,[7] Lerna,[33] and Sesklo[34] in Greece, we find crucibles of the oval type with a handle at the end or side. This type was devised as one way of overcoming the problem of pouring in a Chalcolithic or Bronze Age context. Since the melting point of a pair of copper tongs, such as those from Enkomi in Late Bronze Age Cyprus,[35] is at most 1 084°C, and copper must be poured with at least 20°C superheat, i.e. 1 100°C, there is quite a problem in pouring a crucible containing copper, although this problem is minimized with bronzes. It is probable that the tongs were tipped with clay. The crucibles from Meser[32] (3000 BC), Thermi[6,7] (3000–2500 BC), and Lerna V[35] (2750 BC) (*see* Fig. 10), all show one of the ways of overcoming this problem. This was the moulding of a boss, at the end or side of the crucible, which contained a hole about 1 cm dia. into which a clay-covered rod or stick

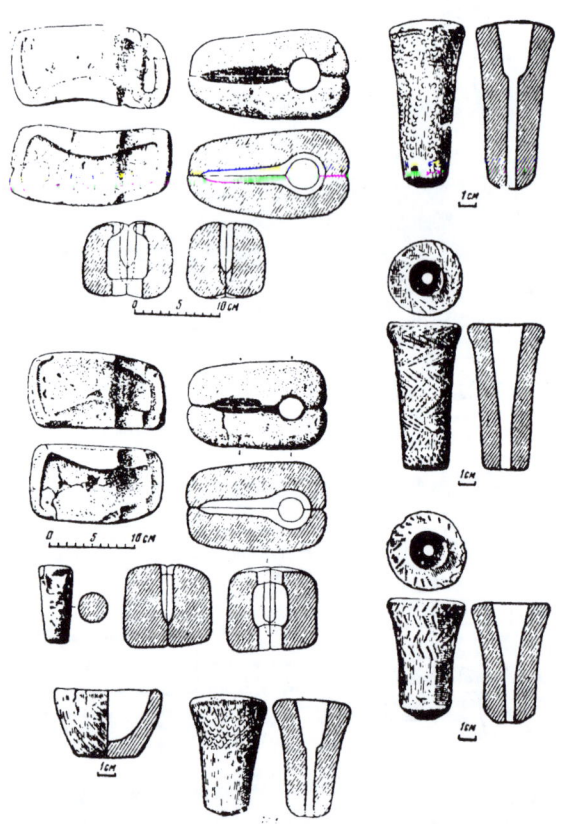

8 *Assemblage of stone moulds, clay tuyeres and a crucible from a bronze-founder's grave at Kalinovka north of Volgagrad USSR*

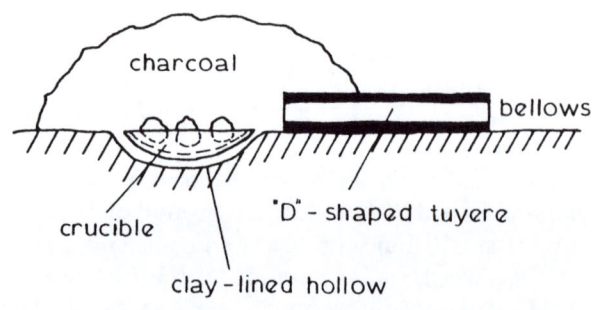

9 *Use of crucible and tuyere from Ambelikou, Cyprus (EBA)*

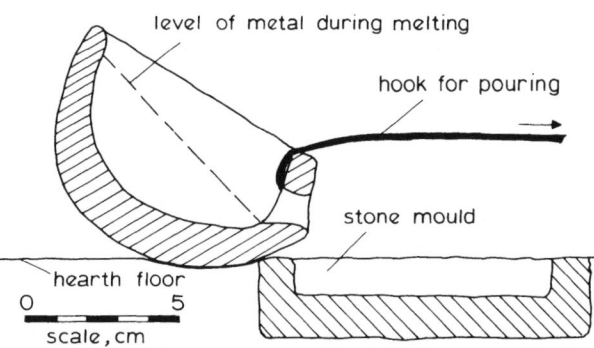

11 Method of using LBA crucible as found at Keos, Greece

10 Examples of crucible with a hole for a handle from Lerna, Greece (MBA) (courtesy of the American School of Classical Studies, Athens)

could be inserted so as to rotate or even raise the crucible for pouring.

Another way of overcoming this problem is that shown by the crucibles from Kea (Keos)[36] and Serabit[37] (1550–1200 BC), which have a hole at the pouring end and rounded bottoms so that they can be rocked. The crucible from Serabit could have held about 870 cc or 7.6 kg of bronze, which could have been poured by rotating the crucible through 40°: the mould would have been embedded in sand in front of the crucible and exposed by pushing away the charcoal from in front, as shown in Fig. 11. The crucible could have been rotated either by pushing with a rod or stick or by pulling on the front bar with a hook. Similar crucibles with side pouring holes have been found at Tell edh Dhiba'i and are shown in the Old Kingdom tomb of Mereruka. These must have had a stopper of some sort, which could be removed allowing direct pouring into moulds without tipping, as Davey has shown.[38]

Some of the Egyptian tomb paintings[39] show the pouring of a thick open-type crucible (such as those from El Argar[40]) by means of withies. The sap in the withies would, for a short time, have prevented loss of suppleness through charring, but one would not have expected withies to have lasted long enough to fill the whole row of about 18 moulds shown in Fig.12. Perhaps there is some artistic licence involved here. But it is possible that it only shows one of several separate casts.

Occasionally, one finds crucibles with feet or a pedestal. One of the earliest examples is from Troy III (2200–2050 BC), and if one had not found them on other sites (such as in Early Christian Ireland) one would doubt their use as crucibles.[41] The Trojan crucible had four legs and contained a deposit of gold and copper carbonate (Fig. 13). Similar crucibles have been found in the Bronze Age site at Dainton in Devon.[42]

The crucible from Balaubash, in Russia,[28] (1500–1300 BC), had a base and it is doubtful whether this would have been considered a crucible unless it had had a deposit. It is often easy to mistake a pottery sherd for a piece of crucible fabric, and the only way to be sure about a crucible is to look for intense vitrification caused by wood ash, together with slag contamination. Sometimes, ordinary pottery sherds can fall into the ash layers of hearths and be vitrified in this way, but it is unlikely that they will contain slag or metallic residues as well.

12 Egyptian crucible held with withies while pouring into moulds; metal is melted in a crucible in a hearth with a forced draught from four foot bellows; withies serve to lift the crucible to the mould in the centre which has a series of runners; one of the men on the right carries an oxhide ingot; from a tomb at Thebes c. 1500 BC (from P.E. Newberry[39])

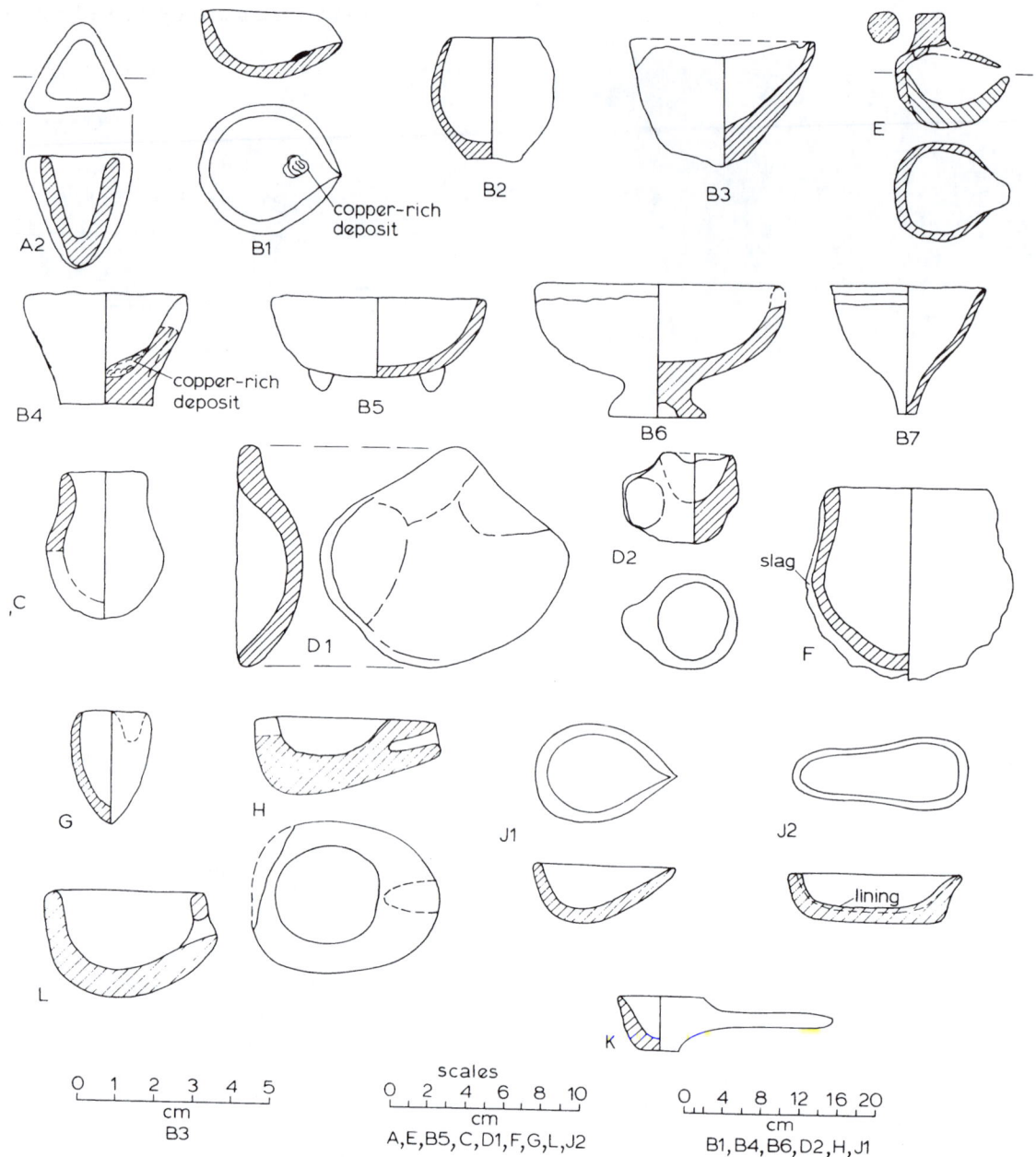

13 *Early crucible types*

A2   Triangular crucible from Mikulcice, Moravia; 8th
      century AD
B1   Hemispherical crucible from Tel Zeror, Israel;
      LBA–EIA
B2   Flat bottomed round crucible from Godmanchester,
      UK; Roman
B3   Conical round crucible from St. Albans (Verulamium),
      UK; Roman
B4   Flat bottomed flared crucible from Tel Quasile, Israel
      (11th century BC); LBA–EIA
B5   Hemispherical crucible with feet from Troy III
      (2100 BC) (dimensions very approximate)
B6   Circular crucible with pedestal base from Balaubash,
      USSR; LBA
B7   Chinese crucible with pedestal base (reverse curva-
      ture); Shang Period (no dimensions given)
C   Necked crucible from Huntsham, UK; Roman
D1   Crucible with horizontal pinch from Kalinovka, USSR;
      LBA

D2   Crucible with vertical pinch from Cullykhan, Scotland;
      LBA
E   Lidded crucible from Dinas Powys, Wales; 7th cen-
      tury AD
F   Bag-shaped or globular crucible from Godmanchester,
      UK; Roman
G   Pointed bottom (interned rim), Corbridge, UK; Roman
H   Socketed crucible from Meser, Israel (3000 BC)
J1   Oval crucible with round bottom from Troy IV
      (2000 BC)
J2   Oval crucible with flat bottom from Scandinavia; LBA
K   Spoon-shaped crucible from Balanovo, USSR; LBA
      (no dimensions given)
L   Crucible from Keos poured through a hole at one end;
      LBA

The greater proportion of early crucibles found are hemispherical or boat-shaped. Some of the circular crucibles have flat bottoms, such as those from Tel Quasile[25] and Timna in Palestine. Later ones are very often triangular or, like modern crucibles, round and slightly flared with flat bottoms (*see* Fig.13). A tentative crucible typology is given in Table 12.

Table 12 Crucible typology

A Triangular or 'D' shaped
    A1   Triangular; pointed bottom – shallow
    A2   Triangular; pointed bottom – deep
    A3   D-shaped
    A4   Triangular – flat bottom
B Circular
    B1   Hemispherical
    B2   Flat-bottom
    B3   Conical
    B4   Flat-bottom – flared
    B5   Flat-bottom – legged
    B6   Pedestal base
    B7   Pedestal base – reverse curvature
C Necked
D Pinched
    D1   Horizontal pinch
    D2   Vertical pinch
E Lidded
F Bag-shaped or globular
G Pointed (Interned rim)
H Socketed
J Oval or boat-shaped
    J1   Round-bottomed
    J2   Flat-bottomed
K Spoon-shaped
L Hole-poured

Crucible Furnaces

It is important to be able to distinguish between smelting and melting, and the composition of slags sometimes makes this possible. As we have seen, smelting slags are essentially ferrous silicates, high in iron and low in non-ferrous metals. On the other hand, crucible slags tend to be high in non-ferrous metal and wood ash and low in iron. There are essentially two types: (*a*) the internal slag, which forms as a result of the reaction between copper and iron oxides in the melt and the constituents of the crucible, mainly silica. This is therefore a complex silicate, but the slag may contain some iron, because the entrapped ferrous silicate smelting slag floats out when the ingot material is remelted and combines with the other silicates. (There may also be residual metal adhering to the walls of the crucible, and the analysis of this gives a clue to the metal being melted). And (*b*) the external slag found on the outside of the crucible caused by a reaction between the wood ash and the clay of the crucible. Sometimes this is modified by the addition of oxidized metal which has spilt down the sides during melting or pouring. Some analyses are given in Table 13 where it can be seen that these slags, though variable in composition, all contain non-ferrous metal and much less iron than the smelting slags. Table 14 shows the results from Israeli sites of remelting raw copper in an inert atmosphere. While no change occurs in the already very low lead content, which is in solution in the copper, the iron content has been markedly reduced.

Crucible melting is often carried out under charcoal and then the atmosphere will be reducing. But if it is oxidizing, a good deal of the alloying elements will enter the slag, as has clearly happened at Serabit where the melt was certainly lead-rich.

The Appearance of Bronze in the Early Bronze Age

The earliest bronzes seem to be those from Mesopotamia, i.e. the early City States which depended upon the deposits of the Anatolian and Persian highlands for copper, and probably trade for their supply of tin. When one examines the chronology of the sites in the Euphrates–Tigris delta region, one is impressed by the agreement in dating for the appearance of true tin bronzes. The first appearance of bronze in all the early sites seems to be between 3000–2500 BC: before this, the metals were pure or arsenical coppers.

The Royal graves at Ur date to the First Dynasty of about 2800 BC, and we find true tin bronzes with 8–10%Sn.[4] An axehead had a typical cored cast structure with the delta phase, and contained blow holes. On the

Table 13 Composition of crucible slags, %

	Chrysokamina (Crete)[67]	Serabit (Sinai)[37]	Abu Matar (Israel)[24]	Tel Quasile (Israel)	Tel Zeror (Israel)	Timna (Israel)
CuO	45·05	Cu 21·7	Cu 47 42	Cu 18	Cu 20	Cu 43
FeO	2·40	Fe 1·9	Fe 9 23	Fe 20	Fe 12	Fe 19
SiO$_2$ Al$_2$O$_3$ } CaO	23·8	37·9	Rest	Rest	Rest	Rest
CO$_2$, H$_2$O } O$_2$	28·75	nil	–	–	–	–
Ni	nil (Desch)	tr.	–	–	–	–
Zn	–	–	<1·0	tr.	nil	<0·1
Pb	–	38·0	Abs.	0·48	0·72	0·62
Sn	–	nil	Abs.	nil	6·4	–
As	–	0·5	–	–	–	–

– = not sought; tr. = trace elements

Table 14 Effect of melting on composition of raw copper (after Lupu and Rothenberg[26])

Specimen no.	As received, %		After remelting, %	
	Fe	Pb	Fe	Pb
1	18·0	0·03	0·07	0·03
2	6·1	0·05	3·40	0·05
3	8·0	0·05	1·34	0·21
4	14·1	0·03	0·40	0·03
5	8·0	0·10	0·09	0·03

other hand, a bolt from the same period was made of worked and annealed copper with final cold working, and objects from Geoy Tepe in Azerbaijan were also free of tin. Later, (c. 2200 BC) the graves at Ur yielded copper and low tin bronzes (with 0–2.4%Sn) which had been worked. These facts suggest that the Anatolian and Persian highlands themselves did not participate in the early use of bronze, and that even in sites like Ur the interruption in the tin supply could lead to a reversion to a bronze containing less tin.

Now let us look at the situation at Troy, to the west of the Anatolian highlands. The occupation of the site of Troy started at about 3000 BC with Troy I and the use of arsenical copper objects such as awls, needles, pins, and fragments of clay moulds. With one exception (a 10%Sn bronze in Troy II), tin bronzes did not appear until Troy III and IV (2200–1900 BC).[43] There is little sign of a reversion, which shows that trading connections were maintained until the end of the occupation of the site (1190 BC). Troy III and IV produced many crucibles and stone moulds.[41] The crucibles were, for the most part, shallow hemispherical types: one had four feet and a deposit of copper and gold. Both open and closed steatite moulds, designed for tools and weapons, were found.

Contemporary with Troy I, and representing the same culture, is Thermi in Lesbos. Excavations at Thermi, besides producing a tin armlet, supplied four bronzes out of a total of 26 analyses of copper-base objects.[6,7] The dating covers the period 2700–1200 BC, and two of the bronzes occur in the later levels. Tuyeres, crucibles and fragments of clay moulds were also found on this site. The Cyclades also yielded a few true bronzes amongst its assemblage of copper and arsenical copper objects. Bronze was used for shaft-hole axes and adze-axes and daggers.

EGYPT

It is not until the Fourth Dynasty (2600 BC) that we begin to see substantial quantities of tin appearing in copper-base objects. By the time of the Middle Kingdom (2000 BC), there are signs of a reversion and a true Bronze Age starts soon after.[3] It is likely that most of the metal used originated in Sinai where Chalcolithic and later sites have been found. The later sites show extensive exploitation of copper-base material and some of the other minerals of the region.[44] Although there are tomb scenes showing blowing by pipes as early as the Fifth Dynasty (c. 2500 BC), we cannot assume that this was a method of melting metal at this time: it would only cause a local-

ized increase in temperature, for brazing or soldering, for example. Either natural draught or bellows must have been used for melting or smelting.

No metallography seems to have been carried out on the bronze artefacts from this region. The Bronze Age artefacts examined both turned out to be arsenical coppers.[45] The first is an axehead, dated to 1800 BC, which contained 1.5%As and 0.2%Sn. The axe was cast but showed signs of hot work, or cold work followed by annealing. The annealed structure had a hardness of 57–90 HB, while a finally cold-worked region had a hardness of 112 HB. The other object was a knife of the 18th Dynasty (1600–1300 BC) with 0.81%As and 0.03%Sn. This also had a hot-worked structure.[46]

Another example of EBA copper is the statue of Pepi I (Sixth Dynasty, c. 2200 BC).[4] Copper was used here because of its malleability, since most of it is made from sheet. Soon after the Middle Kingdom, bronze was used for statues and it is possible that the lost wax or investment process was used for the smaller figurines. Tomb scenes show that closed mould casting methods were employed for doors. Half a stone mould, used for furniture embellishments, is to be seen in the Cairo Museum.

Buhen has provided evidence of Old Kingdom (Third millennium) smelting of copper ores low in iron, but these were smelted with the addition of iron ores as flux. The exact provenance of the ore, which contained up to 0.5% Au, is not known but it may have come from the Eastern Desert.[47]

IRAN

The sites of Sialk,[48] Tepe Hissar,[49] and Hasanlu[50] are all poor in bronze objects and the copper-base alloy levels seem to belong to the non-bronze phases of the archaeological EBA.

Tepe Giyan,[51] on the other hand, besides providing two examples of copper with high nickel content (1–1.35%Ni), has yielded two tin bronzes with 11.4 and 13%Sn, respectively. Geoy Tepe[52] in North-West Iran began to produce tin bronzes by about 2000 BC. A pin containing about 10%Sn was found to be an annealed solid solution with fine grain: it had probably been hot worked. Another pin and a bangle, both with about 5%Sn, were also in the soft annealed condition. On the other hand, a bronze bead showed the alpha–delta eutectoid in a cored solid solution. There was some slag and lead but the outer areas were equiaxed and twinned: it must therefore have been annealed and partly homogenized. The bead's late appearance on the site suggests trade rather than local sources.

The fact that levels at Tepe Hissar,[49] dated to as late as 2100–1800 BC, failed to produce true bronzes but only alloys containing 0.78–2.24%Sn, illustrate the uneven development of bronzes in Iran. At Susa,[53] on the Iraq border, levels as late as 1800 BC failed to produce alloys with a tin content exceeding 1.63%. It is not until the period of the 'Luristan Bronzes' (1500–700 BC) that we see the true flowering of the Bronze Age in Iran. This will be discussed in the next chapter.

THE PERSIAN GULF

One of the possible areas for the supply of copper to the Mesopotamian civilizations has now been located in Oman, and it is almost certain that it is the legendary Magan mentioned in Akkadian and Sumerian texts. This area has produced copper workings going back to the third millennium, the main period being the Umm nan Nar period in the third to second millennia,[54] when rich secondary ores containing 30% copper were smelted. The product was matte and metallic copper, and iron ores were used as flux. The complex product was crushed, and the copper prills separated from the slag and matte and remelted. It is estimated that 2000–4000 t of copper were produced in this area during this period. Some of the analyses show high levels of nickel,[55] and it is possible that this area is one of the sources of the high nickel artefacts often found in the Near East.

SYRIA AND PALESTINE

Amuq, now in Turkey, began to produce bronzes in about 3000 BC (Amuq G). A crucible of this period was found to have a deposit containing 5%Sn, 0.5%As, and 1%Ni. Traces of copper mineral suggest that metal was smelted and bronze produced on the site.[56]

Bronzes from Ras Shamra[57] and Byblos,[58] dated to about 2200 BC, contain tin varying from 2.84% to as much as 18.21%. Many Syrian types, such as the crescentic axes from Jericho and Tell el Hesi,[59] are also recorded from Palestine. A fenestrated axe from Megido, as well as swollen headed pins dated to EB III, may be imports. A metallurgical examination has been made of six artefacts of Palestinian origin, comprising five daggers and one crescentic axehead. The five daggers show the expected transition from arsenical coppers to tin bronzes during the period Intermediate Early Bronze to Middle Bronze Age (c. 2000 BC). The crescentic axe was a full tin bronze with 8.6% Sn and 1.3% Pb; the latter would assist the casting of such a thin-sectioned implement.[60]

THE AEGEAN

The suggestion that it might be from Syria that diffusion of tin bronzes in a westerly direction first took place is supported by certain affinities between Crete and Byblos. Evidence for westerly trade exists in the form of three Byblite daggers, one in Cyprus and two in Crete (2000 BC).[61] There were many common features between Ras Shamra in Syria, Cyprus, and Crete at the end of the EBA (2100–1900), but the trade was certainly not all one way.

Cyprus had much to offer in the way of copper but no tin, and it is not surprising that objects found in Chalcolithic and Middle Cypriot contexts are pure coppers with relatively low As content.[62] Many of these have the odd 0.5% Sn and it is quite possible that some of the ores came from the deposits in the Taurus mountains of Turkey.

When we study the Greek mainland we find the expected arsenical coppers for the Early Helladic period,[63] but bronzes begin to appear at Sesklo in Thessaly.[64] A socketed crucible was found on this site (*see* Fig.13). At Vardavoftsta,[65] in Macedonia, two dilute bronzes were found (2.14% and 3.37%Sn) along with a high-tin bronze

pin containing considerable antimony and nickel. It is unlikely that this site was occupied before 1000 BC.

It is clear, however, that even Late Bronze Age (LBA) sites on the mainland often produced bronzes with very low tin contents so that the figures obtained from analysis for Vardavoftsta are not unusual. Also, complex alloys containing substantial amounts of antimony and zinc are a feature of artefacts of the Greek mainland in the LBA, suggesting that, even as late, as this tin was not always easy to come by and that local sources of antimonial alloys, probably from Macedonia, were substituted.

CRETE

A good deal more is known about Crete[21,66] than about the mainland. The beginnings of metallurgy appear in Early Minoan I (2400–2200) and the main centre seems to be in the Mesara plain in the south. Most of the finds of this period are arsenical coppers, and dilute bronzes (3.14, 3.16%Sn) do not appear until the Middle Minoan I (2000 BC). By Middle Minoan III (1700 BC) Aghia Triada was yielding bronze daggers and knives containing 9.48% and 8.65%Sn.[67] The well-known Cretan Minoan double axes are mostly tin bronze (3–18.0%Sn). Moulds for these and other objects were found at Mallia[68] and on Melos.[69]

While there are some signs of deliberate additions of tin as early as Early Minoan I-II (2500–2200), it was not until Middle Minoan II (1800 BC) that bronze was being used for a wide variety of objects.[70] The standard alloy (10%Sn) was not regularly achieved until Middle Minoan III (1700 BC). By this time trade was capable of bringing both tin and copper, as we see from the two wrecks from southern Turkey.[12,13]

ITALY

The copper-using cultures of Remedello and Rinaldone were giving way to EBA cultures in about 1900–1800 BC. This is shown by the changes of style in the artefacts-cast flanged axes instead of flat axes, often with notches in the butt with which to locate a pin or rivet through the haft (*see* Fig.14). There are signs of trade over the Brenner Pass,[71] but in Tuscany we have deposits of both tin and copper which were undoubtedly used. Of the artefacts analysed from North and Central Italy 44% belong to a group which is relatively high in silver but low in arsenic and antimony.[21] They often contain an appreciable amount of the zinc known to be present in Tuscan ores.[67]

Many hoards have been found in various parts of Italy, which show the increased use of metal in the EBA. An undated bronzeworker's hoard of 35 plano-convex bronze ingots, containing 7.15%Sn and 1.59%Ag, was found in the Val Seriano.[72] Pieces of metal found with it had only 0.42%Sn and 1.4–5%Sb which suggests a very different source, probably Hungarian. Other hoards were found at Noceto near Forli and Montemerano near the Fiora River. At Ripatransome, on the east coast, a hoard of 25 unused solid-hilted daggers was found. This suggests that there was a workshop some distance away, probably in Tuscany.

Most of the Italian EBA axes and daggers contain

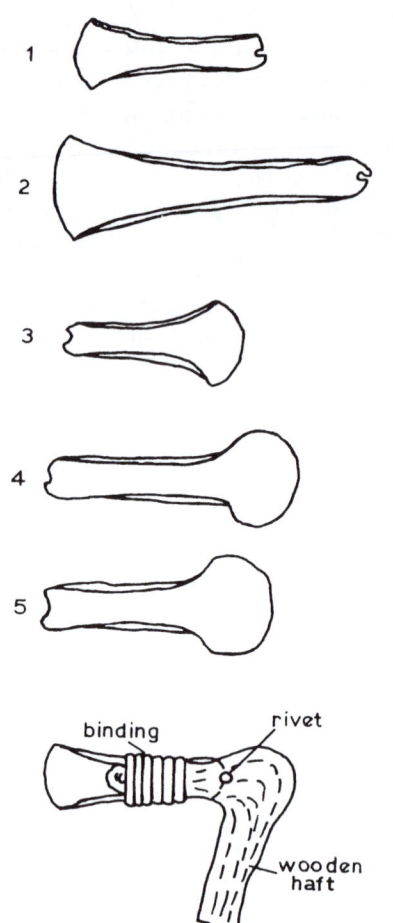

14 Some examples of early Italian flat axes with notched butts showing method of hafting

between 9 and 15%Sn, which appears to indicate that there was no scarcity of tin in this region. Ten out of the sixteen hoards come from the provinces of Siena and Grosseto within 50 km of Monte Aniata where there are copper deposits. Tin ores occur in the Colline Metallifere to the north west. Central Italy was the scene of the rich Apennine culture, and one site of this culture, at Coppa Nevigata on the Adriatic coast, has yielded a copper mould.[67]

While there are signs of contract with the Mycenaean world, which suggests an interest in Italian tin and copper, the chief characteristics of EBA and MBA Italy are those of intense local development.

Sardinia is one of the most important industrial areas of Italy and, besides having several copper and lead deposits, it also has at least one source of tin at Monte Linas, which has been proved in recent times.[73] Unfortunately, very few of the copper-base artefacts are stratified. It would seem that the Chalcolithic period produced a pure copper with little As up to early in the second millennium. After a short EBA a full Bronze Age arrived during the Nuragic period (after 1500 BC) which will be discussed in the next chapter.

THE IBERIAN PENINSULA

This is another area with well-known tin deposits which are now almost exhausted. Tin ores have been found at

Coruna in North West Spain and through the provinces of Pontevedra and Orense to North-East Portugal.[74] Small tin deposits are also known in Murcia and Almeria.[75]

There are many reports of enormous slag dumps in the Iberian Peninsula.[76,77] Most of these are clearly the result of precious metal extraction from pyritic deposits rich in arsenic, lead and antimony but low in copper, such as Minas dos Moros in Portugal.[75] However, the orebody at Rio Tinto in South West Spain carried rich copper deposits (>8%Cu) in secondary enrichment zones, and it is estimated that, in the Roman period alone, 60–70 000 t of metallic copper were produced. There is little doubt that these, and also probably the oxidized zones on the surface, were used in much earlier times.

However, in spite of these resources, no exploitation seems to have taken place until the third millennium BC. Evidence of copper working was found in several settlements where copper and arsenical copper artefacts were produced.[78] The true Bronze Age was not to start until the period 1700–1000 BC, and this is well illustrated at El Argar and other sites in South East Spain,[40] which show the transition from arsenical copper to true tin bronze just as in the British Wessex culture.

The El Argar culture started about 1700 BC and it may have developed out of colonies established by people coming from the Eastern Mediterranean.[79] El Argar is certainly rich in metal and it produced 494 bronzes out of a total copper-base alloy assemblage of 1450 artefacts.[40] The true bronze phase did not start before c. 1500 BC and the bronzes are low in arsenic and lead, as one would expect in the later stages of the EBA. The tin content ranges from 4.6% in a knife, to 13.15% in a bracelet. Moulds of stone and shallow hemispherical crucibles were found, but only a few ingots. One crucible contained an oxidized cuprous deposit which had 5.05%Sn and 0.92%Pb, and there is no doubt that metal was smelted and cast on the site (*see* Tables 15, 16 and 17 for metal, ore and slag analyses).

The principal types of artefacts produced at this time are widely-splayed axes (a mould for these was found at El Argar), wide-hilted daggers, flat knife blades or dirks with organic hilts, halberds, armrings and one long sword blade. Bronze seems to have been used only for the rather thick armrings (which were probably cast), the dirks or knives, and the one very long sword blade: the rest of the artefacts are of copper.

No metallography has yet been carried out so it is not known whether these artefacts have been merely cast or, finally, forged. Most appear to be forged and the mean tin content of the bronzes is about 8%. This is well within the hot forging, or cold working and annealing range. (This seems to represent a typical EBA culture with a rather primitive technique.)

EUROPE

In Central Europe many of the Unetice culture (1800–1500 BC) objects were made from arsenical copper, but we begin to see the beginning of true tin bronzes during this period. The armrings and the 'ingot' rings, from

Table 15 Composition of Iberian bronzes from the El Argar culture

Provenance	Object	Composition, %				Reference
		Sn	Pb	As	Pb	
Campos	Bracelet	7·53				40
Campos	Bracelet	12·39				"
Campos	Bracelet	5·0				"
Campos	Bracelet	13·15				"
C. del Agua	Flat axe	8·26	0·68			"
El Argar*	Rivets	10·00				"
El Argar	Rivets	7·84				"
El Argar	Rivets	6·54				"
El Argar	Knife–dagger	4·60				"
El Argar	Knife–dagger	5·35				"
El Argar	Sword	6·43				"
El Argar	Crucible contents	5·05	0·92			"
El Argar	Ingot	36·21	20·84			"
El Oficio	Rivet (dagger)	10·0	0·0	0·0	0·0	21
El Argar	Axe	6·6	0·85	<0·01	0·32	"
Los Millares	Armring	~10·0	0·16	tr.	tr.	"

* El Argar produced 494 'bronze' objects out of 1 450 copper-base objects
tr. = trace

Table 16 Analyses of some copper ores and concentrates

Element	Sulphide, Wieda (Harz)[1]	Sulphide, Huttenrode[1]	Sulphide, Treseberg[1]	Sulphide, Kamsdorf (Fahlerz)[1]	Sulphide, Velem St. Vid Hungary[1]	Sulphide, Mitsero, Cyprus[103]	Sulphide, Cyprus[35] Concentrate	Oxide, Parazuelos and East Spain[40]	Oxide, Gaudos[67]
Cu	26·0	34·3	30·0	32·04	17·35	2·1, 2·5	22·6	20·5	50 inc. CuO
Ag	1·86	tr.	–	0·22	–	nil	25 g/t	tr.	
Pb	tr.	0·11	–	0·43	29·90	nil	0·02	0·5	
As	0·23	–	tr.	10·19	6·04	nil	0·02	1·21	
Sb	24·11	2·14	0·08	15·05	16·64	tr.	0·02	0·5	
Zn	4·16	0·02	–	3·84	–	0·46	1·5	–	FeO ⎫
Fe	9·32	19·06	–	4·85	1·4	36·51	33	27·8	5·1 ⎬
S	34·64	38·22	–	28·34	28·60	6·29	40	Very little	0·137 ⎭
Sn	–	–	0·08	–	–	–	–	0·1	
Bi	–	–	0·002	1·83	–	–	tr.	–	
Co				2·95	–	–	tr.	–	
Ni								0·3	
SiO$_2$								14·8	27·0
Al$_2$O$_3$									
CaO									

Element	Sulphide, Bocheggiono, Italy[67]	Sulphide, Tepe Hissar[49]	Oxide, Egypt, E Desert Abu Seyal[53]	Oxide, Timna, Israel[26]	Oxide, Burgas, Bulgaria[112]	Sulphide conc. Austrian[108]	Oxide, Egypt, E. Sinai[3]
Cu	11·2	76·1	5·48	41·6	9·0	28·35	3·1
Ag	0·81	–			nil	0·025	–
Pb	0·03	tr.			24·0	0·05	–
As	0·06	–			nil	–	–
Sb	tr.	–			nil	0·27	–
Zn	0·33	–			nil	0·08	nil
Fe	27·01	6·26	Mostly oxide	0·3	8·4	32·04	25·8
S	30·45	17·2	Very little				tr.
Sn	–	tr.			nil	–	–
Bi	0·24				nil	–	–
Co					nil	0·05	–
Ni			0·33		nil	0·7	nil
SiO$_2$				4·92	–	present	
Al$_2$O$_3$				0·33	–		
CaO				1·61			

Table 17 Copper smelting slag analyses, %

Element	Felsberg,[1] Saxony	Kitzbühel,[109] LBA Thick	Kitzbühel,[109] LBA Thin	Apliki, Cyprus[104] LBA	Parazuelos, Spain[40]	Timna II, Israel[26] LBA– EIA	Volo Kastro, Greece[112]	Skouriotissa, Cyprus, LBA[104] I	Skouriotissa, Cyprus, LBA[104] II	Skouriotissa, Cyprus, LBA[104] III	Enkomi, Cyprus[104] LBA	Ras Shamra, Syria[104] LBA
Fe2O3 / FeO	66·6	56·67	44·51	65·6	56·73	43·3	53·5	64·9	69·8	69·4	64·7	65·78
SiO2	16·9	30·25	31·64	21·3	19·7	40·2	7·74	19·42	15·7	24·2	15·6	18·5
CaO	ND	1·85	7·40	1·14	4·06	9·3	0·9	1·22	0·70	–	1·3	2·4
Al2O3	2·65	–	–	7·8	0·34	2·2	–	3·83	1·73	0·28	1·7	5·1
MgO	ND	–	–	tr.	0·54	0·5	2·72	tr.	tr.	–	–	tr.
BaO	3·28	–	–	–	–	–	–	–	–	–	–	–
Cu	2·74	1·08	0·24	0·91	12·1	0·61	2·06	0·7	–	3·6	5·8	1·9
Sn	0·15	–	–	–	0·05	–	0·06	–	–	–	–	–
Pb	tr.	–	–	–	1·8	–	tr.	–	–	–	–	–
Zn	tr.	–	–	–	–	–	0·77	–	–	–	–	–
S		0·54	0·06			0·10	–		1·83	–	3·37	0·26
Sb							0·48	–	–	–	–	–
As							1·52	–	–	0·17	–	–
Bi							0·19	–	–	–	–	–
Ag							2·73	–	–	–	–	–
Ni							0·19	–	–	tr.	tr.	–
P2O5							–		0·12	0·12	0·40	0·27
CaSO4							–		–	2·43	–	–

ND = not detected; – = not sought; tr. = trace

which the former are probably derived, are often true tin bronzes containing 8–10%Sn. In order to work such material a good knowledge of the heat treatment of bronzes would be needed. The flanged axes were undoubtedly cast since the flange depth is so great that it would have been difficult to forge this out of bronze.

In this period one sees the beginning of hoards. These may be of three types: (a) founders' hoards, containing scrap metal and obsolete or worn artefacts hidden in times of trouble; (b) votive hoards and offerings at cult sites; and (c) true 'treasure trove' or hoards of precious metals, also hidden in times of stress. The contents of one group of hoards, from Dieskau[1] (Halle) included axes, halberds, armrings, flanged axes, and neck and spiral rings. Forty percent of the artefacts analysed are of bronze, but most of these are the heavy rings. These hoards are probably votive, and seem to be representative of the objects in fashion at this time. The main source of tin for Southern and South Eastern Europe would have been the Ore Mountains of Saxony-Bohemia.

The emergence of the true EBA in Britain is shown in the Wessex culture (1600–1400 BC), where we have the juxtaposition of two well-defined metal traditions, that of arsenical copper with strong Irish affinities, and a bronze one with Continental associations.

In Britain,[80] evidence of adequate supplies of tin is shown by the high tin content (17%) of the EBA bronzes. The presence of objects in Wessex graves suggests that there were trade contacts over a wide area, and one wonders whether tin was bartered in exchange.

In Western France, and particularly in Brittany,[81] the picture is very much the same. The granites of Brittany carry tin, as do those of the Massif Central. Although the latter is now worked out, at St. Renan in Brittany 500t of concentrate containing 74%SnO2 were extracted in 1962. Again, the same mixtures of arsenical copper and tin bronze are found. For the diffusion of ideas, a Nordic route via the North Sea and the Channel is favoured rather than a direct land route from Germany.[81]

INDIA

The proportions of bronzes from all levels at Mohenjo-Daro (2500–2000 BC), in the Indus valley, is about one-sixth of the total of copper-base artefacts analysed. These cover a range of 3–26.9%Sn and it would seem that tin was sometimes available in the 500 years of occupation on this site.[82]

At Mundigak, in Afghanistan,[83] only one out of a total of five shaft-hole axes analysed contained as much as 5%Sn. We do not know whether this is a representative sample, but it seems probable that it is. The EBA seems to have lasted to about 300 BC in the Ganges Basin and the Deccan.[84] In the former, Harappa-type cultures have produced a few bronzes containing 3.8–13.3%Sn. A site at Jorwe in the Deccan, dated to before 300 BC, produced six very simple flat axes of rectangular form and one copper bangle.[84] The single axe submitted to analysis contained 1.78%Sn, a result very typical of the earliest period of the EBA elsewhere, but one which still begs the question of whether this tin was deliberately introduced.

Considering that India is near the well-known tin deposits of Burma and Malaya, one would not be surprised to see a difference in the use of tin there, and particularly in the quantity used in the EBA, compared with the use made of it in places not so close to known tin deposits. This has not happened, however, and it seems that India was to be overtaken by the Iron Age before reaching a fully developed Bronze Age.

THE FAR EAST

In China, where the LBA Shang period dominates, there is now increasing evidence for a pre-Shang metallurgical tradition, and it would seem that the development of non-ferrous metallurgy in China went through the usual sequence from pure copper to tin bronzes.[85] During this development there were influences from outside, but the high level of ceramic technology was probably responsible for the highly-developed state of metallurgy in the late Shang period.

Chinese history starts with the Shang or Yin Dynasty in about 1500 BC. Before this, we have at least one legendary dynasty, the Hsia, which is supposed to date from about 2000 BC. The great majority of Shang artefacts are of straight tin bronze, and there is little evidence of an earlier Copper Age.

Three spearheads which, although unstratified, closely resemble similar objects excavated on a Shang site in Honan, contain no tin.[86,87] A Chou battle axe (1027–221 BC) was nearly pure copper[88] (*see* Table 18). As all four objects come from a period when bronzes were the norm, one can only assume that these are the usual exceptions in a normal Bronze Age technology. We should note the high lead content in two of them: this is not a feature of the Copper Age but of a developed bronze age when lead was being used as a diluent for copper or bronze.

As would be expected, the majority of the objects from the Shang Dynasty are bronzes with less than 3%Pb[89,90] (*see* Table 19). Lead objects have been found in Shang deposits so there is no doubt that metallic lead was being smelted, but it is unlikely that it was an intentional addition.

It is possible that we can draw the line between an EBA culture and an LBA culture at the time when the Shang Emperor in Honan (Phan-Keng) moved his capital from the south bank of the Yellow River to the famous site on the north bank at Anyang, in the 14th century BC. After this move there was a considerable improvement in technique and an increase in the number of artefacts produced but, so far, no evidence of any change in the composition of these.

An early Shang site near Cheng Chou gives evidence of smelting and casting. A rectangular building contained 12 small, lined pits which were surrounded by pieces of slag, and which may have been melting furnaces. The crucibles have been described as 'bucket' shaped[91] but some have a rather unusual ogival style, in

which the bottom is almost pointed[90] (*see* Fig. 13). A deep layer of slag outside the building seems to be evidence of smelting. If this reconstruction is correct, the elongated base could act as a pedestal or inbuilt stool, which would give several advantages to the heat transfer within the furnace. But in the archaeological museums at Zhengzhou and Beijing one can see tapered furnace linings which suggest heating from within, like the lining from Enkomi in Cyprus (*vide inf.*). At Anyang itself, although copper ore is not known in the area, tin ingots and lumps of copper slag mixed with charcoal have been found.[92]

INDO-CHINA

This comprises the countries of Thailand, Cambodia, Laos and Vietnam. A considerable number of excavations have taken place in recent years in Thailand, and reports of the examination of earlier excavated material from Cambodia have been made. Some of the early work from Thailand gave dates in the third millennium but we now know that many of the sites had been disturbed and more recent work shows that most of the material dates to the second millennium and reflects a developed Bronze Age like that of China. Since Thailand has access to considerable supplies of tin it is not surprising that bronzes were common at the outset, and that little use was made of arsenical coppers.[93]

If we take Lopburi as an example, the extractive technique seems normal, with high iron slags and iron in the copper. But the furnaces seem to be no more than

Table 18 Composition of early Chinese copper artefacts

| Object | Date, BC | Composition, % | | | | Reference |
		Sn	Pb	As	Ag	
Spearhead	1500–1000	tr.	2·90	tr.	0·10	Dono[86]
Spearhead	1500–1000	tr.	22·38	0·80	0·30	Dono[86]
Spearhead	1500–1000	tr.	26·78	tr.	0·05	Dono[86]
Battle axe	1000–221	tr.	tr.	(Cu 98·31)		Fink[88]

Table 19 Composition of Shang bronzes (1401-1122 BC)

| Object | Composition, % | | | | | Reference |
	Sn	Pb	As	Ag	Ni	
Axehead	16·67	tr.	—	—	—	Collins[89]
Ko dagger–axe	4·01	2·59	—	—	0·09	Barnard[90] (Liang & Chang)
Arrowhead	1·83	1·85	—	—	—	Barnard (Liang & Chang)
Knifehandle	3·67	1·03	—	—	—	Barnard (Liang & Chang)
Vessel lid knob	13·07	0·83	—	—	0·06	Barnard (Liang & Chang)
Vessel	20·32	0·05	—	—	—	Barnard (Liang & Chang)
Vessel	17·65	0·09	—	—	—	Barnard (Liang & Chang)
Ornament	16·27	0·22	—	—	0·07	Barnard (Liang & Chang)
Ornament (fish)	16·78	0·06	—	—	—	Barnard (Liang & Chang)

— not determined

reaction vessels (crucibles) from which slag and metal were poured.[94]

SOUTH AMERICA

In view of the proximity of a major tin deposit (south east of Lake Titicaca), one is not surprised that the Copper Age was of relatively short duration and that bronzes were soon introduced along the Pacific coast of the Andes. The principal bronze artefacts in this region in the Bronze Age (c. AD 1000–1540) were pins, chisels or flat axes, and knives (either single edged or of the 'T' shaped tumi type). Examples of these from Machu Picchu (Peru) are shown in Fig. 15. The tin content varied from 2 to 13% but most of the objects contained about 5%Sn.[95] Most of the objects were forged and locally cold-worked, and process would be easier with a 5%Sn alloy. The heavier flat axes with central holes were mostly copper, but most of the lighter tumi knives were of bronze. One piece of sheet metal was found to be 99.79%Sn, which leaves little doubt that the alloys were made by the addition of metallic tin. There seem to be occasional reversions, as though the contact between the coastal and inland tribes controlling tin sources was interrupted. The knowledge of bronze does not seem to have reached Colombia until after the Conquest.

In the Argentine, a crucible containing traces of a 4.84% Sn bronze was found,[96] together with several stone moulds and one of clay for a 'T' shaped axe. Argentine bronzes often contain zinc and bismuth in substantial quantities, but the tin content is generally low (*see* Table 20). It seems that, in some of the objects, the tin content is not intentional: cassiterite has been found at Puna de Atacama and La Rioja[97] and may have been incorporated accidentally.

The transition stage in South America is often represented by complex alloys containing tin and antimony: a pin (*topu*) from a grave in the Queara valley of Peru contained 6.21%Sn, 2.67%Sb, and 1.46%Bi, and there are many other examples of this sort of alloy. As much as 18% Bi has been found in a tin bronze handle from

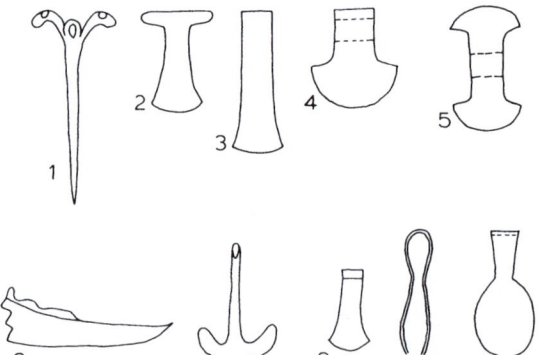

1 lama-headed *Topu* pin from Pelechuco, Bolivia (9 cms long); 2 flat 'T' axe from Pulquina, Bolivia; 3 blade of hoe or chisel; 4 shaft-hole axehead; 5 double axehead; 6 ornamental knife blade; 7 *tumi* knife; 8 knife; 9 tweezers: (mostly after Nordenskiold)

15 Some examples of early South American copper-based artefacts (not to scale) (mostly after Nordenskiold [97])

Table 20 Composition of bronzes from Argentina (after Nordenskiold[97] and De Mortillot[89])

Provenance	Object	Composition, %					
		Sn	Fe	Pb	Zn	Bi	Sb
Calchaqui	Disc	2·58	0·11	0·22	1·65	0·23	—
Calchaqui	Disc	3·07	0·08	1·04	1·15	0·36	—
Calchaqui	Disc	3·46	0·75	0·18	1·01	0·82	—
Luracatao	—	3·03	0·37	0·21	0·94	0·33	—
Parana	—	1·04	—	0·75	0·28	—	—
Guazu	—	0·61	—	0·37	0·49	—	—
Gauzu	—	3·28	—	0·75	0·30	—	—
Gauzu	—	0·77	—	0·17	0·34	—	—
Provenance not given	—	3·04	0·24	0·07	—	—	nil
	—	4·43	0·37	0·12	—	—	tr.
	—	6·91	0·44	0·32	—	—	nil
	—	9·40	0·41	0·32	—	—	nil
	—	13·52	0·17	0·64	—	—	nil

Machu Picchu; as it did not form a grain boundary phase the metal was not embrittled.[98]

A series of Bolivian bronze chisels from the area around Titicaca were analysed[99,100] and found to contain from 4.48–8.92%Sn. Nine bronzes from Peru and Bolivia have been examined, and axes containing 0.7 and 13.42%Sn, respectively, were found to be made of porous cast metal with the lateral ends extended by forging. The low-tin bronze axe had a hardness of 64 HB in the as-cast blade. By forging, the hardness had been increased to 97 and 108 in one of the arms. The high-tin axe had a hardness of 102 HB in the centre of the blade, which had been increased to 150 at the cutting edge and 126 in one of the arms. These should be compared with the results of work on the same type of axe from Egypt, discussed on p. 26. The hoes and tumi knives had also been forged. In the case of a hoe with 5.4%Sn, the hardness had been increased from 82 to 110 HB. The hardness of a tumi knife with 3.4%Sn had been increased from 67 to 128 HB in the shank. No readings were obtained on the cutting edge. Analyses of these and other Bolivian objects are given in Table 21.

Three Peruvian 'T'-shaped axes were found to have a tin content in the range 3.36–12.03%. Two of these came from old Inca settlements near the Pampaconas River

Table 21 Composition of Bolivian objects (after Nordenskiold[96]) and De Mortillot[105])

Provenance	Object	Sn	Pb	Sb	Bi	Zn	S
Samaipata	'T' axe	0·7	0·4	nil	nil	nil	0·5
Peres (Mizque)	'T' axe	3·0					
Covendo	'T' axe	4·67					
Pulquina (Mizque)	Disc	7·03					
Chilon	Chisel	7·12					
Sara	Disc	10·34	—	tr.	tr.	0·28	
Pulquina	'T' axe	13·42					
Caipipendi	—	nil					
Provenance not given	—	5·83	0·63	0·06	—	—	—
	—	6·71	0·17	0·06	—	—	—
	—	7·50	0·11	0·17	—	—	—
	—	9·30	0·14	0·06	—	—	—
	—	10·72	0·28	0·08	—	—	—

and the third from Rosalina.[101] It is interesting to find that the 12%Sn bronze had been either hot worked or cold worked while in the beta phase; if the latter was the case, this must have been achieved by heating above 500°C and quenching. This would have been done several times to obtain the desired shape. None of the axes is very hard, and it would appear that they were finally used in the wrought annealed state.

In conclusion, we can state that these objects are all wrought and therefore typical of the Early Bronze Age cultures and not of the Middle or Late Bronze Age where, in Eurasia, high-tin bronzes (7–15%Sn) were consistently used in the 'as-cast' condition, except for some localized hardening of the cutting edge. True cast bronzes do not seem to have been in use in Central America before the time of the Spanish conquest.[102]

References

1   H. OTTO and W. WITTER: 'Handbuch der altesten vogeschichtlichen metallurgie in Mitteleuropa',1952; Leipzig, Johann Ambrosius Barth (Verlag)
2   G. COFFEY: *JRAI*, 1901, **31**, 265
3   A. LUCAS: 'Ancient Egyptian materials', (revised by J. R. Harris), 1962, London, Edward Arnold
4   C E. ELAM: *J. Inst. Metals*, 1932, **48**, 97
5   SHEREEN RATNAGAR: 'Encounters: the westerly trade of the Harappa Civilization', OUP, Delhi, 1981
6   W. LAMB: *BSA*, 1928–30, **30**, 1
7   W. LAMB: 'Excavations at Thermi, Lesbos', 1936, Cambridge, Cambridge University Press
8   P. STICKLAND: 'The recovery of tin into copper by surface additions of tin-bearing minerals', Under graduate dissertation, Dept. of Metallurgy, Cambridge, 1975
9   HENRIETTA MILES: 'Barrows on the St. Austell granite', *Corn. Arch.*, 1975, (14), 5–81
10   N. D. MEEKS: 'The examination of a sample of a copper ingot found off Plymouth', *IJNA*, (forthcoming)
11   G. F. BASS *et al.*: *Trans Amer. Phil. Soc.*, 1967, **57**,(8), 177
12   G. F. BASS: *AJA*, 1961, **65**, (3), 267
13   G. BASS, D. A. FREY and C. PULAK: 'A late Bronze Age shipwreck at Kas, Turkey', *IJNA*, 1984, **13**, (4), 271–279
14   R. F. TYLECOTE, M. S. BALMUTH and R. MASSOLI-NOVELLI: 'Copper and bronze metallurgy in Sardinia', *JHMS*, 1983, **17**, (2), 63–76
15   K. A. YENER and H. OZBAL: 'Tin in the Turkish Taurus mountains; the Bolkerdag mining district', *Antiquity*, 1987, **61**, 220–226
16   K. A. YENER, H. OZBAL *et al.*: 'Kestel: an EBA source of tin ore in the Taurus mountains, Turkey', *Science*, 1989, **274**, (4901), 200–203
17   D. BRITTON: *PPS*, 1963, **29**, 258
18   P. T. CRADDOCK and D. GALE: 'Evidence for early mining and extractive metallurgy in the British Isles', In:*Science and Archaeology*, Glasgow 1987, (eds. E. A. Slater and J. O. Tate), *BAR Brit. Ser.*, 196, 1988, 167–191
19   V. G. CHILDE: 'Dawn of European civilization', 128,1961, London, Routledge and Kegan Paul
20   S. JUNGHANS *et al.*: 'Metallanalysen Kupferzeitlicher und Frühbronzezeitlicher Bodenfunde aus Europa', (SAM 1),1960, Berlin, Verlag Gebr. Mann
21   S. JUNGHANS *et al.*: 'Kupfer und Bronze in der frühen Metallzeit Europas', (SAM 2), 1968, (3 Parts), Berlin, Verlag Gebr. Mann
22   J. PERROT: *Israel. Excav. J.*, 1955, **5**,17, 73, 167
23   J. PERROT: 'La Préhistoire Palestinienne', 1968, Paris, Letouzey et Ané
24   R. F. TYLECOTE, B. ROTHENBERG and A. LUPU: 'The examination of metallurgical material from Abu Matar, Israel', *JHMS*, 1974, **8**, (1), 32–34
25   R. F. TYLECOTE *et al.*: *J. Inst. Metals*, 1967, **95**, 235
26   A. LUPU and B. ROTHENBERG: *Arch. Austriaca*, 1970, **47**, 91
27   A. P. JOURAVLEV: 'The earliest workshop for the manufacture of copper in Karelia', *Soviet Arch.*, 1974, (3), 242–246
28   M. GIMBUTAS: 'Bronze Age cultures in Central and Eastern Europe', 1965, London, Mouton
29   B. ROTHENBERG: 'An archaeological survey of South Sinai', *Mus. Haaretz Bull.*, No. 11, 1969
30   R. F. TYLECOTE: *Rep. Dept. Antiquities*, 1971, 53, Cyprus
31   MASUDA TSUNA: 'Kodo Zuroku', (ed. C. S. Smith). Burndy Library, Conn. USA, 1983
32   M. DOTHAN: *Israel Explor. J.*, 1959, **9**,13
33   J. L. CASKEY: *Hesperia*, 1955, **24**, 25
34   C TSOUNTAS: 'Dimini e Sesklo', 1908, Athens, Sarkellarios
35   H. W. CATLING: 'Cypriot bronzework in the Mycenaean world', 1964, Oxford, Oxford University Press
36   J. L. CASKEY: *Hesperia*, 1962, **31**, 263
37   F. PETRIE: 'Researches in Sinai', 1906, London, Murray
38   C. J. DAVEY: 'Tell Edh-Dhiba'i and the Southern Near Eastern metal working tradition', In: 'The beginning of the use of metals and alloys', (BUMA), (ed. R Maddin), Zhenzhou, 1986, 21–26
39   P. E. NEWBERRY: 'The life of Rekhmara', Pl.XVIII,1900, London, Constable
40   H. and L. SIRET: 'Les premiers ages du metal dans le Sud-est de l'Espagne', 1887, Anvers
41   H. SCHLIEMANN: 'Ilios, the city and country of the Trojans', 1880, London, Murray
42   S. NEEDHAM: 'An assemblage of LBA metalworking debris from Dainton, Devon', *PPS*, 1980, **46**, 177–215
43   *Bull. HMG*, 1966, **1**, (7), 20
44   B. ROTHENBERG: *Palestine Exploration Q.*, Jan–Jun,1970, 29 pp
45   H. C. H. CARPENTER: *Nature*, 1931, **127**, 589
46   H. GARLAND: *J. Inst. Met.*, 1913, **2**, 329
47   E. S. EL GAYAR and M. P. JONES, 'Old Kingdom copper smelting artefacts from Buhen in Upper Egypt', *JHMS*, 1989, **23**,(1), 16–24
48   R. GHIRSHMANN: 'Fouilles de Sialk', 1935, Paris, Geuthner
49   E. F. SCHMIDT: 'Excavations at Tepe Hissar', 1937, Philadelphia, University of Pennsylvania
50   R. H. DYSON JR: 'Digging in Iran, Hasanlu, 1958', 4,1959, 1–3, Bull. University Museum of Pennsylvania
51   G. CONTENAU and R. GHIRSHMANN: 'Fouilles de Tepe Giyan', 1935, Paris
52   T. B. BROWN and E. VOCE: *Man*, 1950, **50**, 41
53   C. H. DESCH: *Rep. Brit. Assoc.*, 1931, 271
54   G. WEISGERBER: 'Und Kupfer in Oman', *Der Anschnitt*, 1980, (2–3), **32**, 62–66
55   G. WEISGERBER and A. HAUPTMANN: 'Early copper mining and smelting in Palestine', *BUMA*, 1986, (*see* Ref. 38), 52–62
56   R. J. BRAIDWOOD *et al.*: *J. Chem. Educ.*, 1951, **28**,87
57   C. F. A. SCHAEFFER: 'Mission de Ras Shamra (Ugaritica II)', 1949, Paris, Geuthner
58   C. F. A. SCHAEFFER: *JEA*, 1945, **31**, 92

59   K. KENYON: 11th Ann. Report. Univ. London Inst. Arch., 1, 1955

60   K. BRANIGAN, H. McKERRELL and R. F. TYLECOTE: *JHMS*, 1976, **10**, (1), 15–19

61   K. BRANIGAN: *AJA*, 1967, **71**,(2), 117

62   S. SWINY: 'The Kent State University expedition to Episkopi Phaneromeni', *Stud. in Mediterranean Arch.*, 1986, Vol. 4, Pt. 2, Nicosia

63   E. R. CALEY: *Hesperia*, 1949, Sppl.8, 60

64   A. J. B. WACE and M. S. THOMPSON: 'Prehistoric Thessaly', 1912, London, Cambridge University Press

65   W. J. HEURTLEY and O. DAVIES: *BSA*, 1926–7, **28**, 195

66   K. BRANIGAN: *BSA*, 1969, **64**, 1

67   A. MOSSO: 'Dawn of Mediterranean civilization', 1910, London, J. Fisher Unwin

68   F. CHAPOUTHIER and P. DEMARGUE: 'Fouilles à Mallia', 1942, Paris

69   T. D. ATKINSON *et al.*: 'Excavations at Phylakopi in Melos', 1904, London

70   K. BRANIGAN: 'Copper and bronze working in Early Bronze Age Crete', 1968, Lund

71   D. H. TRUMP: 'Central and Southern Italy before Rome', 1966, London, Thames and Hudson

72   C STORTI: *Sibrium*, 1960, **5**, 208

73   R. F. TYLECOTE, M. S. BALMUTH and R. MASSOLI-NOVELLI: 'Copper and Bronze Metallurgy in Sardinia', In: *Studies in Sardinian Arch.*, (eds. M. S. Balmuth and R. J. Rowland Jr.). Ann Arbour, Mich. 1984, 115–162

74   W. R. JONES: 'Tinfields of the world', 1925, London, Mining Publication

75   F. A. HARRISON: *MM*, 1931, **45**, 137

76   J. C. ALLAN: 'Considerations of the Antiquity of mining in the Iberian Peninsula', 1970, R. Anthrop. Inst. Occas. Paper No.27

77   B. ROTHENBERG and A. BLANCO-FREIJEIRO: 'Ancient mining and metallurgy in SW Spain', *IAMS*, London, 1982

78   R. J. HARRISON, S. QUERO and M. CARMEN PRIEGO: 'Beaker metallurgy in Spain', *Antiquity*, 1975, **49**, (196), 273–278

79   B. BLANCE: *Antiquity*, 1961, **35**, 192

80   D. BRITTON: *Archeometry*, 1961, **4**, 40

81   J. BRIARD: 'Les dépots bretons et l'age du bronze atlantique', 1965, Rennes, Faculté de Sciences

82   C. H. DESCH: *Rep. Brit. Assn.*, 1931, 269

83   J. M. CASAL: 'Fouilles de Mundigak', 1961, 2 vols., Paris, Klincksieck

84   H. D. SANKALIA: *Man*, 1955, **55**, (1), 1

85   SUN SHUYUN and HAN RUBIN: 'A preliminary study of early Chinese copper and bronze artefacts', Kaogu Xuebao, 1981, (3), 287–302

86   T. DONO: *Bull. Chem. Soc. Japan*, 1932, **7**, 347

87   T. DONO: *J. Chem. Soc. Japan*, 1932, **53**, 748

88   C. G. FINK and E. P. POLUSHKIN: *Trans. AIMME*, 1936, **122**, 90

89   W. F. COLLINS: *J. Inst. Metals*, 1931, **45**, 23

90   N. BARNARD: 'Bronze casting and bronze alloys in ancient China', 1961, Canberra, The Australian Nat. University and Monumenta Serica

91   W. WATSON: 'China before the Han Dynasty', 1961, London, Thames and Hudson

92   W. C. WHITE: 'Bronze Culture in Ancient China', 1956, Toronto

93   T. STECH and R. MADDIN: 'Reflections on early metallurgy in SE Asia', In: *BUMA*, (see Ref. 38)

94   A. BENNETT: 'The contribution of metallurgical studies to SE Asian archaeology', *World Arch.* 1989, **20**, (3), 329–351

95   C. H. MATHEWSON: *AJS*, 1915, **40**, 525

96   P. RIVET: *JSA*, 1921, **13**, 233

97   E. v NORDENSKIOLD: 'The Copper and Bronze Ages in South America', 1921, Göteburg

98   R. B. GORDON and J. W. RUTLEDGE: 'Bismuth bronze from Macchu Picchu, Peru', *Science*, 1984, **223**, (4636), 585–6

99   M. LOEB and S. R. MOREY: *J. Amer. Chem. Soc.*, 1910, **32**, 652

100   C. MEAD: Anthrop. papers of Amer. Mus. Nat. Hist. New York, 1915, **12**, 15

101   H. W. FOOT and W. H. BUEL: *AJS*, 1912, **34**, 128

102   H. ARSANDAUX and P. RIVET: *JSA*, Paris, 1921, **13**, 261

103   O. DAVIS: *BSA*, 1928–30, **30**, 74

104   J. DU PLAT TAYLOR: *Antiq. J.*, 1952, **32**, 133

105   A. DE MORTILLET: 'Bronzes in South America before the arrival of the Europeans', 261, 1907, Ann. Rept. of Smithsonian Inst.

106   C. H. DESCH: *Brit. Assoc.* 1928, 437

107   C. H. DESCH: *ibid.*, 1932, 302

108   R. PITTIONI: *Arch. Aust.*, 1957, Beiheft I. Supplement

109   R. PITTIONI: *Arch. Aust.*, 1958, Beiheft 3, Studia Palaeometallurgica, In honorem Ernesti Preuschen

110   R. F. TYLECOTE and H. McKERRELL: *Bull. HMG*, 1971, **5**, 37

111   O. DAVIES: *Man*, 1936, **36**, 119

112   O. DAVIES: *J. Hellenic Studies*, 1929, **49**, 89

113   B. ROTHENBERG and A. LUPU: *Bull. Museum Haaretz*, 1967, (9), Pl. X.

114   I. R. SELIMKHANOV: 'Enträtselte Geheimnisse der alten Bronzen', 1974, Berlin, 22

Chapter 4
The Full Bronze Age

The archaeological division of the Bronze Age into Early, Middle, and Late is mainly based on changes in pottery types rather than metal compositions. As has been pointed out, there is little justification for this division from a metallurgical point of view and, while the EBA is mainly concerned with non-bronze compositions, by the MBA we are in a metallurgical bronze age where, with few exceptions, pure and arsenical coppers have ceased to be used and we can now talk for the first time of a Full Bronze Age. The amount of metal increases gradually through the MBA to reach, in most countries, substantial quantities by the LBA.

In the advanced civilizations of Asia Minor the Bronze Age developed in a steady fashion, so that there are no points at which one can say that a marked change in technique or in usage has occurred. However, the appearance of ingots weighing more than 30 kg in Late Minoan times (about 1600 BC) shows a large increase in the scale of operations in the Mediterranean area. In Britain, for example, the Full Bronze Age coincides with an enormous increase in the quantity of metal artefacts, especially socketed axeheads. Naturally, mass production techniques were used to supply the demand. In China, this stage was shown by an increase in the size of objects, as seen in the late Shang and Chou bronzes.[1]

One important change in technique is discernible. Whereas EBA objects required forging and hammering to give them their final shape, most LBA objects were cast in double (two part) moulds and needed very little, if any, final finishing by hammering. There was now more attention to detail, with the production of finer objects and more accurate castings: an addition of lead was often made to increase the fluidity of the metal, an innovation of the LBA. The lead content is an indication that an object belongs to this, and not an earlier period. Analyses show that Chou bronzes from China contain additional lead, while those of the (earlier) Shang period are virtually without lead (see Chapter 3, Table 19): and in South East Britain we find that most cast objects contain 5–10%Pb.

There is plenty of evidence that casting had reached a high degree of technical skill: moulds of stone and bronze have been found in Asia Minor and Europe, and the use of a core for casting socketed objects had been widely applied, as well as casting by the lost-wax process. In China, piece moulds were used for large objects, as can be deduced from the magnificent ceremonial vessels which required as much as 1400 kg of metal. An increase in the number of artefacts, and the numerous hoards found in all areas of the Old World, are testimony to a flourishing metal industry.

Furnaces and Smelting Techniques
As far as smelting technique is concerned we have no more examples of copper-smelting furnaces than in earlier periods. One assumes, perhaps because of lack of evidence, that little advance was made in this sphere and that the copper itself was produced mainly near the mines and in small units. The large oxhide-shaped ingots, weighing 30–38 kg, are an exception to this, and one might assume that China was producing ingots of comparable size. Britain and other European countries seem to have been producing plano-convex ingots weighing only 2–4 kg from small furnaces. It was not until the Roman period that the ingot size reached 20 kg.

For details of furnace technology we must rely on work in the Palestine Negev. In this area, which is now known to have had Egyptian connections, the excavated furnaces date from the 14 to the 12th century BC and it is reasonable to assume that this type of furnace is typical of those used in the LBA in the more developed areas of Asia Minor.[2]

The Negev furnaces in the Wadi Arabah have been dug into the sandy soil of the desert and are of roughly cylindrical shape, about 60 cm dia., and the same in height (see Fig.16).[3] The edges have a stone kerb and the bowls are lined with a limy cement. The ground in front of the furnace has been cut away to give access for tapping slag, and the furnaces were blown through inclined tuyeres of a complicated form (see Fig.17). Near the furnaces are the remains of large discs of slag, weighing about 50 kg, with a central hole. Many ends of tuyeres with holes of about 2 cm dia. were also found.

It is clear from experiments made on these furnaces that the slag was tapped first, leaving lumps of copper in the bottom.[4,5] These lumps were then removed from the furnace, remelted out of contact with slag but with a fresh charge of charcoal, and cast into the desired ingot shape, whether it be plano-convex or oxhide. Taking into consideration the quality of the ore, the weight of the slag discs, and the size of the furnace, it may be calculated that the weight of the ingot would have been about 3–4 kg. This weight is fairly typical for LBA ingots. It is a fair assumption that these ingots supplied the Egyptian markets, although only oxhide shaped ingots are depicted on the tomb paintings.

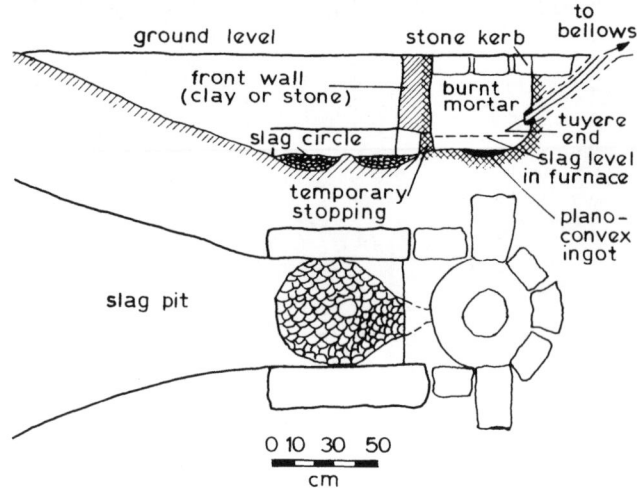

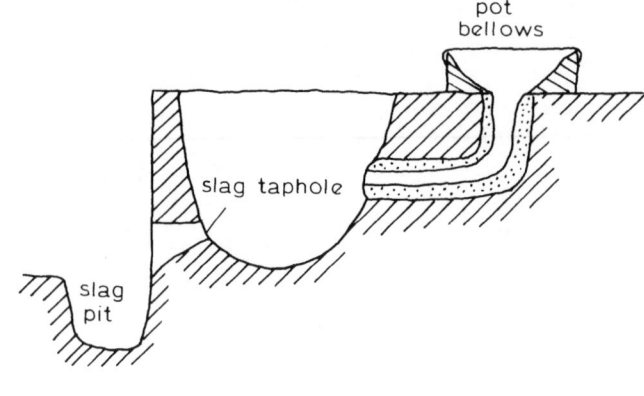

18 Method of use of elbow-type tuyere found at Apliki, Cyprus

16 Copper smelting furnace from Timna, Israel (after Rothenberg and Lupu[2])

A Late Bronze Age settlement at Apliki in Cyprus showed evidence of metalworking,[6] and there is no doubt that smelting had been carried out in the area where copper ore was being mined as recently as the late 1930s. The ferruginous sulphide ore (*see* Table 22) was heap-roasted using silica as a flux. The slag heaps contained many pieces of discarded furnaces: some of the tuyeres were 'D' shaped as in the EBA site at Ambelikou, and one tuyere had a right-angled bend (*see* Fig. 18).[7] The slag analyses are given in Table 17.

The copper sulphide deposits in the Austrian Alps have been worked since at least 1200 BC, and the remains of furnaces and slag heaps have been found on many sites.[8,9,10] The area of an individual smelting site averages 100–150 m² and the size of the slag dumps about 30 m³. The two furnaces on a site in the Mitterberg suggest an above-ground cylindrical furnace of about 50 cm i.d., with a slag tapping pit in front. This may be one of the earliest sites known to have possessed sulphide ores (*see* Table 16). Two types of slag were found (*see* Table 17):

one type occurred in the form of flat cakes about 30 cm dia. and was extremely inhomogeneous, and contained some of the original charge.[11] This slag seems to have been tapped in a very viscous state and contains drops of matte (mixed copper and iron sulphides) about 5 mm dia.[12]

It appears that this badly-settled rich slag, to use a modern term, was crushed to release the matte, as is shown by the dumps of finely broken up slag. It must be remembered that matte does not separate as easily as metallic copper from slag as the specific gravity of the matte is much closer to that of slag. Two grades of matte were found: one containing 35–40%Cu and the other 60–65%Cu. A few discs of an even richer matte, 3–5 mm thick, were found, but this was rare.

The matte must have been completely roasted to copper oxide and the final smelting of the oxidized matte, to which silica must have been added, resulted in a very homogeneous slag in the form of thin discs about 3–6 mm thick and 20 cm dia. This was probably obtained in the same type of furnace and tapping pit as used for matte smelting. The result was an impure plano-convex ingot of 'blister' copper containing about 98%Cu and weighing about 6 kg.

Table 22 Composition of Cyprus ores (after Du Plat Taylor[6])

Element	Content, %			
	Paphos	Kalavassos	Mitsero	Sha
Fe	13·50	29·75	21·00	23·74
Cu	2·56	12·80	0·52	0·92
S	7·89	32·00	22·48	51·31
SiO₂	53·3	18·00	40·07	0·19
As	0·01	tr.	0·02	0·03
Ni	nil	nil	nil	nil
Pb	0·03	0·01	0·03	tr.
Zn	nil	nil	0·35	nil
Ag	0·02	<0·01	<0·01	<0·01
Sb	0·03	0·03	} 0·03	tr.
Sn	0·04	0·06		
Al₂O₃ } CaO } MgO } TiO₂ }	Rest	Rest	Rest	Rest

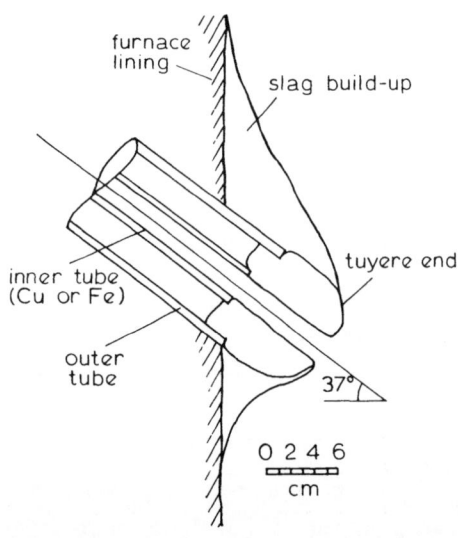

17 Type of inclined tuyere reconstructed from remains found at Timna, Israel (after Rothenberg and Lupu[2])

It is estimated that about 20 000 t of raw copper have been produced in this area over a period of one thousand years, up to the end of the Urnfield period (*c*. 600 BC). Slag from the early smelting sites was used to temper potter's clay in settlements far removed from the mining centres. It seems that the use of sulphide ores at about this time was not unusual. We now have similar evidence for Bronze Age copper mining from the British Isles.[13] An Irish mine was producing low-grade sulphide ores with the aid of stone tools at about 1300 BC, i.e. in the Irish MBA. We now know that Bronze Age mining was carried out at Cymystwyth and at Great Orme's Head, both in Wales. The ^{14}C dates are as follows:

St. Lorenzen, Austria (VRI 1657)[14] 1060∓100 BC
Gaishorn, Austria (VRI 720)[14] 690∓ 90 BC
Cwmystwyth, Wales (Q-3078) 1560–1420 BC(Cal.)
Gt. Orme's Head, Wales (HAR 4845) 1300–1000 BC(Cal.)
Mt. Gabriel, Ireland (Grn-13980) 1620–1510 BC (Cal.)

So far there is no evidence for the same smelting techniques in the British Isles as in the Mitterberg. The low level of iron in the British Bronze Age plano-convex ingots points to a primitive smelting technique using very high-grade copper oxide minerals without iron fluxing.[15]

Ingot Types

We do not know the exact form of the smelted ingot material from the Copper Age, but it may have been plano-convex or bun-shaped. The ingots of the LBA are usually flat on the top surface and hemispherical at the bottom and would have been a natural shape to make in a bowl furnace (*see* Fig.19). But some, such as those from Funtana Jana in Sardinia have more of a truncated cone shape.[16] It seems increasingly likely that all these were cast from a remelting furnace into suitably shaped moulds, as it does not seem possible to make such shapes direct from a smelting furnace. But the plano-convex ingots may form in the bottom of the smelting furnace as a result of the liquation of the smelted copper from the super-incumbent slag. Metallic copper has more than twice the density of this slag and, in due course, falls through the slag to form the ingot. When the smelting of one charge is complete it can be allowed to solidify and the brittle slag broken away, revealing the copper ingot. The more convenient process, however, is to drain off the slag in the liquid state at the end of the smelt and then to withdraw the ingot of raw copper after it has solidified. The melting point of this slag would have been about 1 150°C, and that of the copper below 1 084°C.

Analyses of plano-convex ingots of the LBA show most of them to be of pure copper,[17,18,19] although small bronze ingots were made for forging into cauldrons and bowls, examples of which have been found containing 8%Sn. The metallographic examination of a copper ingot from Cornwall shows it to have solidified in the furnace and not to have run into a mould near the furnace.[18] The practice in Palestine in the 14–12th century BC was similar.

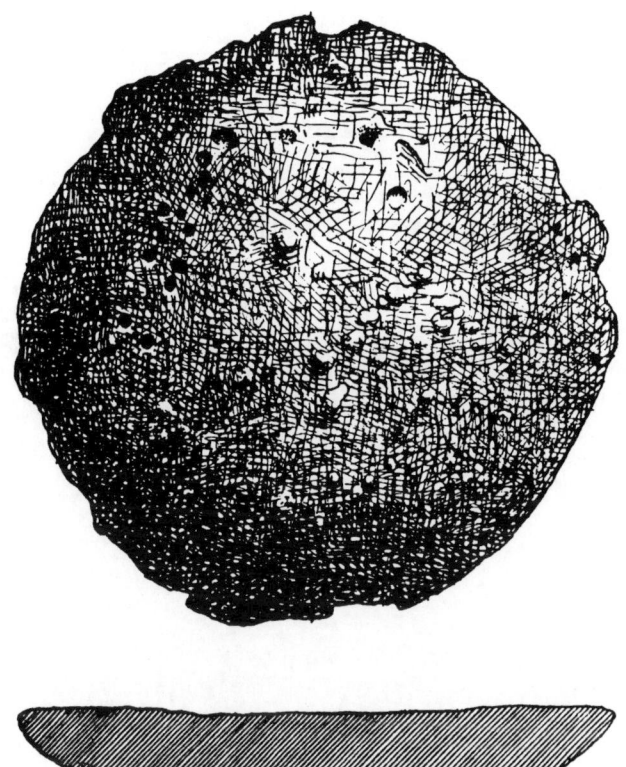

19 Example of an LBA plano-convex ingot of pure copper (145 mm wide)

The plano-convex ingots coming directly from the furnace have not been refined in any way, but are already relatively pure. The Cornish LBA examples show that they contain considerable amounts of sulphur and oxygen, as would have been expected, and are therefore like modern impure (blister) copper; the absence of iron suggests that they were made by a process not using iron oxides as flux.

One of the most interesting types of ingots are those of oxhide or double-axe shape, current in the Eastern Mediterranean during Minoan times (*see* Table 23, and Fig.20). There is no doubt that they were traded over a wide area and they are depicted on Egyptian tomb paintings of *c*. 1450 BC.[20,21] The 19 oxhide ingots from Agia Triada in Crete, which might have been imported from Cyprus, were found to be of almost pure copper with 0.445%S.[22] The ingots weigh about 30–40 kg with a thickness of about 4 cm.[23] They show the typical gassy appearance of blister copper: the top surface is roughened by the blisters while the bottom surface, although smoother, shows cavities owing to gas evolution from the metal or from the moisture in the clay moulds. The sides are the smoothest of all and the corners tend to be thicker than the rest of the ingot. The edges show a 'rim', resulting from the initial rapid solidification against a cold mould wall.

Two underwater wrecks off the south coast of Turkey have revealed a large number of these ingots,[24,25] 34 in the Cap Gelidonya wreck and over 200 in the later Kas wreck, weighing 25 to 30 kg. Experimental work has shown how these ingots may have been made. Metal would first have been smelted in furnaces of the type

Table 23 Provenance and typology of copper oxhide ingots

Provenance	No.	Date, bc	Weight, kg	Type (Fig.20)	Ref.
Crete					
Agia Triada	19	LMI-II	27-40	1	72
Kato Zakro	6	1600-1500	-	1	72
Paleokastro	1	-	-	-	72
Sitsia	1	-	-	-	72
Mochlos	1	-	-	-	72
Knossos	1	-	-	-	72
Tylissos	1	LMI-II	26.5	1	72
Cyprus					
Enkomi (BM)	1	1200	37.1	3	72
Nicosa (Enkomi)			39.2(38.7)	3	72
Los Angeles (")	1		32.1		
Turkey					
Cap Gelidonva	<40	1200	19-22.3	2	24
Kas	c.200	14th cent.	27	2	25
Antalya	1	1600-1400	25.7	1	72
Greece					
Mycenae	1	LHIII	-	2	72
Kyme	11	-	12-17.6	1	72
Sicily					
Canatello	1	-	-	-	72
Sardinia					
Serra Ilixi	3	-	27.1-35.6	1/2	72
Ozieri	2	-	-	-	16
Bulgaria					
Sozopol	1	-	29	3	Dimitrov[73]

Several other sites have yielded fragments

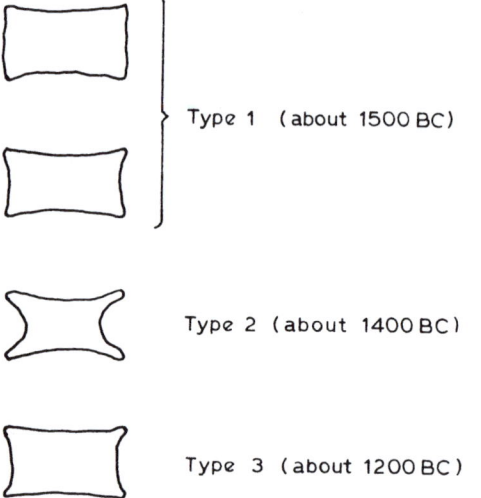

Type 1 (about 1500 BC)

Type 2 (about 1400 BC)

Type 3 (about 1200 BC)

20 *Examples of Minoan copper ingots of oxhide shape (after Buchholz[76])*

shown in Fig. 21 and the superincumbent slag tapped off to the right as shown. The lumps of metal inside would be remelted with additions of other copper metal such as scrap or plano-convex ingots, with charcoal in the same furnace. It would then be possible to tap the metal, now free of slag, into sand or stone moulds as shown on the left of Fig. 21. A stone mould of the right size and shape for these ingots has been found at Ras Ibn Hani in Syria.[26]

Melting and Casting

Furnaces are necessary for remelting pieces of ingot and adding tin and other alloying metals. We know from the Chalcolithic site at Beersheba that this sort of operation can be done in large cylindrical clay furnaces or, as in Late Bronze Age Palestine, in square stone 'boxes'. All that is required is a container for charcoal, so arranged as to provide air from below or from the sides, as in a brazier, or for it to be blown by a tuyere connected to bellows. Reducing conditions are now no longer necessary and as much air as is needed may be blown in. A crucible, with or without a lid, would have been filled with pieces of ingot and alloying additions and inserted into the centre of the furnace somewhere near the bottom (*see* Fig.9). While we have a good collection of tuyeres from many periods and sites, we know little about the bellows. These can be extremely simple, like the goatskin bag bellows used by African blacksmiths today, or the pot bellows depicted on Egyptian tomb paintings and found in the Sudan, where they are dated to the first 500 years AD.[27,28] These paintings are probably inaccurate, since we should not have expected a shallow open crucible to have been put on top of a fire but within it where, of course, it would not have been seen by the artist.

An interesting collection of tuyeres came from the LBA levels of the Palestinian site at Tel Zeror[28] (*see* Fig.22). None of these was complete: they were all fired red and black and one had a deposit of greenish slag, so it must have been used; Fig.22*d* shows the bellows-end of a long tuyere of 7 cm o.d., with a hearth-end of 12 mm dia. This is about twice the size of those from Kalinovka (p. 22). The others have closed ends with 5 mm dia. holes leaving them at angles between 25° and 30°. These are unique and must have been used for some oxidation operation, either cupellation of lead or the flapping (refining) of copper. In this process, a very intense air blast is directed upon the metal itself (*see* Fig.23). There is no doubt that bronze was being made on the site, and a piece of a lead pig was found in a later Hellenic level.

Moulds for casting may be of stone, clay, copper or bronze, and examples of all types have been found from early periods onwards. Two moulds from Mallia, Crete,[29] dated to the Middle Minoan period c. 2000 BC, are of the open type consisting merely of a matrix carved into an irregular piece of 'rough stone'. Another mould from the

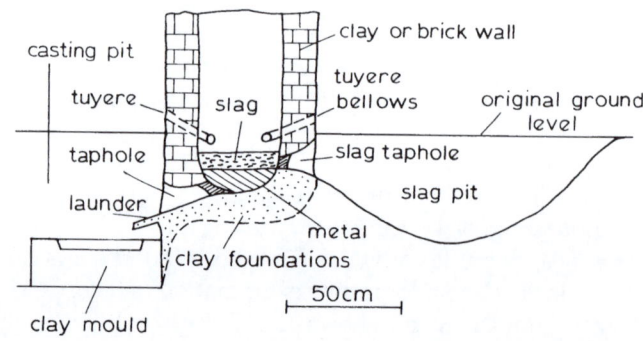

21 *Type of smelting furnace capable of producing Minoan oxhide ingots*

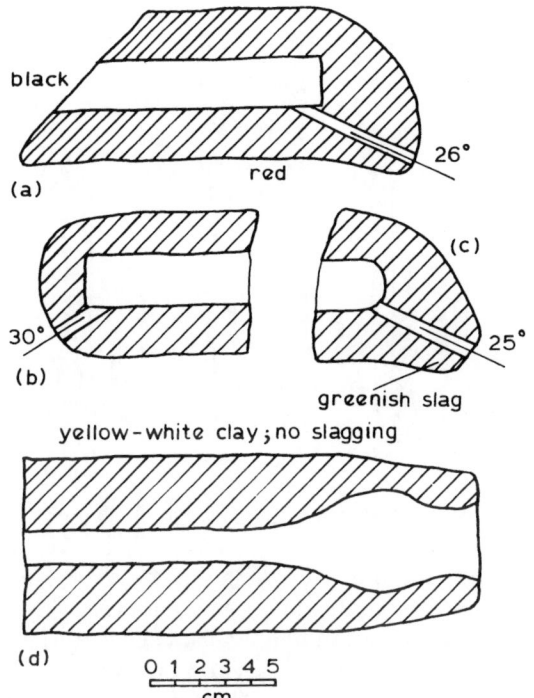

22 Types of tuyeres found at Tel Zeror, Israel

same site has matrices on all four sides of the block but seems to have been intended for use as an open mould, and is not one of a pair of similar moulds, like some British and Irish examples of the LBA (*see* Fig.24). The matrices seem to have been intended for casting blanks for awls, chisels, and perhaps knives, and would have needed considerable forging to give them their final shape.

Steatite was used whenever available because it could be easily carved, and we have parts of the two-half moulds of steatite for a double axe from Mallia. The Minoan double axe (a shaft-hole axe) represents a met-

allurgical transition between the Copper Age and the EBA, since some examples analysed are of impure copper and others are bronzes (3.71–18%Sn). The Mallia double-axe mould has an enclosed core print, a device for supporting the claysand core that makes the hole for the shaft. The mould is in very bad condition and seems to have continued in use after fracture. The two halves would have been located correctly with the aid of dowels and held together by means of a stone packing in a pit. Such a mould would have had to be preheated to prevent it cracking from thermal shock when the hot metal was poured in: it must have been filled through the side with the two halves held with their parting line vertical.

Fragments of another double-axe mould were found on Melos (Fig.24), together with many fragments of rough pottery vessels with nodules of bronze adhering to their undersides. These were undoubtedly the remains of a casting installation.[30] The mould was made of coarse-grained micaceous stone although here the core prints were not enclosed but open, as in the bronze moulds for socketed axes from Britain. Here also, the metal was poured through a hole from the side in the parting line of the mould.

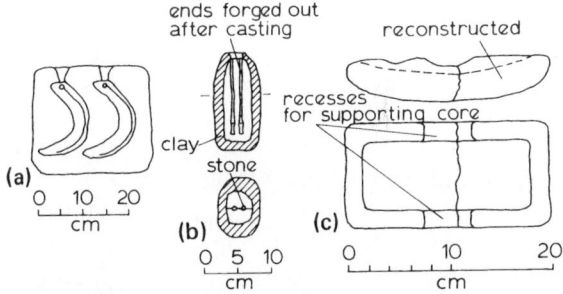

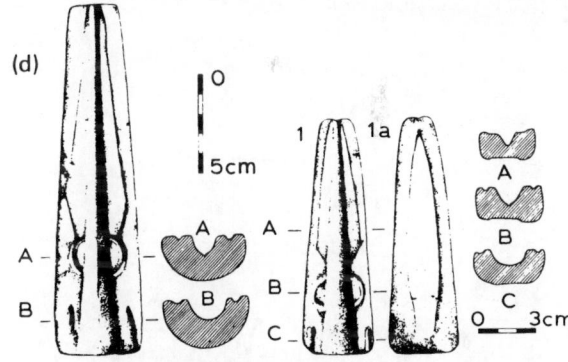

a mould of sandstone for two sickles from Trisko-lupy, Bohemia: this was closed by a flat plate with matrixes for runners only; *b* stone mould for two pins from Zvoléneves, near Prague: the two halves were held together by a clay envelope (after Drescher); *c* reconstruction of half a mould for double axes of micaceous sandstone from Phylakopi, Melos: the core for the socket rested in the recesses in the centre and the mould would be closed by an identical half (based on Atkinson); *d* two half-moulds for looped-spear-heads and a knife dagger from Ireland: both spearhead moulds were probably cast from the tip and the clay cores supported by means of the two sets of hollows in the bases (after Coghlan and Raftery)

23 Method of use of diverted type of tuyere in copper refining

24 Some examples of stone moulds of LBA type

A good many of the stone moulds were for trinkets, i.e. a multitude of small articles of jewellery or votive objects. Most of them are made of steatite. One shows rather more than the technique of use;[31] the variety of objects made with this one mould is so great that it suggests the smith who used it must have travelled a circular route round much of Western Anatolia in about 2200 BC. For this reason, it would have been important that the moulds weigh as little as possible and all surfaces, including the sides, were used. This is clearly shown by the trinket moulds from Poros and one of the earlier moulds from Mallia, referred to above.

More massive semi-finished objects, such as discs, were also cast into stone moulds.[32] Such a set was found on the Welsh hillfort of Dinorben in a 3rd–4th century AD context (*see* Fig.24). The 39 cm dia. discs were probably used for the production of cauldrons of Bronze Age type.

Clay moulds are so fragile that there are not many examples from this period, but sites such as Dainton in Devon are now producing fragments of clay moulds and crucibles. Copper and bronze moulds, on the other hand, have survived in plenty. Their use often comes as a surprise to those who find it difficult to believe in the sophisticated techniques of early metalworkers. The composition of the mould usually reflects the metal cast into them, and it is not surprising that we have few from the Copper Age considering the difficulty of casting copper, with its high melting point. However, we have a copper mould from Coppa Nevigata in Italy which contained no tin but 0.5%Ni and Co, 0.3%S, and some iron.[22] In Britain, bronze moulds date from the Middle Bronze Age (1400–1000 BC): the examples from the LBA are made of leaded bronze, as were the artefacts cast in them (*see* Figs.26 and 27).

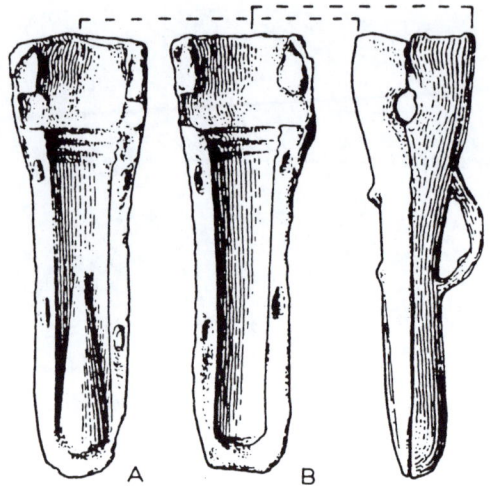

26 *Examples of bronze moulds with trunnion cores*

Bronze moulds are somewhat heavier than stone moulds, although the difference is not very great as the metal moulds have thinner walls. They would not have suffered from thermal shock and therefore would generally have lasted longer than stone. Metal moulds deteriorate because of distortion and the welding of the hot metal to a badly dressed mould. Nothing is known about early mould dressings but it is possible that a smoky flame from burning oil was used. Subsequently, the two halves would have been heated to about 200°C. Registration of the two half-moulds is sometimes brought about with dowels and holes, sometimes by external marks, and in the case of some of the British bronze moulds, by allowing one half to fit inside a flange cast round the other.

In the case of both stone and bronze two-part moulds, the running of the metal into the mould must have been assisted by clay runner bushes stuck on to the sides of the assembled mould, clay/sand cores would have provided the socket.[36]

Many arrowhead moulds have been found and one twopiece stone mould from Susa (3000–1000 BC) was intended for four tanged arrowheads.[34] Another, from Mosul,[35] is a very sophisticated multipart bronze mould for socketed arrowheads with a core (*see* Fig.25) and is dated to 700–600 BC. A tanged arrowhead from Marlik Tepe, Iran (about 1000 BC), had started as a cast rod, and the two lobes had been forged out of the bar.[37]

There is now little doubt that metal was cast into the stone moulds. This has been confirmed by finding traces of copper-base metals on the surfaces of stone moulds.[38] But, when it comes to metal moulds, the situation is not always so clear. While the British bronze moulds shown in Fig.27 have undoubtedly been used for direct casting, it is possible that some bronze moulds have merely been used for wax casting. The delicacy of the ingate system used in the bronze socketed axe mould from Orviedo, Spain suggests that this mould and core was not used for direct casting of metal.[39] The mould from Mosul shown in Fig.25 may also be an example of the use of lost-wax casting.[35]

Lost-wax casting involves the making of wax patterns in pattern moulds, or carving them, and investing them

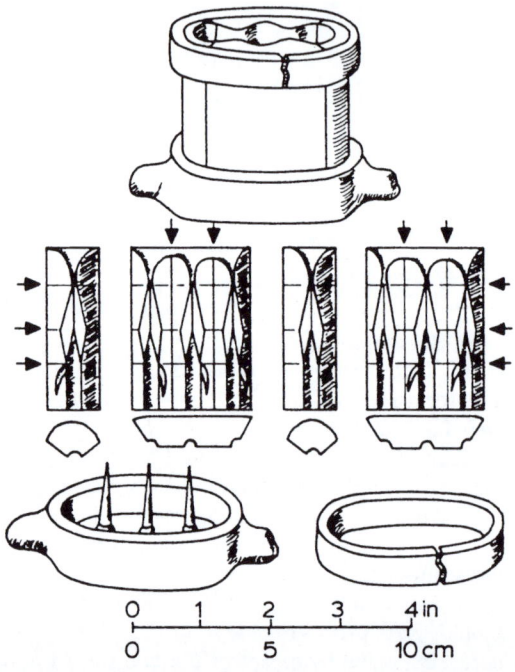

25 *Bronze arrowhead mould from Mosul, Iraq (after Maryon[35]; courtesy of The American Journal of Archaeology)*

made of dried clay or wax with a clay core. Each individual mould-piece is mated to the next ones above and below, and to those at the sides, by means of mortices and tenons. When the mould has been built up brick-by-brick and dried, it can be taken apart, the pattern removed, and the mould reassembled.[40] The clay pattern may be scraped down to give the metal thickness required, or a new core of a rough but smaller shape may be made and baked. There is no evidence for the latter but, if used, it would have preserved the original pattern for the making of new moulds.

The moulding material was the normal loess of North China and consisted of sand, limestone and a little clay. It was not ideal as a moulding material because of its limited plasticity. Most of the material cast into it would be 15–20%Sn bronze, with or without lead, which has a comparatively low melting point of about 900–950°C.

To ensure that the casting had the designed wall thickness, metal spacers called 'chaplets' were positioned in the space between the mould and the core. These can be seen in a bronze *hu*-vessel of the Western Han Dynasty (202 BC–AD 9) as square pieces of metal with a lighter corrosion product. They should, in fact, be made of a metal of slightly higher melting point than the metal being cast so that they undergo slight surface fusion and thus bond to the new metal without losing too much of their strength.

In stone or metal moulds, the cores are usually positioned by an extension which is located in a specially prepared cavity (print) in the mould. A two-piece clay mould for a shaft-hole axe from Kalinovka in South Russia (2000–1800 BC) was found, together with a core consisting of a bar or peg which was located in the prints in the two halves of the mould[41] (*see* Fig.8). In European bronze moulds for socketed axes or gouges, the clay core forming the socket is sometimes held in place with a 'trunnion' (*see* Fig.26). In other cases the core is held in position by extensions of the mould which act as a clamp (*see* Fig.27). In the Mosul bronze mould for three socketed arrowheads, the three cores for the sockets are part of the bottom part of the mould and stand vertically like three spikes (*see* Fig.25).

Fabrication and Joining

Another of the outstanding features of the LBA is the wealth of tools made for wood, stone and metalwork. Socketed hammers, gouges, chisels, punches and awls were used in quantity. Two-handed saws, probably of bronze and several being as long as 1.5 m and 23 cm wide, have been found in Crete.[42]

The techniques of working and decorating had been well developed. Bowls and vases were made by sinking, i.e. with blows on the inside of the vessel, or by raising, i.e. by blows on the outside of the vessel, both done with special hammers and anvils. Finishing by turning was certainly known by the EIA but there is no evidence for earlier usage (*see* Fig.28).

Decoration can be applied by repoussée, in which the design is in relief and the background depressed by means of the hammer or punches. Alternatively, the pattern is worked from the back into a soft base of

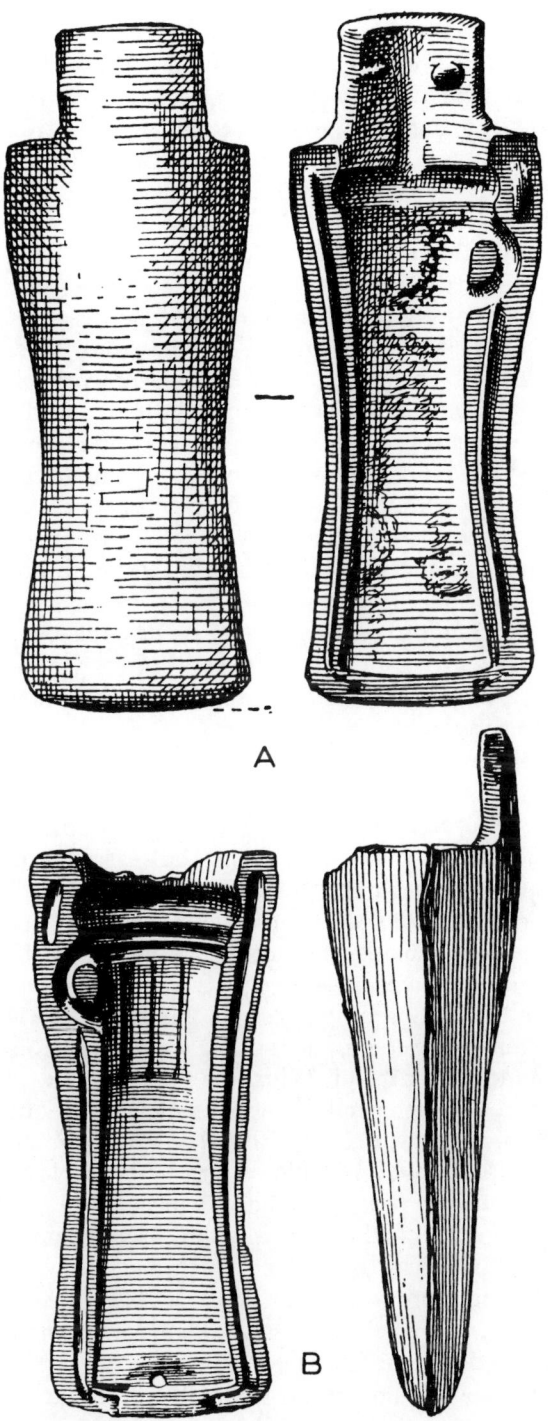

27 *Examples of bronze moulds with clamp cores*

in a sand-clay mixture which is dried. The wax is then melted and run out, and molten metal is poured into the preheated mould. This process is glibly referred to by many authorities, but incontrovertible evidence of its use before the Iron Age is still lacking.

In China, a 'piece-moulding' technique was used and it probably originated from a similar process used for the decoration of pottery. This process was later introduced into Europe and is the basis of the decoration in the well known Roman *terra sigillata* or Samian ware. Piece-moulding consists of making an impression of the pattern or model by impressing square or rectangular blocks of wet clay upon it until the whole pattern is enveloped in individual 'bricks'. The pattern could be bitumen or lead (*see* Fig.28). Details may be applied by the graver or the tracer. The former tool is pushed by the hand and cuts a groove, producing a small curl of swarf ahead of the tool which is removed. The outline may be completed and emphasised by means of short narrow marks with the chisel like tracer which is hammered on to the metal.[43]

Joining was carried out either by cold (forge) welding, which was probably only applied to gold, or by soldering. The latter was in general use for gold and silver work by the fourth millennium in Asia Minor. One of the earliest processes was the joining of gold by copper, or copper by gold. Certain copper minerals were applied to the area to be joined under reducing conditions, so that they were reduced to copper which alloyed with the gold and produced a low melting point solid solution which acted as the solder.[44] Fine gold granules would have done the same to copper. Later on, high-tin bronzes with low melting points were used to join low-tin bronzes with their higher melting points. Copper or copper–silver alloys could be used to join silver.

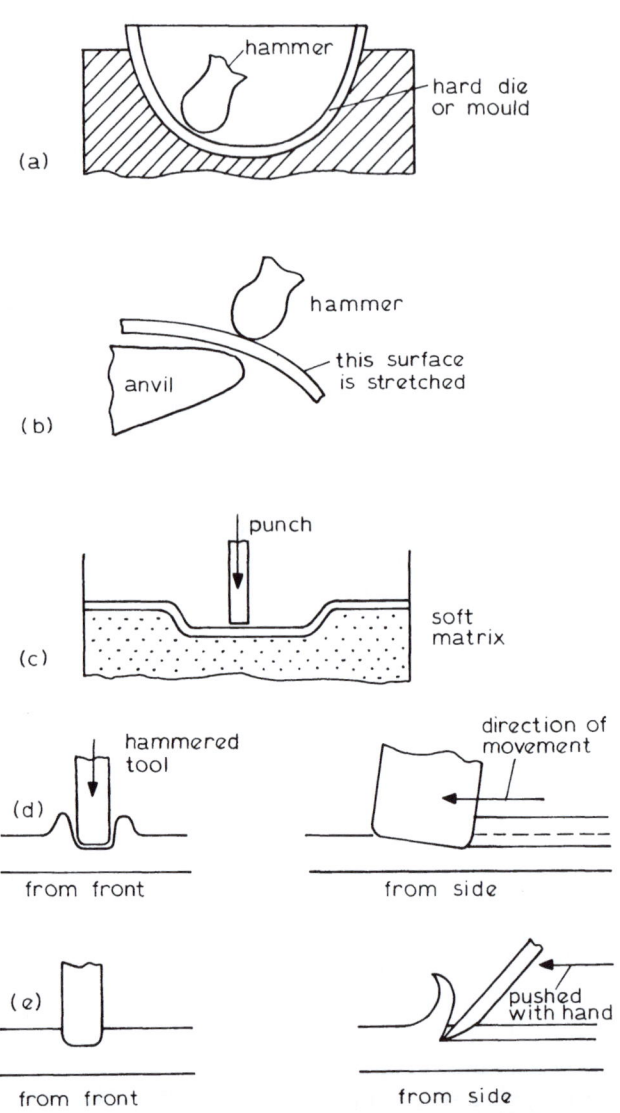

a sinking; *b* raising; *c* répoussée; *d* tracer; *e* graver or scraper

28 *Shaping and decorating techniques used in prehistoric times (after Lowery et al.[43])*

The Spread of Bronze Age Techniques

In Iran, the flowering of the LBA is best shown in the so-called 'Luristan Bronzes'. Luristan is an area in South West Iran, in the Zagros mountains, which borders on Mesopotamia and which could be expected to be one of the main sources of minerals for this region. Although there was a local industry from about 3000 BC which went through the normal metallic sequence, its zenith was reached in the period 900–650 BC when the knowledge of iron was gradually being introduced to this area.

Most of the Luristan bronzes have no known archaeological provenance.[45] But sufficient work has now been done to identify the types produced and the area they came from. Most of the artefacts come from plundered cemeteries and cist graves, and many are horse trappings. Although undoubtedly LBA in period, few of the bronzes are leaded, showing that this tendency is by no means universal.[46] Most cast bronzes contain between 4 and 13%Sn.

The standard sword of the LBA was a one piece flat-hilted sword usually made of leaded bronze.[47] In the Near East, and in Luristan in particular, there were variations on this involving the 'casting-on' of a separate hilt to a simple tanged bronze blade.[48,49] Later, when iron was coming in, the hilt was cast-on to a tanged iron blade.[48] It is believed that the cast-on bronze hilt on a bronze blade was derived from the bronze-hilted iron blade and it is very probable that there was some overlap.[48] Considering the difficulty of making iron hilts for some of the early iron blades it is not surprising that the bronze hilt was often preferred.

In the casting-on process,[49,50] the bronze blade was first made by casting, and the tang was then surrounded by a mould for the hilt. This was probably a two-piece clay mould with runners and risers. In at least one case the tang was covered with a thin clay–sand 'core' and the cast-on hilt only came into contact with the blade near the shoulder, producing only a mechanical joint. In another example the blade and tang were made from a bronze with less than 5%Sn which had been worked, while the cast-on hilt was an 11.5%Sn bronze. As in many other cases the cutting edge had been finally hardened by cold hammering.[51]

The casting-on process is capable of infinite elaboration and was used in China for attaching legs to vessels. A projection was cast into the original vessel, a new mould was then built on to the vessel and over the projection so that the new cast metal leg interlocked with, and sometimes fused on to, the projection.

In India, the Indus civilizations of Harappa and Mohenjo Daro came to an end at about 1750 BC, probably due to an invasion from the north.[52] To the east, on the Ganges, many copper-base hoards dated to 800 BC have been found. Some objects are of arsenical copper and others of tin bronze which show Caucasian affinities. This period ended about 200 BC with the coming of the Iron Age.

Perhaps the greatest advances were made in the Far East. One might say that China exploded into the LBA. Certainly she made up quickly for her apparently late arrival on the metallurgical scene. Shang sites antedating the Anyang period (before 1400 BC) were discovered at Erhlikang near Chengchou in 1953. Elsewhere, in Chengchou, in Hantan in Hopei, and at Anyang itself, bronze weapons and tools, *ting*-tripods, and many other artefacts have been found that belong to this earlier period.[53] These early bronzes are comparatively crudely made, few in variety and simple in design. Later in the Anyang period, there was a widening in the sphere of Shang influence and trade and it was therefore possible to obtain copper and tin from a larger area; this coincided with an improvement in technique.[54]

By the time of the Western Chou Dynasty (770 BC) there is no doubt at all that we are in a Late Bronze Age. The artefacts are large and there is a great quantity of them; the tin content is more consistent and the casting technology is typical of the LBA. A bronze cauldron found in Anyang in 1946 weighed 1 400 kg (3 100 lbs) and was about 1 m across.[1] Most of the compositions of this period were leaded (*see* Tables 24 and 25) and the objects were often cast in clay piece-moulds which were made from individual baked clay pieces; these were made from a baked or dried clay pattern on to which the detail had been inscribed or carved. Pewter (lead–tin alloy) objects are also known from early Chou times.

It is probable that low-shaft furnaces were used for smelting. From the Ming period (AD 1368–1644) we have the remains of a 10 m runner channel that might have been used for casting large objects in a pit, like the technique used in medieval Europe for bell casting.[1] We also know that two-handled ladle or crucible furnaces were in use at this time. But we really have no evidence to show how the Chou *ting*-cauldrons were cast.

The modern copper mining complex at Tonglushan shows extensive evidence of LBA mining and smelting.[55] Material dates to the Spring and Autumn period (770–

Table 25 Composition of Chinese LBA artefacts (after Dono[75])

| Element | Content, % | | | |
	No.1 Halberd	No.2 Dagger	No.3 Socketed axes	No.4
Cu	78·68	82·21	68·65	82·18
Sn	18·40	9·34	9·70	12·24
Pb	1·71	7·50	19·83	2·60
Fe	0·30	0·15	0·08	0·08
As	tr.	tr.	tr.	tr.
Ag	0·14	0·06	0·03	0·03
Au	tr.	tr.	tr.	tr.

479 BC) and bronze tools were replaced by iron ores later in the sequence which continued into the Han dynasty (second century AD). The ore was smelted in oval-hearthed shaft furnaces (hearth dia. 68 × 27 cm) with two tuyeres. The metal produced was high in iron due to the use of ferruginous fluxes.

Coins of various shapes were being made by Chou times and these tended to have a low tin content: some contained appreciable amounts of nickel (*see* Table 26). The coins were cast in clay moulds fired in kilns by Han times (208 BC–AD 220). Some late Chou coins analysed by means of X-ray fluorescence were found to have high tin (11.5–13%Sn) content, and a lead content as high as 30–36%. These indicate considerable segregation. It is clear that coiners were using cheap high-lead compositions with good fluidity.

Bronze mirrors were introduced during the Eastern Chou period (770–475 BC) and were very popular in the Han period (208 BC–AD 220).[56] Here, the composition is reversed, i.e. they consist of high-tin bronze (speculum) with low lead, the tin content being 24–31%, and the lead content 0–9%.

Table 24 Composition of Chou bronzes (1100-221 BC)

| Object | Composition, % | | | | | | |
	Sn	Pb	Ni	Zn	Sb	As	Ref.
Armour	9·44	2·33	0·01	0·07			1
Hemisphere	11·67	11·33	0·01	0·09			"
Halberd	12·84	0·26	tr.	0·05			"
Bell	13·27	nil	0·05	0·06			"
Bell	16·88	0·04	nil	0·93			"
Horse bit	14·44	8·37	0·08	0·04			"
Kuei	16·70	13·65	0·06	0·13			"
Bell	17·45	8·45	0·03	0·08			"
Spear cap	17·71	11·45	0·08	0·05			"
Ting	18·21	8·65	0·08	0·15			"
Sword	19·84	–	2·47		3·8	0·55	77
Chisel	4·60	tr.	–	–	–	–	74
Buckle	8·19	5·1	–	–	–	–	"
Head	15·36	10·92	–	–	–	–	"
Tripod urn	8·19	9·79	–	–	–	–	"
Spearhead	15·76	5·30	–	–	–	–	"
Spearhead	0·94	18·04	–	–	–	–	"
Beaker	12·89	12·64	–	–	–	–	"
Dish	2·05	40·89	0·10	0·13	0·37	0·02	
Vessel	20·30	4·60	–	–	–	0·1–1·0	76

Table 26 Chinese bronze coin compositions

Ref.	Type	Date, BC	Composition, %			
			Sn	Pb	Ni	As
77	Knife Coin	770−249	1·66	55·41	1·03	0·6
"	Knife Coin	770−249	2·12	48·60	1·63	1·72
"	Knife Coin	770−249	2·12	47·32	5·08	3·28
"	Knife Coin	770−249	6·76	21·25	3·04	3·88
"	Knife Coin	522	12·5	36·0	0·18	−
"	Knife Coin	412	11·5	30·0	0·21	−
"	Knife Coin	400	13·0	17·0	0·20	−
"	Knife Coin	c. 300	3·0	27·0	0·17	−
"	Knife Coin	c. 300	6·0	24·0	0·22	−
"	Knife Coin	c. 300	9·25	25·5	0·10	−
"	Spade Coin	250−221	3·5	23·0	0·21	−
"	Spade Coin	340−325	1·8	62·0	0·25	−
56	Spade Coin	722−481	9·92	19·3	0·35	

The Han period arrowheads were of a more normal composition with 10.8%Sn, but they must have been made from tetrahedrite ore as they also contained 8.0–8.7%Sb, 2.0–3.7%Ni, and a trace of silver. The bronze sword was also introduced during the late Chou period (770–475 BC). There are signs of similarity with European Hallstatt types, which maybe indicative of west–east diffusion.[57] The compositions were not unlike the Han arrowheads mentioned above.

The Eastern Chou period saw the beginning of the Iron Age. Cast iron cauldrons have been dated as early as 512 BC, but bronze types lasted for a considerable time and cast iron moulds were probably used for casting some bronze artefacts.[58] Other evidence of overlap is provided by the existence of long-shafted pickaxes, which have a core of iron around which bronze has been cast. Arrowheads with iron stems and bronze points have also been found. It appears that, while cast iron was soon used for agricultural tools, the need for a less brittle metal for weapons meant the continued use of bronze in this sphere, while the conversion of cast iron to wrought iron and steel was being mastered .

Zinc was beginning to appear in bronzes in Han times, but true copper–zinc alloys (brasses) did not appear until post-Han times, i.e. after AD 220. There is no evidence that the Chinese used brass at the same time as the Romans in Europe, but the appearance of zinc, together with nickel, in copper-base alloys makes it seem likely that the principal source of the zinc (and the nickel) was a complex Cu–Ni–Zn ore. It was, without doubt, this ore which in later times gave rise to a ternary alloy of this composition known as *paitung* or *paktong*.

Other Metals

GOLD

The attention given to copper alloys is, of course, justified by their position in ancient economics. But to judge from museum exhibits one would have thought that gold and silver were of equal importance to bronze. Naturally, these metals were used for artistic purposes, and because of their corrosion resistance have great lasting properties. Thus, gold can be used again and

again, so that much of the gold one sees in museums has probably been re-used many times

The production of gold was almost entirely a question of organization of manpower and not of technique since the metal occurs mostly in native form and no smelting is required. The wealthy civilizations of Asia Minor, the Fertile Crescent, and particularly Egypt, could afford to employ workers to mine the gold wherever it was known to occur. Furthermore, such countries had big armies and could seize it from others or obtain it in the form of tribute. So the gold changed hands, as indeed it does today for rather different reasons.

The only techniques involved were those of mineral dressing, i.e. grinding the quartz rock fine enough to release the small particles of gold, and then washing away the fine and less dense rock with a controlled stream of water. In some cases, the particles of gold were caught in a fleece because of its oily and fibrous nature, and we have the stories of Jason and others to support the use of this technique in early times. In Nubia and Afghanistan the bottom stones of grinding mills can be seen. These are of the hemispherical edge-runner type known to have been used for olive mills in Roman times and their dating is probably much later than the period under consideration. Naturally, nuggets would be used when found and these still come to light in unexpected places. One weighing 610 g was found near Dublin in the 18th century,[59] and one weighing 62.2 kg was found in Brazil in 1983.[60]

Once the gold dust was concentrated it would have been melted down in small crucibles and cast into trinkets or ingots. By the MBA in Britain the natural gold–silver alloy was modified by the addition of copper.[61] Analyses have shown (*see* Fig.29) that the amount of copper added to gold depended on its silver content. It was, therefore, probably added to improve the colour. Gold with more than 20%Ag would have given a whitish colour (electrum), which does not seem to have been as popular as the more reddish colour of pure gold. Additions of copper, however, would have hardened the gold and made it less easy to work.

Gold, with its easy workability, was an ideal material

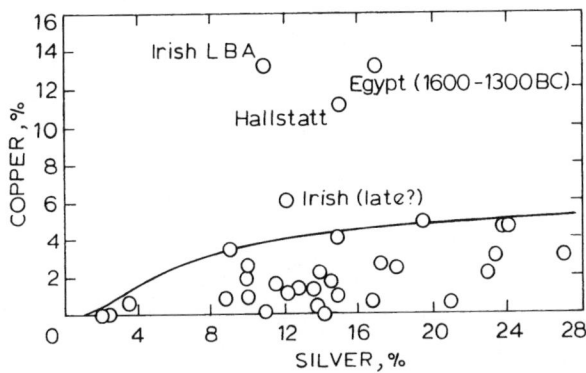

29 *Silver and copper contents of early gold artefacts as evidence of the use of copper in colouring the whiter natural golds or electrums*

for ornaments, and the craftsman could lavish his skill on the making of beautiful objects. Much of the early gold work is made from very thin sheet or leaf. This was made by hammering out a small ingot into a sheet. This sheet was cut into squares which were placed on top of each other with thin animal skin in between. The resulting packet was hammered down again to make the sheets even thinner and the skin layers prevented the sheets from sticking to each other. This process was repeated many times until each packet of gold consisted of thin leaves a few hundredths of a millimetre thick.[62] The figurines of the bulls found at the Royal cemetery at Ur were covered with gold varying in thickness from 0.5 to 2 mm. Decorations on ornaments and swords consisted of repoussée, filigree, and granulation. We also find much inlay, such as in the Mycenaean daggers, where gold is used with great effect against a black background to provide colour contrast.

SILVER AND LEAD

Although native silver does occur, it is comparatively rare, and a good deal of early silver was derived from the cupellation of lead.[63] The cupellation processes involves the oxidation of the lead to litharge (PbO) at temperatures of the order of 900–1000°C. This is usually done in a saucer-shaped hearth with the fuel on top (wood or charcoal), with a large pair of bellows positioned on the lip of the saucer. The hearth is often made of easily available bone ash (burnt bones) and the PbO is poured off, absorbed by the hearth, or vapourized to leave the unoxidized silver. The sign of the use of such a process is the presence of cakes of litharge or the remains of the cupels.

This process must have been known by about 2000 BC, judging from the finds of fine silver objects found at Ur, Maikop, and many other sites. Silver and lead bars were found at Troy I and II (2500–2000 BC), and the lead content of the silver suggests the use of cupellation.[64] Three silver daggers were found in a tholos tomb in Crete belonging to the Early Minoan period[65] (2200–2000 BC). One dagger was analysed and contained 71%Ag, 27.5%Cu, and 0.78%Sn. This was a Copper Age period in Crete and alloying with silver in this proportion would have given a very high-strength weapon, superior to

any weapon of this sort at the time. One feels, however, that these artefacts must have been used for ceremonial purposes.

There are large quantities of prehistoric slags near the copper mines of Rio Tinto[66,67] and Tharsis[68] in Spain which contain small amounts of silver but very little copper, and it is now believed that these represent slags resulting from the smelting of the high-silver layers at the base of the gossan.[66] The silver values have been concentrated by a leaching process into argentojarosites containing 0.20%Ag, which are now almost entirely worked out but which were extensively smelted for their silver content in LBA and perhaps Roman times. Exactly how the silver was recovered is not at present known, but the slags are essentially fayalite like other slags and contain some lead (1.37%). It is possible, therefore, that lead ores were added to the argentojarosites to collect the silver, and the silver metal recovered from the high-silver lead by cupellation. Litharge has been found at Tharsis which is said to date from the 'Pre-Phoenician' period. The lead remaining in the slags from the LBA silver mines at Rio Tinto contained about 0.06%Ag, equivalent to 600 g/t, which indicates that the silver content of the lead produced was at least equal to that of the famous silver mines of Laurion in Greece, of a later date.[69]

Lead itself was not of much value in the LBA, but was often used for the objects produced in trinket moulds which may have had cult or votive significance.[70] The socketed axes found in Brittany in great numbers contained 45%Pb, the remainder being copper, and would have been useless as working tools.[71] Some of these axes were found laid out in a circle and it has been assumed that they must have been votive. A large full-size lead axehead has been found in a bronze mould from Cambridge. This raises the question of whether bronze moulds had been used for the production of lead patterns for investment and casting. In view of the large numbers of miniature lead axes found this seems unlikely, and one must assume that lead axes and copper–lead axes were used for ritualistic or votive purposes.

References

1   N. BARNARD: 'Bronze casting and bronze alloys in ancient China', 1961, Canberra, The Australian National University and Monumenta Serica
2   B. ROTHENBERG and A. LUPU: Bulletin no.9, Museum Haaretz, 1967, Tel Aviv
3   R. F. TYLECOTE *et al.*: *J. Inst. Metals*, 1967, **95**, 235
4   J. F. MERKEL and R. F. TYLECOTE: 'Experimental casting of an oxhide copper ingot', Paper given at the 22nd symposium on Archaeometry, Bradford, 1982
5   J. F. MERKEL: 'Summary of experimental results for LBA copper smelting and refining', *MASCA. J.*, 1983, **2**, 173–178
6   J. DU PLAT TAYLOR: *Ant. J.*, 1952, **32**, 133
7   R. F. TYLECOTE: 'From pot bellows to tuyeres', Levant, 1981, **131**, 107–118
8   E. PREUSCHEN: 'Copper', 1966, Hamburg, Norddeutsche Affinerie
9   R. PITTIONI: *Arch. Aust.*, 1958, 29–32, (3), 19, et seq

10 K. ZSCHOCKE and E. PREUSCHEN: 'Deas Urzeitliche Bergbaugebiet von Muhlbach-Bischofsofen', Wien, 1932

11 C. EIBNER, 'Kupfererzbergbau in Osterreichs Alpen', In: *Sudösteuropa zwischen*, 1600 und 1000 v. Chr., (Ed. B. Hansel), *Prähistorische Archaologie in Sudosteuropa*, Berlin, 1982, **1**, 399–408

12 R. PITTIONI: *Man*, 1948, **48**, 120

13 P. T. CRADDOCK and D. GALE: 'Evidence for early mining and extractive metallurgy in the British Isles', In: *Science and Archaeology*, Glasgow, 1987, (eds. E. A. Slater and J. O. Tate), *BAR Brit. Ser.*, No. 196, 1988, 167–191

14 H. PRESSLINGER, C. EIBNER *et al.*: 'Ergebnis der Erforschung Urnenfelder-zeitliche Kupfermetallurgie in Paltental', Berg und Huttenmannische Monatshefte, 1980, **125**, (3), 131–142

15 P. T. CRADDOCK: 'Bronze Age metallurgy in Britain', *Curr. Arch.*, 1986, Feb. 106-108

16 F. LOSCHIAVO: 'Copper metallurgy in Sardinia during the Late Bronze Age: new prospects on its Aegean connections', In: 'Early Metallurgy in Cyprus', (ed. J. D. Muhly, R. Maddin *et al*), Nicosia, 1982, 271–281

17 R. F. TYLECOTE: 'The Prehistory of Metallurgy in the British Isles', *Inst. Metals*, London, 1986, 18

18 R. F. TYLECOTE: *Corn. Arch.*, 1967, **6**, 110

19 C. DORFLER *et al.*: *Arch. Aust.*, 1969, **46**, 68

20 H. H. COGHLAN: 'Notes on the prehistoric metallurgy of copper and bronze in the Old World', 1951, Oxford, Pitt-Rivers Museum

21 P. E. NEWBERRY: 'The life of Rekhmara', 1900, London, Pl. XVIII

22 A. MOSSO: 'The dawn of Mediterranean civilization', 1910, London

23 H. W. CATLING: 'Cypriot bronzework in the Mycenaean world', 1964, Oxford, Oxford University Press

24 G. F. BASS *et al.*: 'Cape Gelidonya; a Bronze Age Shipwreck', *Trans. Amer. Phil. Soc.*, 1967, **57**, (8), 177

25 G. F. BASS, D. A. FREY and C. PULAK: *IJNA*, 1984, **13**, (4), 271–279

26 J. and E. LAGARCE, A. BOUNNI and N. SALIBY: 'Fouilles a Ras Ibn Hani', *Acad. des Inscript. et Belles Lettres*, Nov. 1983, 249–290

27 C. J. DAVEY: 'Some ancient Near Eastern pot bellows', Levant, 1978, 11, 101–111

28 M. KOKHAVI and K. OBATA: *Israel Explor. J.*, 1965, **15**, 253; 1966, **16**, 274 (these tuyeres were examined and drawn in Tel Aviv with the kind assistance of Prof. Ben Rothenberg)

29 F. CHAPOUTIER and P. DEMARGUE: 'Fouilles à Mallia', 1942, Paris

30 T. D. ATKINSON *et al.*: 'Excavations at Phylakopi in Melos', 1904, London

31 J. V. CANBY: *Iraq*, 1965, **27**, 42

32 G. GUILBERT, Dinorben Hillfort, North Wales. Pers. Comm. 1980

33 S. NEEDHAM *et al.*: 'An assemblage of LBA metal working debris from Dainton, Devon', PPS, 1980, **46**, 177–215

34 R. DE MECQUENEM: *Mét. et Civil.*, 1946, **1**, (4), 77

35 H. MARYON: *AJA*, 1961, **65**, 173

36 H. HOWARD: 'An axe core from East Kennet', *Wilts. Arch. Nat. Hist. Soc. Mag.*, 1983, **77**, 143–4

37 R. F. TYLECOTE: *Bull, HMG*, 1972, **6**, 34

38 D. A. JENKINS: 'Trace elements analyses into the study of ancient metallurgy', *Aspects of ancient mining and metallurgy*, Bangor, 1986, 95–105

39 R. J. HARRISON: 'A Late Bronze Age mould from Los Oscos (Prov. Oviedo)', Madrid, Mitteil, 1982, **21**, 131–139

40 C. S. SMITH: 'Metal transformations', (eds. W. W. Mullins and M. C. Shaw), 1968, New York

41 M. GIMBUTAS: 'Bronze Age cultures in Central and Eastern Europe', 1965, London

42 These can be seen in the Museum of Antiquities, Heraklion, Crete

43 P. R. LOWERY: *PPS*, 1971, **37**, (1), 167

44 H. MARYON: *AJA*, 1949, **53**, (2), 99

45 P. R. S. MOOREY: *Iran*, 1969, **7**, 131

46 P. R. S. MOOREY: *Archaeometry*, 1964, **7**, 72

47 K. R. MAXWELL-HYSLOP: *Iraq*, 1946, **8**, 1

48 K. R. MAXWELL-HYSLOP and H. W. M. HODGES: *Iraq*, 1964, **26**, 50

49 J. BIRMINGHAM *et al.*: *Iraq*, 1964, **26**, 44

50 H. DRESCHER: Der Uberfanggguss, 1958, Mainz, RGZM

51 J. BIRMINGHAM: *Iran*, 1963, **1**, 71, et seq.

52 SIR M. WHEELER: 'Early India and Pakistan', 1959, London, Thames and Hudson

53 SUN SHUYUN and HAN RUBIN: 'A preliminary study of early copper and bronze artefacts', Kaogu Xuebao, 1981, (3), 287–302

54 N. BARNARD and SATO TAMOTSU: 'Metallurgical remains of ancient China', Nichiosha, Tokyo, 1975

55 *HUANGSHI MUSEUM, CHINESE SOC. MET., et al.* (eds.): 'Tonglushan; a pearl among ancient mines', Cult. Relics Publ. House, Beijing, 1980

56 M. CHIKASHIGI: *J. Chem. Soc.*, 1920, **117**, 917

57 J. NEEDHAM: 'The development of iron and steel technology in China', 1958, London, The Newcomen Society

58 LU DA: *Vita pro Ferro*, 68, 1965, Schaffhausen, Festschrift für R. Durrer

59 G. A. J. COLE: 'Localities of minerals in Ireland', 1922, Dublin, HMSO

60 The *Guardian*, 22 Sept. 1983

61 C. F. C. HAWKES: *Archaeometry*, 1962, **5**, 33

62 J. H. F. NOTTON: 'Ancient Egyptian gold refining', *Gold Bull.* 1974, **7**, 50-56

63 R. J. FORBES: 'Studies in ancient technology', 193, vol.8, 1964, Leiden, Brill

64 H. SCHLIEMANN: 'The city and country of the Trojans', 1880, London

65 S. XANTHOUDIDES: 'The vaulted tombs of the Mesara, 1924, London, Liverpool University Press

66 J. C. ALLAN: 'Considerations of the antiquity of mining in the Iberian peninsula', Occ. paper No.27, 1970, London, RAI

67 B. ROTHENBERG and A. BLANCO-FREIJEIRO: 'Ancient mining and metallurgy in SW Spain', *IAMS*, London, 1982

68 S. G. CHECKLAND: 'The mines of Tarshish', 1967, London, Allen & Unwin

69 A. BLANCO and J. M. LUZON: *Antiq.*, 1969, **43**, 124

70 S. P. NEEDHAM and D. R. HOOK: 'Lead and lead alloys in the Bronze Age', In: *Science and Archaeology*, Glasgow, 1987, (eds. E. A. Slater and J. O. Tate) *BAR. Brit. Ser.* No. 196, 1988, 259–274

71 PITRE DE LISLE: *RA*, 1881, **42**, 335

72 H. G. BUCHHOLZ: 'Minoica—Festschrift zum J. Sundwall', 921, 1958, Berlin; and *PZ*, 1959, **37**, 36, et seq.

73 B. DIMITROV: 'Underwater research along the south Bulgarian Black Sea coast in 1976-77', *IJNA*, **8**, (1), 70–79

74 W. F. COLLINS: *J. Inst. Metals*, 1931, **45**, 23

75 T. DONO: *Bull. Chem. Soc. Japan*, 1932, **53**, 748

76 R. J. GETTENS: *J. Chem. Educ.*, 1951, **28**, 67

77 C. F. CHENG and C. M. SCHWITTER: *AJA*, 1957, **61**, 351

Chapter 5
The Early Iron Age

It is generally accepted that the Iron Age started in Asia Minor, where iron-using people have occupied the area from about 2000 BC. During the Bronze Age copper ores would have been smelted with the aid of iron fluxes, with a distinct possibility of iron being reduced in the bottom of the furnace. This would have caused the furnace bottoms to contain much slag and ductile iron, as can be seen in more recent copper smelters in Central Iran.[1] This possibility could have occurred anywhere in the LBA and there is no reason why the peoples of Asia Minor should have made use of it before anyone else, except for the fact that they had had a longer acquaintance with copper smelting.

As has been pointed out in Chapter 1, there are occasional references to iron in earlier periods, but the pieces of metal referred to have been found to be meteoric, or single artefacts that might well be out of context, or else the result of accidents of the type mentioned above. One example is the plate from the pyramid at Gizeh.[2] It is alleged that this was well stratified when found and therefore must date to c. 2750 BC. Certainly its composition and structure is not typical of later wrought iron with fayalite slag stringers, and it may well be an example of early ironmaking.

It is not known for certain who began to make iron intentionally, and in quantity. It could have happened in Anatolia where, at Alaca Hüyük, we have one of the earliest man-made iron daggers. It seems that the supply of man-made iron in the second millenium BC was small and spasmodic, but it gradually increased until it was being used on quite a large scale for weapons in about 1200–1000 BC. The slow development of iron was repeated elsewhere in the Iron Age world and, for a long time, bronze was to continue to serve for many of the applications of metal.

Because of its rarity, iron was at first used in small items of jewellery. Apart from this, its early use seems to have been for dagger blades. Cast tin bronzes and leaded bronzes are not sufficiently ductile to survive a forceful blow without breaking, and this must have been a severe disadvantage. When iron blades made their appearance the advantages would have been immediately obvious. A sword that bends may be straightened, but one that breaks is useless. So we see the Luristan smiths producing bronze-hilted iron blades in about the 11th century BC. One such blade from Marlik Tepe, Iran, had a cast hilt of 10–12%Sn bronze. The blade, which was really an inhomogeneous low-carbon steel, was harder than a cast tin bronze and far more ductile.[1]

The hilts of the Luristan iron swords are very primitive, and it is clear that joining separate pieces of iron to fabricate a hilt was a real problem to the early smiths.[3] The same problem can be seen in the joining of the pieces that go to make the miniature headrest found in Tutankamun's tomb (1400 BC).[4] It seems that the smiths were able to make only small pieces of iron, and that these were joined to other similar pieces without sufficient welding heat or deformation. We find that iron was beginning to appear in quantity by the 8th century BC. In Sargon II's Palace at Khorsabad (720–705 BC) 160 t was found, much of it in the form of double pointed square-section bars.[5] (see Fig.30[2]). No doubt this was tribute and came from a wide area, much of it from outside the Assyrian kingdom. This shape was to become common over much of Europe in the succeeding five centuries, when bars of various types were used as articles of commerce (see Fig.30).

The Expansion of Ironworking

If we assume that the knowledge of ironworking was incubating somewhere in the Anatolian–Iranian region during the period 1500–1000 BC, its spread across parts of Europe, Asia, and North Africa in the following five centuries can be explained. Soon after 1000 BC it was penetrating the coast of Palestine: the Philistines had iron in the 11th century BC[7] and we see its appearance at Gerar,[7] while Galilee and Egypt were still in the Bronze Age. There is little doubt that the Phoenicians had it as well and spread it to the Western Mediterranean and Carthage. From here it probably spread to Nigeria, where the Iron Age Nok culture was smelting iron by about 400–300 BC.[8,9] It entered Greece by about 900 BC and seems to have reached Egypt via Greek or Carian traders, as there is real evidence of smelting from the emporium of Naukratis[10] where iron slag was found in 7th century BC levels. In the Sudan, iron smelting started in about 200 BC and this knowledge could have come via Greek mercenaries into Egypt or through Mesopotamia, South Arabia, and Ethiopia.[11] Central and East Africa probably received their knowledge of iron smelting about AD 500 from Nigeria with the migration of the Bantu tribes. This route finally ended in South Africa about AD 1000.

Meanwhile, Greek or Anatolian influences were making themselves felt in Etruscan Italy and Spain. By 800 BC, Hallstatt Europe was converted, and the knowledge reached Britain in about 500 BC. At the same time, the movement in an easterly direction from Iran had

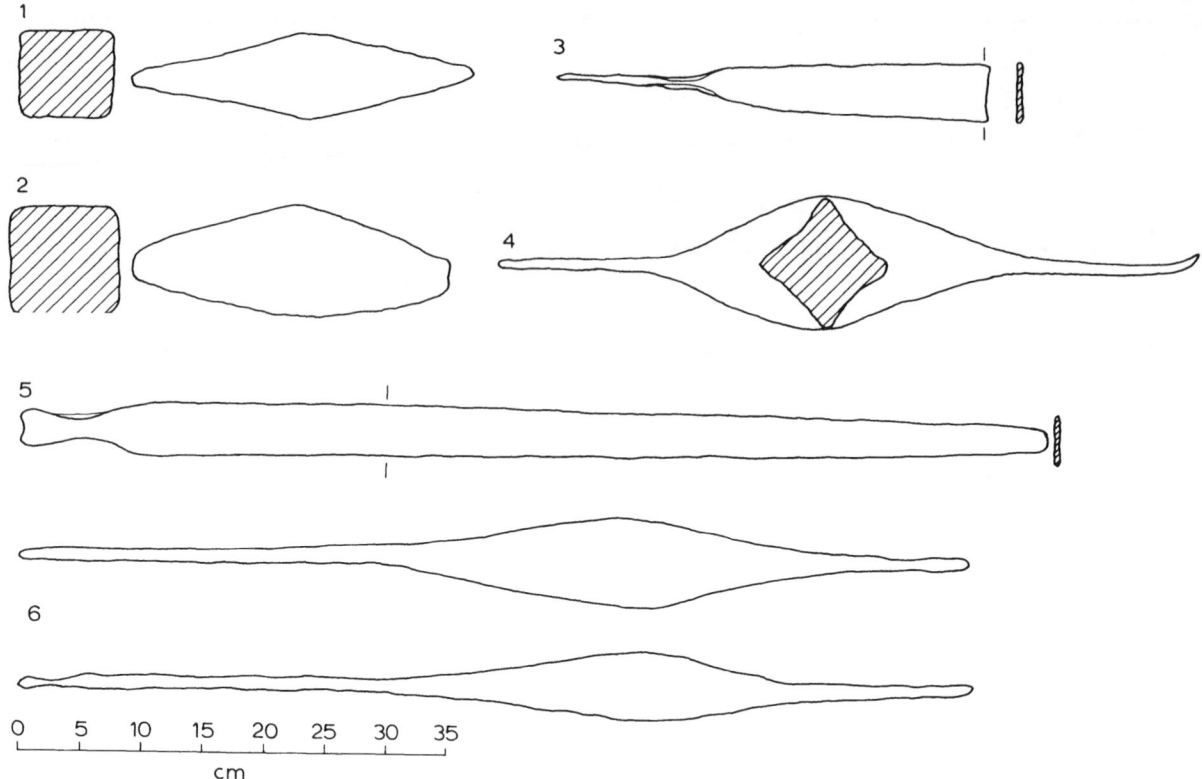

1 Spitzbarren from Mt/ Lassois, France, weight, 5 200 g (after France-Lanord[5]); **2** Stumpfbarren from Strasbourg, weight 6 540 g (after France-Lanord[5]); this is similar to those found at Khorsabad; **3** La Tene, Switzerland, weight about 700 g; **4** Spindle-shaped bar from Portland, UK (after L.V. Grinsell: 'The archaeology of Wessex'.1958, London); **5** Sword-shaped bar from Bourton-on-the-Water, UK (weight 560 g); **6** Spindle-shaped bar (two sides) from Neu Ulm, West German, weight 5 260 g (after Radeker and Naumann[34])

30   Some examples of iron bars used as articles of trade during the EIA

reached India, and possibly China, in about 400 BC. China was to be the one country where iron technology was to take a different course, i.e. towards cast iron rather than wrought iron, and it is therefore possible that ferrous metallurgy evolved independently from the highly sophisticated non-ferrous techniques of the region.[12] The same sort of accident that produced recognizable ductile iron at the bottom of a copper-smelting furnace could, under certain conditions, have produced cast iron and the Chinese would have realised that they had found a bronze substitute.

Technology

The technology of ironworking divides into two sections, smelting and hot forging. While the latter could have been known first and practised on meteoric iron, there is no evidence for this: all the working of meteoric iron has been done by cold hammering without heating.

Pure iron has a melting point of 1 540°C, and this temperature could not be obtained until the 19th century AD. So, all early wrought iron was produced in the solid state by chemical reduction of iron ore to solid, almost pure iron at about 1 200°C, with the aid of charcoal. The reduced iron was removed as a clod or 'bloom', which was a mixture of solid iron, slag, and pieces of unburnt charcoal. In some cases this lump was broken up and the small pieces of iron were separated by hammering; these could be distinguished from the rest

because they were ductile and would flatten on hammering. These were then welded up into a larger piece by heating them in a smith's fire followed by hot hammering. In some cases the bloom consisted of coherent iron and could be smithed in one piece. In other cases, the bloom was too large and had to be cut into smaller pieces which were individually smithed.

The product of the bloomery process can be very heterogeneous, with areas of high and low carbon and variable amounts of such elements as As and P. This gives rise to uncertainty when we come to examine the artefacts to determine the level of technique at a particular time. When we see relict areas of steel in a corroded iron object can we assume that the whole artefact was intentionally made of steel? When we see martensite in the structure can we assume that the quenching necessary to produce it was intentional or merely done by the smith to cool it?

If the ratio of fuel to ore is large and the bellows are efficient, the iron can be made to absorb so much carbon that it forms an alloy of iron and carbon or 'cast iron', which melts at 1 150°C and forms pools at the bottom of the furnace. These liquated lumps could have been broken up and remelted in a crucible in a hot smithing fire, and cast like bronze. It seems that early people in Asia Minor and Europe occasionally made cast iron by accident, but only the Chinese appreciated the advantages and made it regularly. Even so, it would not fulfil all the applications required of iron: wrought iron had to be made either by conversion of cast iron in a smith's fire or in the direct, European, manner with a lower fuel/ore ratio.

There is no doubt that the heterogeneity of bloomery iron was well known to early smiths, and that the high-carbon areas could be separated from the softer iron. Also it would be noted that an increase in the fuel/ore ratio would produce more of the harder iron. What is not clear, however, is how common was the knowledge that quenching such a steel into water from the right temperature was a method of hardening it further.

Furnaces

The main problem in deciding what sort of furnace was used in any period is the fact that, in most cases, only the base of the furnace remains. This has meant that many talk of 'bowl' furnaces in which the diameter is much the same as the height when excavated. We do not often know the original height but it is clear from experiments that the height/diameter ratio does not need to exceed 2:1 to get iron smelting conditions and that, with proper manipulation, smelting can be carried out with ratios less than this.

The bowl furnace is usually thought of as the simplest type of furnace. This is often no more than a hole in the ground or rock into which air from bellows can be directed through a tuyere with a short, probably cylindrical-shaped, superstructure of clay above *(see* Fig.31). The broken ore and the charcoal are mixed together or charged in layers on to a hot charcoal fire. The maximum temperature should be at least 1 150°C. This type of furnace has no outlet for slag, and the slag runs down to the bottom forming a cake or furnace bottom or, in some cases, just small rounded particles or 'prills' of slag *(see* Fig.32). The bloom remains above the slag, and after the process is completed the clay superstructure is broken away, the bloom removed, and the furnace cleaned out.

This type of furnace was superseded in Roman times by the 'developed' bowl furnace, which looks very similar to the Timna copper-smelting furnace *(see* Fig.16). This may be evidence that iron smelting was not invented by advanced copper smelters but by more primi-

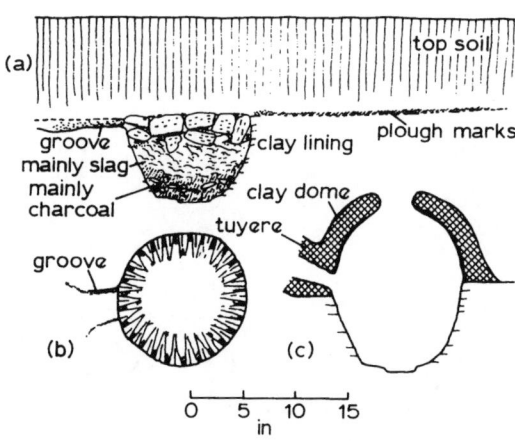

a section; *b* plan; *c* conjectural reconstruction

32 An example of an EIA bowl furnace from West Brandon, County Durhan (from G. Jobey[115])

tive copper smelters, or even by a completely new group who did not even know the technique of smelting copper. Whatever the explanation, the Early Iron Age in Europe was typified by the non-slag tapping bowl or shaft furnace, while the art of tapping slag used in the developed bowl furnace was not introduced into Europe until Roman times.

The shaft assists the maintenance of reducing conditions which is even more important for iron than for copper. The smithing furnace does not need reducing conditions and, although a hole in the ground can be used, it is not necessary, as can be seen from primitive forges still in use in developing countries.[13] All that is needed is a tuyere held down by a stone and long enough to keep the bellows from scorching. A pile of charcoal is then ignited with air from the goatskin bellows. The smith places his piece of iron in the charcoal near the mouth of the tuyere and a good heat can easily be obtained (1 200°C). With wrought iron, most of the work can be done by cold hammering and annealing at 700°C, and many primitive smiths work this way today.

There are hardly any remains of furnaces known from Asia Minor or Persia. The remains of a low shaft furnace measuring 0.4 m dia., with walls of 3–4 cm thick, and a height of 1 m, were found in an Islamic context at the site of Arsameia near Yenikale in Turkey; much slag in the form of furnace bottoms surrounded them.[14]

Further to the east at Sirzi, near Malatya, a large deposit of slag was found in a valley[15] *(see* Table 27). It was not free-flowing at the working temperature of the furnace and yet had a very high iron content which suggests the use of a rich ore *(see* Table 28).

Iron Ores

Table 28 gives some analyses of iron ores known to have been used in early periods. Up to now the only iron ores that were thought to be suitable for ironmaking were the oxides and the carbonates. But we know that sulphide nodules are readily available in some areas, e.g. in the south of England, and that many of them have been oxidized naturally to give high grade oxidized nodules or limonites.[16]

31 The lining of an EIA bowl furnace from Chelm's Combe, Somerset, showing tuyere at top; width about 300 mm

Table 27 Typical Early Iron Age iron smelting slags

Provenance	Date	Composition, %							
		FeO	Fe_2O_3	SiO_2	CaO	MnO	Al_2O_3	MgO	P_2O_5
UK, Maiden Castle, Dorset[103]	AD 24–25	53·00	22·87	15·95	2·75	tr.	1·47	0·45	0·40
Turkey, Sirzi[15]	7th century BC	55·65	13·96	8·60	5·16	7·18	1·89	4·87	1·95
Austria, Noreia[101]	700–600 BC	48·26	24·29	14·78	2·13	1·29	3·65	0·90	0·20
Austria, Noreia[101]	400 BC	55·39	12·62	24·48	1·99	2·35	2·54	1·43	0·15
Austria, Noreia[101]	Late La Tène	55·72	10·33	20·72	3·85	2·38	1·96	1·85	0·40
Czechoslovak M. Zehrovicich[102]	La Tène	51·63	20·08	18·37	1·73	0·46	0·60	–	1·83

The sands on the Turkish shore of the Black Sea are often black and magnetic and, like sands from Japan and Thasos in Greece, contain appreciable iron.[17] The Japanese sands have been very successfully used in the Tatara process but difficulties have been met with in some cases, e.g. in Virginia,[18] and the sands have had to be blended with other ores. Some of the sands contain appreciable Mg and others Ti.[19]

Iron from pre-Roman Britain can contain up to 1%P and 0.6%As, which have a marked hardening effect on the object.[20] Occasionally, as at Bogaskoy, iron arsenide is found, suggesting that such an ore or residue was used for ironmaking.[21] Experience in the use of arsenides would have been known from copper smelting, and it is possible that arsenical iron ores were used to produce Fe–As alloys which are known from several sites.[22] Naturally, some degree of pre-roasting would have been necessary to decrease the As content before reduction.

There are also sources of low nickel-iron ores. We decided in Chapter 1 that the level of Ni in meteorites exceeds 5% and that Ni levels below this most probably come from Fe–Ni ores. Greece today is a producer of Fe–Ni alloys from its own indigenous ores, and it is therefore not surprising that we find a number of Fe-Ni objects among Greek artefacts. For example, Varoufakis[23] has analysed a number of rings of Mycenaean date, and Photos[24] has carried out primitive smelting experiments on Greek ores and examined Fe–Ni objects. Nickel tends to segregate, as it does not easily diffuse in iron, so that there are high Ni and low Ni regions in the artefacts, which make forging difficult.[24]

The Hellenistic site of Petres in Macedonia has given slags and a discarded bloom showing that Ni ores were smelted but, perhaps, not very successfully.

Composition and Structure of Iron Artefacts
Unlike bronze, the composition of iron is not much of a guide to the type of ore used, or to its provenance. Iron, being a relatively base metal, will take up any more noble elements, such as nickel or copper from the ore, but will leave behind in the slag elements such as manganese, chromium and zinc. Recent work has shown some value in determining the amount of nickel and copper in objects from one area,[25] but this is not of much value as a general aid to provenancing, as the different types of iron ores are so common. In Britain, almost every county has or had a local source of iron ore and some had sources of more than one type. The elements which are of some value in assessing the provenance within a small area or the manner of manufacture of iron implements are: C, P, S, Ni and Cu. The carbon content depends on technique, but phosphorus usually comes from the ore. Sulphur comes mainly from the ore since coal was not used for smelting until recent times and the sulphur content of charcoal is very low. Nickel and

Table 28 Analyses of main types of iron ore (%)

Element	Hematite, Cumberland[104] UK	Magnetite, Japan[105]	Limonite, Forest of Dean[104] UK	Weathered siderite, Northants[104] UK	Siderite, Kent,[106] UK	Hematite, Sirzi, Turkey[15]	Bog Iron Ore, Jutland[107]
FeO	–	22·89	–	–	42·08	–	1·30
Fe_2O_3	84·47	64·75	90·05	64·62	6·85	68·35	59·70
SiO_2	6·95	9·14	1·07	13·52	6·46	–	17·90
CaO	0·25	1·18	0·06	0·90	3·87	–	1·60
MnO	0·22	0·96	0·08	3·91	2·32	5·93	14·00
Al_2O_3	–	0·34	–		2·64	–	1·00
MgO	0·41	0·69	0·20	0·25	1·76	–	0·20
P_2O_5	0·03	0·02	0·09	2·15	0·65	0·11	2·94
H_2O	8·48	–	9·22	14·60	–	–	–
CO_2	–	–	–		32·70	–	–
SO_3	–	0·07	–	–	0·11	–	0·20
Alkali					(FeS$_2$)		0·37

copper also usually come from the ore, but there are cases where pieces of meteoric iron containing 7–10%Ni have been incorporated into EIA and later objects.[26,27,28] In the cases where the nickel comes from the ore, it never seems to occur in amounts exceeding 4%. The solubility of copper in iron is very limited (about 4% under normal conditions), but there is the possibility of picking up small amounts of other metals from fluxes. The use of additions to render the slag more fluid was not common in this period but additions of lime and manganese ores were sometimes made, and such additions might contain small amounts of other metals.

In view of these problems most of the work on the examination of artefacts has concentrated on the structure and manner of fabrication of the object, rather than on its composition. The poor corrosion resistance of iron has meant that very few of the earliest objects are left in a sufficiently good state to give much information. Indeed, it is often difficult to decide whether an early artefact was originally metallic or merely a piece of ore or mineral.

Examination of some of the early iron finds from Asia Minor, i.e. Troy III (*c.* 2300 BC), showed two objects to contain 2.44 and 3.91%Ni, respectively, which is well above the level of nickel likely to be obtained from nickeliferous iron ores. These artefacts probably consisted of smithed material, in which meteoric iron was mixed with smelted iron. Tell Chagar Bazaar V provided a fragment of rust of about this period.[29] From Geoy Tepe D (2000–2300 BC) we have, however, a white cast iron bloom and it is difficult to explain this by suggesting that it is an import,[29] because white iron was too brittle for use and was normally only produced for conversion to wrought iron. Although the bloom was much rusted there is no doubt about its composition, which includes 3.51%C, 0.45%P and 0.16%S. This is probably the result of a mistake by an early smelter who used too high a fuel/ore ratio. By about 1200 BC, Geoy Tepe was producing a recognizable iron sword hilt and other iron artefacts of indisputable validity. An iron sword hilt for a bronze blade was also found at Yorgan Tepe (1600–

1200 BC),[30] and it is clear that these objects belong to a period when iron was used more as a precious metal for decoration and its rarity value, than as a metal with special physical properties.

From about 1200 BC iron blades were making their appearance with cast-on bronze hilts.[30,31] The blade material was very often of steel rather than iron (*see* Table 29). This probably has no special significance and arises from the heterogeneous nature of the bloom, which is a mixture of reduced iron of varying carbon content, slag and charcoal.

Unfortunately, we have no raw material before the 8th century BC bars from Sargon II's Palace at Khorsabad, which were found to be made of good soft iron free of nickel and manganese.[32] These measured 30–50 cm long by 6–14 cm thick, weighed from 4 to 20 kg and are similar to the spindle-shaped bars from the Hallstatt and La Tène iron ages of Europe (500–1 BC). As expected, the European bars were found to be very heterogeneous. Their carbon content ranged from 0 to 0.85%. Some contained charcoal powder in fine fissures and it is clear that they had not been heated longer than necessary for their shaping.[33,34] They contained much slag, and in

Table 29 Analyses of Near Eastern iron

Element	Content, %			
	Luristan sword[35]	Spearhead Deve Hüyük (6th century)[39]	Luristan sword, (7th century)[37,38]	Philistine sword (1100 BC)[4]
C	0·067	0–0·6	0·3	0–0·8
Mn	<0·01	–	<0·17	0·01
Ni	–	0·21	0·024	0·10
Cu	–	–	0·04	0·01
Cr	–	–	<0·015	–
Si	0·23	–	–	0·01
P	0·04	–	Scale; $P_2O_5 = 0·8$	0·01
S	0·002	–	(Sn = 0·015)	(As = 0·052)
Hardness, HV		153–108		Cold worked

Table 30 Composition of iron blooms and other semifinished products

Object	Provenance	Date	Composition, %						Reference
			C	Si	Mn	P	S	Others	
Pointed bar		La Tène	0·44	–	0·1	0·04	0·012		101
Bloom	Siegerland		0·23	–	tr.	0·30	tr.		"
Currency bar	Noreia		0·12	–	0·02	0·017	0·004		"
21 bars (mean)	Rhein-Pfalz		–	0·24	0·04	0·37	0·025		"
Currency bars	UK	La Tène	tr.	0·09	nil	0·69	–	0·23Ni	108
			0·08	0·02	nil	0·35	–	–	108
			0·02–0·8	0·2	0·05	0·35	0·014	0·05Ni	108
			0·06	0·11	tr.	0·954	0·014	tr.Ni	108
Bloom	Wookey Hole, Somerset	1st century BC	0·74	0·61	nil	0·15	–		108
Cast iron bloom	Hengistbury, Hants.	1st century AD	3·49*	0·38	tr.	0·18	0·035		109
Cast iron bloom	Siegen		2·78*	0·05	nil	0·29	nil	0·21Cu	57

* Mostly combined, i.e. as Fe_3C

some cases, unreduced ore particles were trapped in the slag *(see* Table 30).

There is no doubt that these represent smelted metal with the minimum of forging. The size almost certainly shows the range of bloom size being produced, and the largest (20 kg) from Khorsabad indicates that by the end of the eighth century BC some very large furnaces must have been in use. It is difficult to believe that these were bowl furnaces and it is probable that some of the blooms were the products of induced-draught furnaces similar in size to recent African ones[13] *(see* Fig.33). The European bars provide no evidence that they were made by welding up small blooms.

The Structure of Implements and Weapons
While the composition of early iron has relatively little significance when compared with that of copper-base artefacts, the structure gives some idea of the manner of fabrication and the heat treatment. Unlike bronze, iron –carbon alloys can be hardened by the rapid cooling brought about by plunging the red-hot metal into cold water or brine — a process known as quenching. There is, however, no evidence for the intentional use of this hardening process during the early phases of the EIA, and one must assume that it was unknown at this time or else not considered beneficial. Quenching a high-carbon steel results in a brittle structure and is usually followed by a low-temperature treatment, known as tempering, which restores some of the lost ductility. Much the same properties as those of a quenched and tempered steel can be obtained by cold working and annealing at much lower temperatures and it would seem that much early iron and steel was alternately cold forged and annealed at 600–700°C, as is much African material today.

The hilt end of an 11th century blade from Marlik was of the latter type.[1] The mean carbon content was 0.1–0.2%, and therefore the hardness was not much higher than that of bronze. The disc pommel of a 9th–7th century BC sword contained 0.5%C and had been fairly

rapidly cooled after forging at about 900°C.[36] The blade of the Luristan sword at Toronto was also a 0.2%C steel with a structure that showed that it had been fairly rapidly cooled from 900–600°C, and held at about 700°C for some time.[36] It was evidently finally forged while cooling down to ambient temperature, as the structure shows deformation markings which are only obtained by smithing with very heavy blows at low temperature.

The hilts of the 11th century Luristan swords in the British Museum show that the knowledge of welding was limited. The smith had riveted the various parts of the hilt to the blade tang.[36] The blade and tang of one of them were of high-carbon steel, and the structure showed that it had been forged or heated in the range 700–800°C. The carbon content of the blade was very variable, and the cutting edges had been heavily decarburized by heating in an oxidizing environment. This shows a clear lack of technical knowledge.

A seventh century BC Luristan sword consisted of many pieces of mean carbon of 0.3%, but it had been air-cooled from 1000°C and forged between 850–700°C.[37,38] The socketed spearhead consisted of 50 layers of surface-carburized material formed by repeated forging and folding over like flaky pastry. This must have been done at a relatively low temperature (700–800°C), otherwise the carbon would have diffused. The maximum carbon was 0.6%, the mean about 0.1%, and it contained 0.32%Ni. The hardness was 108–153 HV—no harder than that of a bronze. A Philistine (1000 BC) blade varied in carbon content from 0 to 0.8% and had been smithed at a very low temperature (less than 800°C), after being annealed at 800–900°C[6] *(see* Table 29).

All these implements show the structure expected from most primitive forges in use today, and also show the lack of any structures which would have resulted from quenching and tempering. Quench-hardening is most likely to show at the edges of cutting tools and other artefacts and may be of very limited extent, so that it will be the first material to suffer from corrosion. However, the conclusion that quenching was not often used is supported by the material from Hallstatt and La Tène Europe, much of it in better condition because of its more recent date, and also by the still more recent African evidence.

It is of the greatest interest, therefore, that we see the first evidence for quench-hardening in Egypt considering the alleged backwardness of that area.[40] A lugged axehead, dated to 900 BC, had never been used as it was covered with a thin layer of magnetite from the last heating. The carbon content varied from zero in the centre to 0.9% at the blade edge. The whole axe had been quenched from a temperature of 800–900°C, giving a hard martensitic cutting edge. This edge had been tempered by the conduction of heat from the thicker parts of the axe, which had not been cooled to ambient temperature before removal from the quenching liquid. The final hardness, therefore, varied from 70 HB away from the edge, to 444 HB at the edge itself. The result was a first-rate axe correctly heat-treated to the hardness one would expect from an axehead today. A second axe of the same period had been used and was much corroded. The

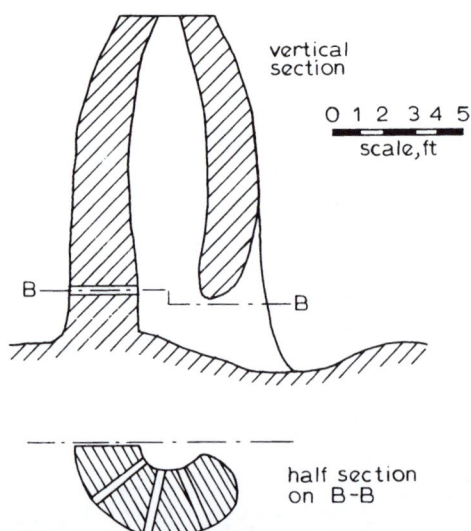

vertical section

0 1 2 3 4 5
scale, ft

B———·——·——B

half section on B–B

33 A shaft furnace from Togoland; African Iron Age (after Hupfeld[63])

remains of the edge showed that this blade had also been quench hardened, although the hardness was not high owing to the low carbon content of the remains. It is therefore clear that, while some smiths knew the art of quench-hardening, it was not widely practised either in the Near East or in Europe during the pre-Roman Iron Age.

The Spread of Ironworking Techniques

Greece

The rise of the Iron Age in Greece occurred towards the end of the 11th century BC.[41] There is little doubt that the knowledge spread from Anatolia, although some people believe it was introduced from the north. While we have many finds of weapons, mainly from rich cemeteries, the early material is in bad condition. Even so there is little doubt that most of the material was made in Greece itself from local ore; bundles of iron rods or 'spits' were clearly used as raw material and currency, just like the 'currency bars' of other countries. Spindle-shaped bars of the Khorsabad type were also found.

Most of our evidence dates from the Classical and Hellenistic periods (after 500 BC). From a Classical vase we have a good representation of what is almost certainly an ironsmelting furnace in a smithy,[42] (*see* Fig.34). It is a shaft furnace blown with bellows and closely resembles a Hungarian furnace from Gyalar[43] (*see* Fig.35). In front of it is a smith working a bloom or piece of iron on an anvil.

The cauldron-like object on the top of this furnace presents a problem. It could be that it has water in it to keep it cool, and that it serves as a lid or seal to reduce the draught of the furnace while the smith is using it for forging. It could alternatively be a number of stones on top of the shaft before it turned a right angle into a horizontal flue, as shown in the Hungarian Gyalar furnace. (*see* Fig.35).

Amongst material examined are iron clamps from the Parthenon.[44] These were set in lead and can be found all over the Near East, dating from the 5th century BC.[41,45] They are of typical inhomogeneous iron made, in many cases, by forging out a single iron bar and then bending it and welding it into the required shape. By this period iron was used for the mining tools used by the Laurion miners and for all plate armour and weapons. There are many literary references to the quench-hardening of Greek iron but, so far, insufficient artefacts have been examined to show how effective it was.

Palestine and Egypt

The diffusional route to Egypt seems to have been along the coast of Palestine, and evidence of this is supplied by workshops at Tel Zeror (Hedera)[46] and Gerar near Gaza,[47] and a 12th century Philistine sword from a grave.[6] It is well known that the Philistines reached the Iron Age before the Israelites and this may be due to contacts with the north, or even actual migration after the break-up of the Hittite Empire. At Gerar, a number of smithing furnaces, dated to between the 12th century and 870 BC, were found along with iron knives which were dated to 1350 BC.

In the Nile Delta, Greek or Carian traders established 'emporia'. At one of these, Naukratis,[10] deposits of iron-smelting slag, iron ore, and a spindle shaped currency bar were found. Iron is not well represented in the 20th Dynastic pyramids and the earliest iron finds are dated to about 580 BC. Near Tanis,[47,48] also in the Delta, iron slag was found together with iron artefacts and the debris of non-ferrous metalworking—the ores could have come from the Sinai peninsula. Very little industrial material

34 A shaft furnace from Greece (6th century BC) depicted on a vase showing a smith forging a bloom in front with the bellow behind (after Blumner[42])

35 *Hungarian shaft furnace of uncertain date; note the flat stone closing the shaft which could well be that shown on top of the Greek furnace in Fig.34 (Crown Copyright, Science Museum, London)*

has been found in Central Egypt and this has led to the belief that the Egyptians were somewhat backward in the Iron Age. It is unlikely that they were actually without the knowledge of ironworking but, considering the unrest during this period, it is possible that the metal production was left to foreigners. This can only be settled by further excavations in Central Egypt.

At Thebes, Petrie found a group of 23 tools dating to the Assyrian invasion of 667 BC. This comprised saws, chisels, a rasp, centre bits, a file, gouges and a sickle, together with a bronze helmet of Assyrian type. Three of the tools were wrought iron, two of the others had been case-carburized on the cutting edges and then quenched. Two were homogenized 0.2% carbon steels quenched to give a maximum hardness of 487 HV.[49]

PENETRATION ALONG THE MEDITERRANEAN

One of the earliest Iron Age civilizations in the Mediterranean is that of the Etruscans, and a good deal of work has been done on weapons and implements from this area of Italy. Urartian affinities can be seen in bronzes in Etruscan tombs,[50] but of course, contact with Anatolia may have been second-hand. In any event, in about 1000 BC, a new group of people with a knowledge of iron appeared in North West Italy: after this, there was much contact with other Mediterranean peoples, and by 750 BC the Etruscan civilization had appeared, supported by the Tuscan mineralized area.[51] Many of their weapons have been examined and found to be typical inhomogeneous steels which show no sign of quench-hardening.[26,27] A spearhead was found to contain a lamination of 70Fe–30Ni, which was almost certainly meteoric and incorporated unknowingly by the smith into his assemblage of iron from various sources.[27]

Phoenician traders were establishing their town of Carthage in the 8–9th century BC, and it is almost certain that it was the spread of ideas via the trans-Saharan route that introduced iron to West Africa and, in particular, to the Nok culture of Nigeria. It appeared there

in about 400 BC, well before it was known anywhere else in sub-Saharan Africa. One is tempted to wonder whether Nigerian tin travelled northwards along this route and whether some of the Phoenician tin trade was supplied from Nigeria.

Early Spanish swords[52] show the same structures as those from other EIA areas and the Spanish contribution to this period is primarily on the non-ferrous side, which will be discussed below.

Excavations of the pre-Roman (Punic) levels at Carthage have shown much evidence of smithing but non of iron smelting.[53] It would appear that this was done in the more wooded areas of the hinterland such as Sbeitla. The type of tuyere used consisted of two tubes coming together at the hearth end, and entering it at an angle of 30°. This type is also known from Phoenician sites in Spain.

THE SPREAD OF IRON ACROSS EUROPE

It is from Europe that most of our information on the EIA comes. The two great type-sites are Hallstatt in Austria and La Tène in Switzerland. The first is the earlier site and typifies the impact of the Iron Age on Central Europe from Asia Minor, probably via the Danube. The cemetery at Hallstatt dates from the 8th century BC and the weapon types found show close affinities with the earlier Bronze Age types. This tendency is a typical and perhaps natural one, but the translation of a cast bronze socketed axe into a forged iron copy shows how tradition has triumphed over reason. We have similar cases today where metal prototypes have been translated into polyethylene without sufficient change in design.

Austria has provided many examples of early furnaces. These may be classified into three types; (a) the bowl furnace, which we seem to have at Huttenberg together with a roasting hearth (Fig.36); (b) the shaft furnace which may, like that from Togoland, have induced draught (see Fig.33); and (c) the domed furnace. Only the latter is new and, so far, peculiar to Europe and the British Isles (see Fig.37).

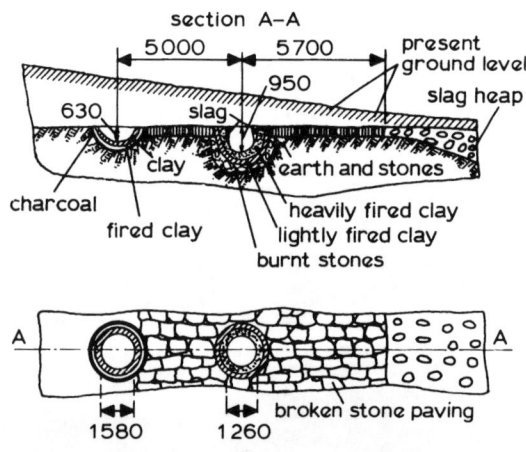

36 *Early Iron Age bowl furnaces from Huttenberg, Austria; dimensions in mm (after Coghlan[114])*

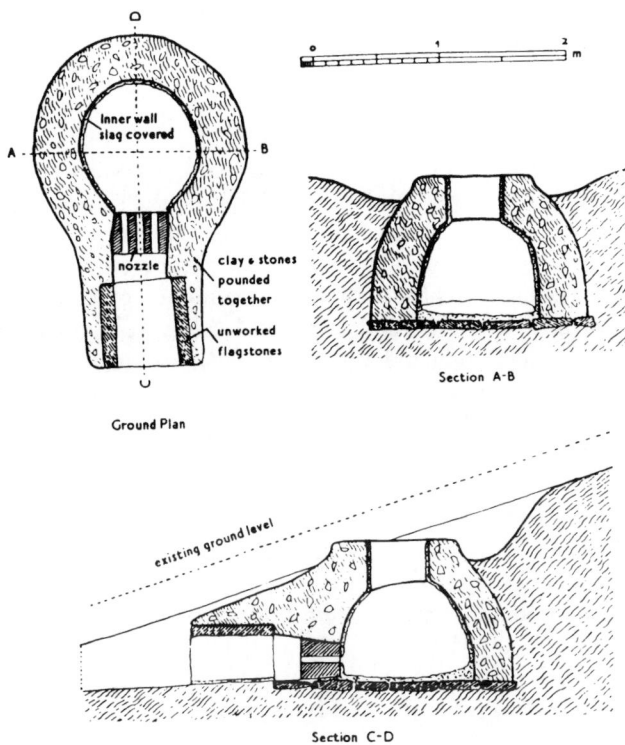

37 Domed furnace from Engsbachtal, Germany (after Coghlan[114])

The shaft furnaces at Lölling and at Feisterwiese in the Austrian Erzberg have been used as models for smelting experiments with a forced draught.[54] But the Lölling furnace had multiple tuyeres and seems to have been blown by induced draught.[55] This type of furnace was found in the Siegerland, in Germany,[56] and experiments have been made with it. The third type of furnace is the domed furnace from Engsbachtal, also in the Siegerland.[57,58] This is an unusual type, but a similar one was found at Levisham in North Yorkshire. This is also dated to the pre-Roman Iron Age.[59] It is almost certain that the induced-draught shaft furnaces were not slag tapping, but yielded a bloom consisting of pieces of iron slag and residual charcoal which had to be separated by hammering. The bowl and domed furnaces would have been blown by forced draught, but the earliest of these would also have been non-slag tapping. It was the ability to tap slag which distinguishes the earlier Iron Age furnaces from the later proto-Roman furnaces of Europe.

Recent finds in Wales have produced iron smelting furnaces dated to the pre-Roman Iron Age.[60] Within the walls of the hillfort of Bryn y Castell are the remains of a snail-shaped smithy. Outside the walls were a number of non-slag tapping low-shaft smelting furnaces; a reconstruction of this furnace was used for a number of smelting experiments on bog ores.[61] The local ore gave metal containing as much as 1%As.

AFRICA

The iron industry penetrated into sub-Saharan Africa by way of North Africa to Nigeria, and through Egypt to the Sudan. The EIA phase has persisted in Africa until the present day but is fast dying out. Luckily, in the last few years, anthropologists have obtained a great deal of information on ironworking which has been used to interpret the archaeological evidence from more northerly areas.

The furnaces used by the Nok culture of Nigeria (400–200 BC) were shaft furnaces of quite a large diameter, greater than 30 cm (*see* Fig.38). They were blown through short tuyeres by forced draught and do not seem to have produced tap slag.[11] But a number of discarded saddle querns had been re-used for breaking up the mixtures of metal and slag. The furnaces had thin clay walls built above slag pits cut into the natural soft rock, and in this sense bore considerable resemblance to the shaft furnaces of Jutland and North Germany used during the first few centuries AD.[59,62]

The pieces of iron extracted by crushing the slags could be consolidated by collecting them together and making them into a ball which was covered with a heavily grogged clay, to protect the iron from the oxidizing atmosphere of the forging hearth. The ball was heated to a welding heat (1 200°C) and hammered, whereupon the brittle clay envelope was broken and the welded metal could be forged into a plate. Such a process was carried out near Jos in Nigeria to give a plate of iron with 0.1%C and a hardness of 137–167 HV5, good enough for a hoe blade.[63,13]

The furnaces used in Africa recently belong to a large number of types, from small bowl furnaces of Kordofan, in the Sudan, and the central Sahara, to the 3.35 m induced draught furnaces of Togoland.[63] Some of the induced-draught furnaces had more than 100 tuyeres. One type of bellows-blown furnace used by tribes in the Mandara Hills and the Nigerian plateau uses a long single tuyere which goes down the centre of the shaft like a proboscis, terminating just above the hearth.[64,65] Presumably, an area with an EIA tradition lasting 2 500 years will have developed more types of furnace than an area with a much shorter tradition.

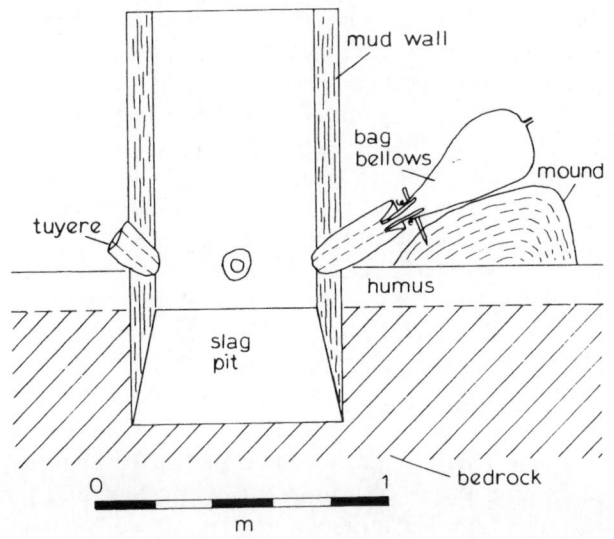

38 Low shaft furnace from near Jos, Nigeria; African EIA — 300BC

The metal from the Nok culture shows an extraordinary degree of purity and freedom from slag inclusions, as would have been expected if the raw bloom had been carefully broken up in the cold state in order that iron might be extracted, instead of the whole mass having been forged at a high temperature, as was generally the case in medieval Europe. The smithing technique was typical of the EIA, with a large number of artefacts showing evidence of long periods in the temperature range 600–750°C. In no case had quench-hardening been used. Even today a traditional African smith would not quench-harden the blade of a socketed axe he had made from a piece of 0.6%C rail steel of European origin. The need for stronger steels was satisfied merely by an increase in the carbon content: the blooms of iron produced at Oyo, Nigeria, in about 1910 contained 1.67%C.[66]

It is reported that cast iron was being made in crucibles in East Africa.[67] As late as the 1970s the Bari ironworkers of Bilinyan in the south Sudan were making cast iron by reducing fine iron ore in lidded crucibles. It is possible that this arises out of an isolated and late contact with India, or even China.

INDIA

Until recently, parts of India were also in a pre-industrial state and it is reasonable to assume that techniques in these areas have changed little since the inception of the Iron Age. The knowledge of iron seems to have come from the north to the valleys of the Indus and the Ganges, and this is usually ascribed to Aryan settlers. Sites in the north of India, dated to 800 BC and later, have produced a small number of iron objects showing close affinities with those of the later phases at Sialk in Iran. Other sites have produced iron slag,[68] and at Ujjain the remains of a furnace were discovered which belonged to a period not earlier than 500 BC.[69]

There is plenty of evidence to show that in about AD 400 the area had a similar technological level to that of Asia Minor and Europe. The famous pillars of Delhi and Dhar date from about AD 300 and, like the columns from the temple at Konarak, are similar in structure to the beams from bath houses in the Roman period. At Besnagar there are wedges that are said to date from 125 BC and which are possibly of Greek origin.[70,71]

Indian bloomery furnaces are, for the most part, shaft furnaces with internal heights of up to 3 m, all forced draught, and most of them slag tapping. The only natural draught furnace in this region was found in Burma: it had 20 tuyeres and was 3.2 m high. In the Central Provinces at Tendukera, developed bowl furnaces of the Catalan type were in use in the 1850s.[13]

One interesting Indian development is the melting of steel in crucibles. Bloomery iron or inhomogeneous steel and charcoal were put into crucibles which were sealed and heated for four hours in a hearth with a forced draught. This is an improvement of the African process of heating pieces of iron in a clay envelope and then forging them. The Indian product was usually a homogeneous carbon steel with 1–1.6%C known as 'wootz' and exported to the West. Perhaps it is synonymous with Damascus steel, since this is probably the town through which it entered the West in medieval times.

CHINA

China seems to have entered the Iron Age in about 600 BC and could therefore have derived the knowledge from Asia Minor. But the fact that China's Iron Age started with cast iron suggests that it might have had independent origins, or else the Chinese were quick to appreciate the value of their first accidental piece of cast iron and turn it to good account. As elsewhere, the Iron Age came slowly and one of the first uses of iron was for 87 cast iron moulds for hoes, sickles and chisels (475–221 BC); some of the implements were undoubtedly cast in bronze. According to literary sources, iron cauldrons were being cast in 512 BC, and the fifth century saw the beginning of iron weapons. These were not made of cast iron but of wrought iron, and one must infer either that the Chinese knew how to convert cast iron to wrought iron, or that they worked the direct process when necessary.[72,73]

The first recorded blast furnace explosion was in 91 BC, and this was followed by many others.[73] From this, there is little doubt that cast iron was being produced on the scale of the bronze, which was used for *ting*-cauldrons weighing as much as 1 400 kg as early as the Shang dynasty (c.1000 BC).[74] We have no archaeological evidence of blast furnaces before the Han period (200 BC), when the furnaces were 1.4 m i.d. at the top and were built of refractory bricks. Thick-walled clay tuyeres were used, 1.4 m long, 28 cm o.d. at the furnace end, and 60 cm at the other. The charge was charcoal and a ferruginous sand which seems to have been a mixture of ground hematite and magnetite sand.

So far there is no archaeological evidence for the bloomery process in China but from Han times there is a long tradition of crucible manufacture of steel, as in India.[74]

The need for a more malleable metal such as wrought iron or steel definitely existed and there are references to 'a hundred refinings' which is taken to mean the number of layers in a forging. During a more recent period (1940–50) cast iron was converted into wrought iron by a hole-in-the-ground hearth which was filled with cold or molten cast iron and blown with an intensive blast of air. It would seem that this is a traditional Chinese process which might well date back to the Han period.[75]

Most of the iron- or steelmaking processes were blown with air from fan of piston-blowing machines which were often water driven. They were very much more efficient than the bag, pot and concertina bellows of the rest of the iron-using world and it would appear that it is to their mechanical supremacy that the Chinese metallurgists owe their success.

From the structure of early Chinese artefacts[12] we see that the moulds of the Warring States and Han periods (475 BC–AD 24) were made of white cast iron, while a coffin nail of the same period consists of wrought iron. Two Han white iron castings contained 4.19 and 4.32%C, respectively, and were low in phosphorus, sulphur and silicon: these are cold-blast charcoal irons made from low-phosphorus ores.[76] What was even more surprising was the discovery of a spade or shovel of blackheart malleable iron, which was doubtlessly made by heating white iron in a forge at 900–1 000°C. A steel sword of the Han period (206 BC–AD 24) had a martensitic structure, so we can safely assume that the production of wrought iron and steel and the quench-hardening of steel was well understood by AD 24.

Non-Ferrous Metals in the Early Iron Age

The principal differences between this and the earlier periods are the introduction of zinc in copper-base alloys, and the complex alloying now seen in gold work. Zinc is an element that is usually absent in copper artefacts of the Bronze Ages but its mere presence cannot be used as evidence of forgery. Zinc occurs as an impurity in many copper minerals, as in some of those from Cyprus and Ireland. While its volatility is well known, it is no worse than that of arsenic, so if zinc-containing ores were used we should expect to find some zinc introduced into the metal. Bronzes dated to between 1800 and 1400 BC from Palestine and Cyprus contained about 3%Zn, which was almost certainly accidental. We begin to see the presence of zinc in Chinese bronzes from Han times, i.e. after 220 BC (*see* Table 31 and Fig.39).

Brass did not appear in Egypt until about 30 BC but after this it was rapidly adopted throughout the Roman world. It was made by the calamine process, in which additions of zinc carbonate or oxide were made to copper and melted under a cover of charcoal to give reducing conditions. No zinc metal has been found in an EIA context, but one or two pieces have been found in Roman Athens.[77]

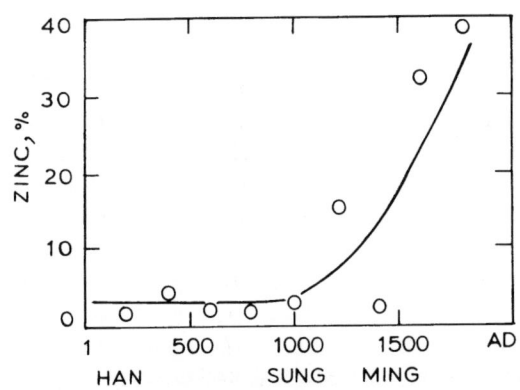

39 *Increase in the zinc content of Chinese bronzes over the period 220 BC to AD 1600 (after N. Barnard[74])*

In Chapter 4, reference was made to the additions of copper to restore the natural colour of gold to alloys lightened in colour by silver existing as an impurity (i.e. electrum). Analyses from Iron Age Britain[78,79] show that large additions of both copper and silver were being made to gold, so that its purity was reduced to values as low as 30% (i.e. 9 carat).

Silver was, as we know from the Greek workings at Laurion (600–25 BC), a product of lead mining in which the noble metal was extracted by the cupellation process.[80,81] Since noble metal coinage was introduced into Europe and the Near East in the seventh century, a good deal of this silver and some of the gold went into currency although, as we see from Persian (Achaemenid-Sassanian) material, much went into art metalwork. We have no means of knowing the exact source of this metal: while a good deal was the result of conquest, some no doubt was obtained from Persian lead ore deposits. The silver content of the ore from Laurion was as high as

Table 31 Analyses of later Chinese bronze artefacts

Refer-ence	Object	Period	Composition, wt-%					
			Sn	Pb	Ni	Sb	As	Zn
110	Mirror*	Tang	23·11	2·5				
"	Mirror	Han	24·56	1·47				
"	Mirror	Han	26·97	1·65				
"	Mirror	Han	24·16	2·06				
"	Sword	Han	19·25	1·93				
111	Arrowhead	Han	10·82	—	3·68	8·0	tr.	
"	Arrowhead	Han	16·15	—	2·04	8·69	tr.	
"	Mirror	Han	25·07	9·12	—	tr.	tr.	0·16
"	Mirror	Han	26·6	5·30	0·63			0·20
"	Mirror	Han	26·5	5·30	—	1·08	—	5·00
"	Mirror	Han	24·3	5·60	—	—	—	—
"	Mirror	Han	24·1	3·97	1·83	—	2·63	—
"	Mirror	Han	25·4	0·61	—	—	—	6·73
"	Mirror	Han	28·7	5·05	—	—	—	—
"	Mirror	Han	30·7	0·69	—	—	—	—
"	Mirror	Tang	27·6	5·40	—	—	—	—
"	Mirror	12th century AD	11·6	20·5	0·34	2·35	0·13	—
"	Mirror	19th century AD	3·3	7·6	0·35	1·81	tr.	12·96
"	Mirror	19th century AD	5·6	7·7	3·17	—	—	23·3

* Few alpha dendrites in a fine-grained delta matrix; a few Pb globules and a little cuprite in the alpha phase

1 200–4 000 g/t (0.1–0.4%), which is the highest recorded in the pre-Roman period from the Near East and the Aegean.

After the cupellation of lead to extract the silver, lead can be recovered from the litharge at additional cost. Since little litharge was found at Laurion it must be assumed that this was done, either at Laurion or elsewhere. The latter is more likely; fuel would have been a considerable problem at Laurion[82] and as much as possible would have been reserved for the first smelting and cupellation. However, there is some evidence that timber in the form of pinewood was brought into Laurion for metallurgical purposes. Lead recovered from the litharge was cast into 15 kg ingots and sold cheaply.[83] The only uses that we know of in Greece for the lead so recovered was for sealing the iron cramps used in classical times for joining masonry, for anchors and for cladding ships' bottoms.[41] A small amount was used domestically and we see examples of its use for water pipes in the Acropolis and at Delos for rainwater pipes. But the intensive use of lead for plumbing was yet to come with the Roman period.

The copper ingots of this period seem not unlike the plano–convex ingots of the LBA. A wreck off the French coast at Rochelongue near Agde produced 800 kg of copper ingot and bronze objects.[84] The plano-convex copper ingots varied from 100 kg to 11 kg and reached 25 cm dia. There were also some tin ingots. This hoard was dated to the sixth century BC and could have belonged to a travelling merchant like that involved in the LBA wreck off Cap Gelidonya.

Two types of tin ingot have been found in the Mediterranean, a large plano–convex ingot with Greek inscriptions and a smaller purse-shaped form. The latter are Roman and date from 42–48 AD. The plano-convex ingots found off the isle of Bagaud in the Hyères group are similar in shape, but not in size, to the unstratified British ones.[85] These date from 125–75 BC and the ship carried 2–3 t of tin ingots as well as other cargo. There were about 100 tin ingots averaging 25 kg each which makes them a good deal larger than the British 10kg ones. They contain some stamped markings in Greek such as a coin-like figure of Hermes, and a rectangular cartouche— ARISTOKR—on their flat sides. The legend written around the head of Hermes appears to contain the words translated as HYPO-KELTON which could be 'low celts' or 'southern Celts'. These could be British or Iberian, and the voyages of Pytheas would strongly suggest the former. [86]

The Begining of Metallurgy in Japan
Japan probably acquired the art of metalworking from China via Korea, and the Japanese people were in a LBA phase when they left the mainland for the Japanese Islands at a date estimated to be between 600 BC and AD 200. Previously, the Islands had been occupied by the paleolithic and neolithic Ainu.

The Japanese had reached the EIA by the time of the construction of the dolmens in Kyushu in the seventh century AD, and our knowledge of bronze working dates from this time. The dolmens contain bronze castings which are mainly arrowheads, mirrors, and bells, but very little iron.[87] The bronze arrowheads and swords have been cast in stone two-piece moulds—a typical LBA technique. The mirrors are often copies of Chinese designs. By AD 708, large bronze castings were being made in sand or loam moulds.

There is little doubt that the Japanese had considerable knowledge of advanced blowing machines such as the 'Tatara', possibly obtained from China. A segmented cupola furnace was used for melting bronze in the 19th century. This was perhaps the type used for casting the large bronzes known from China of Chou and later times and is therefore worth describing here.

A furnace consisting of four segments would be as much as 2.5 m high: the metal was tapped from a taphole into a launder or ladle.[88] Smaller pit-type crucible furnaces, blown with a forced draught, were also used. The crucible consisted of porcelain enclosed in fireclay. For moulding temple bells and other large objects a wax pattern was used. The wax was moulded on to a hollow clay core and more clay was then moulded on to the wax. After this, a charcoal fire was lit under the mould, the wax melted out and the clay mould dried and hardened.[87,88,89]

The copper was smelted in a simple bowl hearth with an inclined tuyere. In the 18th century, the copper was refined by flapping (oxidation) and purities of 99.24–99.80% were obtained. The principal impurities were lead, silver, and iron. Such metal was often alloyed by the addition of a pseudo-speiss, which was a product of the earlier desilverization of the copper by lead: the composition of the speiss is given in Table 32. This is the first attested example of the deliberate alloying of copper with antimony and arsenic, and may be responsible for the content of these metals in some of the bronzes shown in Table 33. It was a prescribed addition to bronze coinage in 1764 and was almost certainly used for the Buddha of AD 1614, for improving fluidity and lowering the melting point. Furthermore, it assists the production of a metal with a dark grey colour, which is used for colour contrast with silver in decorative objects.

Coins and Coining
The EIA saw the introduction of coins. The first gold coins came from Lydia c.550 BC, but early Greek silver coins were being produced from Laurion lead slightly earlier—about 580 BC—in Corinth and in Aegina. Perhaps the best known gold coinage is that of Philip of Macedon (350 BC): this was widely copied and a debased form of the design was used on gold coins as far away as Britain.

While the earliest gold coins would undoubtedly have been made from natural gold, it would not have been long before alloying was resorted to in order to increase the hardness and wear resistance of the natural golds with low silver content. The Gallo–Belgic coins contained about 70%Au and 10%Cu, the latter being an intentional addition. Some British coins of the pre-Roman period contained only 47%Au with 40%Cu, which suggests an intentional devaluation. Carthaginian electrum coins show a tendency for the copper content to

Table 32 Composition of pseudo-speiss (after Gowland[88])

Composition, %	
Cu 72·7	Sn 0·93
Pb 8·53	Fe 0·13
As 11·37	Ag 1·33
Sb 4·27	S 0·33

increase with the gold content: this must have been done to increase the hardness.[90] Most of the coins contained 30–50Ag (*see* Table 34). This certainly shows that silver additions were normally made since natural gold usually contains less than 20%Ag. There was a tendency for the gold content to be higher on the surface owing either to inverse segregation (if cast), or to differential corrosion.

It is clear that there were wide variations of compositions in the gold and silver contents of the lead ores in the Athenian mines, both horizontally and vertically. Early silver coins from the period 196–169 BC contained 0.0076–0.33%Au and 0.035–5.3%Cu and the coins were uneven in size and quality.[91] In the middle period (168–132 BC) the style was good and the dies well cut: the gold content varied between 0.09 and 0.25% while the copper content varied between 0.05–0.4%. Microscopic examination of Athenian silver coins of the period 158–132 BC

shows some segregated lead and traces of iron, calcium and gold. In the late period (131–87 BC) the composition was more uniform, the gold content varying from between 0.25 and 0.5% and the copper content from between 1.5 and 5.5%, and this suggests deliberate adulteration: also, the minting technique was poor. It is possible that the higher gold content indicates a decline in home production, and the use of imports. In any event, the presence of gold shows that no attempt was being made to recover the gold by chemical separation (parting). We have no evidence of the manner in which the blanks or flans were made in the Near East, but we know from the innumerable coin dies found in Western Europe that the method of making noble metal blanks was to weigh carefully the required amounts of the noble metal and to place them in hollows in a baked clay plate.[92] After the metal had been put into the hollows the clay plates were covered with charcoal and heated to the required temperature to melt the metal. Owing to the effect of surface tension, the result of this operation was to form a spherical globule which could either be flattened and then coined, or merely coined with a punch. The fifth century Persian silver *syglos*, which weighed 5.4 g, was made in a similar way, after which it was struck on an engraved die with a square punch somewhat smaller than the coin.[93]

Table 33 Composition of Japanese bronzes

Object	Date	Composition, %						Reference
		Sn	Pb	Ni	Sb	As	Others	
Temple bronze		2·58	3·54				3·71 Zn	88
Coins	AD 1835–1870	8·26	8·74	–	–	0·18		"
Coins	1863	3·21	11·22	–	0·49	1·5		"
Cannon	18th century	12·68	3·32	–	–	–		"
Mirror	17th–18th century	0·58	3·19					"
Mirror	Recent	23·64	0·13					"
Arrowhead	7th century	2·46	0·56			0·50		111
Arrowhead	7th century	3·20	0·28			0·03		"
Mirror	5th century	22·5	8·99	1·52				"
Mirror	11th century	9·3	5·7	0·88	1·55	1·68		"
Buddha	3rd century	9·93	–	0·79	2·60	1·68		"
Sword	(Chikuzem)	14·13	1·32	2·93	4·93	tr.	0·71 Au	"
Buddha	1614	4·32	15·33	–	–	1·14		88

Table 34 Precious metal coinage alloys of the Early Iron Age

Mint and type	Date	Composition, %				Reference
		Au	Ag	Cu	Pb	
Phillip of Macedon, Gold stater	4th century	99·7	0·3	–	–	112
Greek, Terina, stater	5th century	0·09	95·32	1·42	2·19	94
Greek, Corinth, stater	4th century	tr.	94·12	4·01	0·57	94
Gallo-Belgic, A		69·02	22·83	8·15	–	113
British (QC)	40–20 BC	57·30	16·40	23·90	–	113
Carthaginian (Panormus)		60·80	36·30	2·30	–	90
British (Verica)	AD 10–50	72·2	7·60	17·20	–	113

Greek silver coins containing 93–99%Ag, of the period 500–300 BC, were struck in various ways. In some cases the cast blanks were struck directly, in others the blanks were worked and annealed before striking.[94] Athenian coining dies of the period 430–332 BC were usually made of cast high-tin bronze (22.5%Sn) and were therefore reasonably hard.[95] The as-cast hardness of such a composition would have been about 270 HV and would no doubt have hardened during use. In Greece, two types of coin die were known, the first consisting of a die into which a design was cut and which was intended to be placed on a flat plate, and the second was a punch. An example of the first is a 6th–7th century BC bronze block in the Ashmolean Museum in Oxford which measures 12.5 × 2.8 × 1.0 cm, and has designs cut on each of four sides in such a way that the coin flans must have been hammered on to them. It is possible that the designs were first cut with a gem cutter's wheel, but later they were cut with a number of iron or steel punches. Of the 46 Greek and Roman dies known, only eight are of iron.[96]

In the case of low-value bronze and brass coins a casting process is quite feasible. The Indian bronze coins, like the bronze 'tin money' of Britain and the bronze coins of Chou China, were cast in moulds.[97] The 'links' or runners on the British coins suggested that they were cast in strings like sausages, while the Indian coins of about 100 BC were cast in a radial manner rather like that used for counterfeit coins of the Roman period in Europe. There is little doubt that the normal method of making coinage which had been established before the Roman period was that of striking prepared flans and not that of casting. Casting was considered a much inferior technique that no self-respecting European coiner would consider satisfactory.[98] The method of striking a prepared blank was to remain the standard for the next 2 000 years.

The tendency for arsenical coppers to produce silvery layers by the process of inverse segregation seems to have been exploited in the Carthaginian currency of the third century BC. The designs were struck on to cast flans of arsenical copper.[99]

Some of the Bactrian coinage of the second century BC issued in North Afganistan is known to contain 16–22%Ni. It would seem that copper-nickel ores were available in this area and that there was no intention of making use of the nickel contents for reasons such as surface whitening, as impure copper was often used as well.[100]

References

1   R. F . TYLECOTE: *Bull. HMG*, 1972, **6**, 34
2   EL SAYAD EL GAYAR and M. P. JONES: 'Metallurgical investigation of an iron plate found in 1837 in the Great Pyramid at Gizeh, Egypt', *JHMS*, 1989, **23** (32), 75–83
3   K. R. MAXWELL-HYSLOP and H. W. M. HODGES: *Iraq*, 1966, **28**, 164
4   H. MARYON: *WMF*, 1955, **23**, 383
5   M. A. FRANCE-LANORD: *RHS*, 1963, **4**, 167
6   C. BOHNE: *ibid.*, 1967, **8**, 237
7   SIR F. PETRIE and B. QUARITCH: 'Gerar', British School of Archaeology in Egypt, *1928; Nature*, 1927, **120**, 56
8   B. FAGG: *World Arch.*, 1969, **1**, 41
9   R. F. TYLECOTE: *Bull. HMG*, 1968, **2**, (2), 81
10   SIR F. PETRIE: 'Naukratis, Pt. II', 1886, Third Memoir, London, Egypt Exploration Fund
11   R. F. TYLECOTE: *Bull. HMG*, 1970, **4**, (2), 67
12   LU DA: *Acta Met. Sin.*, 1966, 9, 1 (Trans. Durrer Festschrift, 1965, 68)
13   R. F. TYLECOTE: *J. Iron Steel Inst.*, 1965, **203**, 340
14   F. K. DORNER *et al.*: *JDAI*, 1965, **80**, 88
15   H. G. BACHMANN: *Arch. Eisenh.*, 1967, **38**, 809
16   R. F. TYLECOTE and R. E. CLOUGH: 'Recent bog iron ore analyses and the smelting of pyrite nodules', *Offa*, 1983, **40**, 115–118
17   R. F. TYLECOTE: 'Iron sands from the Black Sea', *Anat. Studies*, 1981, **31**, 137–139
18   HENRY HORNE: 'Observations on sand iron', *Phil. Trans. Roy. Soc.*, 1763, **53**, 48–61 (Lowthorp's abridgement)
19   E. PHOTOS, H. KOUKOULI-CHRYSANTHAKI and G. GIALOGOU: 'Iron metallurgy in E Macedonia; a preliminary report', In: 'Craft of the Blacksmith', (eds. B. G. Scott and H. F. Cleere), Belfast, 1987, 113–120
20   R. M. EHRENREICH: 'Trade, Technology and the Ironworking Community in the Iron Age of S. Britain', *BAR Brit. Ser.*, 144, Oxford, 1985
21   J. D. MUHLY, R. MADDIN *et al.*: 'Iron in Anatolia and the nature of the Hittite iron industry', *Anat. Studies*, 1985, **35**, 67–84
22   J. PIASKOWSKI: 'Das Verkommen von Arsen in Antiken und frumittelaltterlichen Gegenstanden aus Renneisen', *Z. f. Arch.*, 1984, **18**, 213–226
23   G. VAROUFAKIS: 'Investigation of some Minoan and Mycenaean objects', In: *Fhuhestes Eisen in Europa*, Festschrift für W. U. Guyan, Schaffhausen, 1981, 24–32
24a E. PHOTOS, R. F. TYLECOTE and P. ADAM-VELENI: 'The possibility of smelting nickel-rich lateritic ores in the Hellenistic settlement of Petres', In; 'Aspects of ancient mining and metallurgy', (ed. J Ellis-Jones), Bangor, N. Wales, 1988, 35–43
24b E. PHOTOS: 'The question of meteoric versus smelted Ni-rich iron; archaeological evidence and experimental results', *World Arch.*, 1989, **20**, (3), 403–421
25   LENA THALIN: *Jernkontorets Annaler*, 1967, **151**, 305
26   C. PANSERI and M. LEONI: *Met. Ital*, 1966, **58**, 381
27   C. PANSERI: *Sibrium*, 1964–6, **8**,147
28   J. PIASKOWSKI: *J. Iron Steel Inst.*, 1961, **198**, 263
29   T. B. BROWN: *Man*, 1950, **50**, (4), 7
30   R. MAXWELL-HYSLOP: *Iraq*, 1946, **8**, 1
31   H. DRESCHER: Der Überfangguss, 192; 1958, RGZM
32   C. DESCH: *Brit. Assoc.*, 1928, 437
33   O. KLEEMANN: *Arch. Eisenh.*, 1961, **32**, (9), 581
34   W. RADEKER and F. K. NAUMANN: *Arch. Eisenh.*, 1961, **32**, (9), 587
35   F. K. NAUMANN: *Arch. Eisenh.*, 1957, **28**, 575
36   H. MARYON: *AJA*, 1961, **65**, 173
37   J. LECLERC *et al.*: *RHS*, 1962–3, **3**, 209
38   E. SALIN (ed.): *ibid.*
39   H. H. COGHLAN: 'Metallurgical analysis of archaeological materials', 1959, Wenner Gren Found. Congress
40   SIR H. CARPENTER and J. M. ROBERTSON: *J. Iron Steel Inst.*, 1930, **121**, 417
41   R. PLEINER: 'Ironworking in ancient Greece', 1969, Prague, National Technical Museum
42   H. BLÜMNER: 'Technologie und terminologie der

gewerbe und künste bei Griechen und Romern', 152, 1886–7, vol.4, Leipzig

43   Exhibited in Budapest in 1897; model in Science Museum, South Kensington

44   C. J. LIVADEFS: *J. Iron Steel Inst.*, 1956, **182**, 49

45   R. PLEINER: 'The beginnings of the Iron Age in ancient Persia', 1967, Prague, National Technical Museum

46   K. OBATA and M. KOKHAVI: *IEJ*, 1965, **15**, 253; 1966, **16**, 274

47   SIR F. PETRIE: 'Tanis, Part I, 1883–4', 1885, London, Second Mem. Egypt Exploration Fund

48   SIR F. PETRIE: 'Tanis II, Nebesha and Defenneh', 77, 1888, London, Fourth Mem. Egypt Exploration Fund

49   A. R. WILLIAMS and K. R. MAXWELL-HYSLOP: 'Ancient steels from Egypt', *J. Arch. Sci.*, 1976, **3**, 283–305

50   K. R. MAXWELL-HYSLOP: *Iraq*, 1956, **18**, 150

51   A. R. WEILL: *Rev. Met.*, 1957, **54**, 270

52   H. H. COGHLAN: *Sibrium*, 1956-7, **3**, 167

53   R. F. TYLECOTE: 'Metallurgy in Punic and Roman Carthage', In: *Mines et Fonderies antiques de la Gaule*, (ed. C. Domergue), CNRS, Paris, 1982, 259–278

54   H. STRAUBE: *Arch. Eisenh.*, 1964, **35**, 932

55   K. KLUSEMANN: *MAGW*, 1924, **54**, 120

56   J. W. GILLES: *Arch. Eisenh.*, 1957, **28**, 179

57   J. W. GILLES: *Stahl u. Eisen*, 1936, **56**, 252

58   J. W. GILLES: *ibid*, 1957, **77**, 1883; 1958, **78**, 1 200

59   Anon: *Bull. HMG*, 1970, **4**, (2), 79

60   P. CREW: 'Bryn y Castell hillfort; a late prehistoric ironworking settlement in NW Wales', In: 'Craft of the Blacksmith', (eds. B. G. Scott and H. F. Cleere), Belfast, 1984, 91–100

61   P. CREW and C. J. SALTER: 'Comparative data from iron smelting and smithing experiments', In: 'From Bloom to Knife', (eds. E. Nosek *et al.*), Cracow, 1988

62   W. WEGEWITZ: 'Nachr. Niedersachsens Urgeschichte', Special Report No.26, 1957

63   F. HUPFIELD: *Mitt. a.d. deutschen Schutzgebieten*, 1899, **12**, 175

64   RENE GARDI: 100 Al. Lloyd's Register of Shipping, 1959, (4), 32

65   H. SASSOON: *Man*, 1964, **64**, 174

66   C. V. BELLAMY: *J. Iron Steel Inst.*, 1904, **66**, (2), 99

67   M. RUSSELL: Pers. Comm. Aug. 1981

68   H. WALDE: 'Durrer Festschrift', 71, 1965

69   N. R. BANERGEE: 'The Iron Age in India', 1965, Delhi, Munshiram Manoharlal

70   S. C. BRITTON: *Nature*, 1934, **134**, 238, 277

71   SIR R. HADFIELD: *J. Iron Steel Inst.*, 1972, **85**, (1), 134

72   J. NEEDHAM: *Technol. Cult.*, 1964, **5**, (3), 398

73   J. NEEDHAM: 'The development of iron and steel technology in China', 9, 1958, London, Newcomen Society

74   N. BARNARD: 'Bronze casting and bronze alloys in ancient China', 1961, Canberra

75   D. B. WAGNER: 'Dabiesham', *Scand. Inst. Asia Stud. Monog. No. 52*, Copenhagen 1985

76   G. W. HENGER: *Bull HMG*, 1970, **4**, (2), 45

77   M. FARNSWORTH *et al.*: 'Hesperia', 1949, Suppl.8, 126

78   R. R. CLARKE: *PPS*, 1954, **20**, (1), 27

79   J. E. BURNS: *ibid.*, 1971, **37**, (1), 228

80   C. E. CONOPHAGOS: 'Le Laurium Antique', Athens, 1980

81   C. E. CONOPHAGOS: 'La technique de las coupellation des Grecs anciens au Laurion', In: *Archaeometry*, 25th Int. Symp., Athens, 1986, (ed. Y. Maniatis), 1989, Amsterdam, 271–289

82   R. J. HOPPER: *BSA*, 1953, **48**, 200

83   C. E. CONOPHAGOS: 'Thorikos V', 1968, Gent, Comitée des Fouilles Belges en Grece

84   A. BOUSCARAS: 'Epaves de VIe au IIIe S. Av. J-C.; Le site des bronzes de Rochlongue', In: *Archéologie Sous-Marine*, Maison des Traquieros, Perros-Guoirec, 1986, (Exhibition Catalogue), 41–42

85   LUC LONG: 'L'Epave antique Bagaud 2', VI *Congress Int. de Arqueologia Submarina*, Cartagena, 1982, 93-98, (and Pers. Comm.)

86   C. F. C. HAWKES: 'Pytheas - Europe and the Greek explorers', 8th J. L. Myers memorial lecture, Oxford, 1975

87   W. GOWLAND: *Arch.*, 1897, **55**, (2), 439

88   W. GOWLAND: *J. Roy. Soc. Arts*, 1895, **43**, 609

89   W. GOWLAND: *J. Inst. Metals*, 1910, **4**, 4

90   H. A. DAS and J. ZONDERHUIS: *Archaeom.*, 1964, 7, 90

91   M. THOMPSON: *ibid.*, 1960, **3**, 10

92   R. F. TYLECOTE: *Num. Chron.*, 1962, **2**, 101

93   C. S. SMITH: 'A history of metallography', 1960, Chicago, Chicago University Press

94   C. F. ELAM: *J. Inst. Metals*, 1931, **45**, 57

95   G. F. HILL: *Num. Chron.*, 1922, **2**, 1

96   D. G. SELLWOOD: *ibid.*, 1963, **3**, 217

97   F. C. THOMPSON: *Nature*, 1948, **162**, 266

98   G. C. BOON and R. A. RAHTZ: *Arch J.*, 1966, **122**, 13

99   S. LA NEICE and I. A. CARRADICE: 'White copper; the arsenical coinage of the Libyan revolt, 241–238 BC', *JHMS*, 1989, **23**(1), 9–15

100   M. COWELL: 'Analyses of the copper-nickel alloy used for Greek Bactarian coins', In: *Archaeometry*, (ed. Y. Maniatis), Athens, 1989, 335–345

101   B. NAUMANN: Die Altesten Verfahren der Erzeugung Technischen Eisens, 1954, Berlin

102   R. PLEINER: 'Zaklady Slovanskeho Zelezarskeho Hutnictvi', 1968, Prague, Czechoslovak Academy of Sciences

103   R. E. M. WHEELER: 'Maiden Castle, Dorset', 1943, Res. Rpt. No. 12, Soc. Ants. London

104   J. D. KENDALL: Iron ores of Great Britain and Ireland, 1893, London

105   K. KUBOTA: 'Japan's original steelmaking and its development', 1970, 8, Int. Co-op Hist. Tech. Committee, Pont à Musson

106   D. W. CROSSLEY: personal communication

107   R. THOMSON: personal communication

108   R. F. TYLECOTE: 'Prehistory of Metallurgy in the British Isles', *Inst. Met.*, London, 1986, 144

109   J. P. BUSH-FOX: 'Excavations at Hengistbury Head, Hants, in 1911–12', Report. No. 3, Soc. Ant. 1915, London

110   W. F. COLLINS: *J. Inst. Metals*, 1931, **45**, 23

111   M. CHIKASHIGI: *J. Chem. Soc.*, 1920, **117**, 917

112   J. HAMMER: *Zeit Numismatik*, 1908, **26**, 1

113   S. S. FRERE (ed.): 'Problems of the Iron Age in Southern Britain', 308, 1958, Occ. Paper No. 11, Inst. Arch. London

114   H. H. COGHLAN: 'Early iron in the Old World', 1951, Oxford, Pitt Rivers Museum

115   G. JOBEY: *Arch. Ael.*, 1962, **11**, 1 Early Iron Age 41

Chapter 6
The Roman Iron Age

It might seem difficult to justify the term 'Roman Iron Age' for a chronological period of the world history of metals, since it appears that the main contribution of the Roman Empire to world technology was not one of originality but one of organization. It would seem that the effect of the Roman Empire was the widespread dissemination of the best techniques that existed anywhere in the Romanized world. This dissemination was not limited to the Roman area but affected the Iron Age tribes living on its periphery, who both traded in metals with the Romans and used them for defending themselves against the Imperial power.

However, there was an enormous increase in the scale of the industry: whereas the amounts of slag found on pre-Roman iron-smelting sites are measured in kilograms or hundreds of kilograms, the slag heaps of the Roman period are usually measurable in hundreds of tonnes. It is probable that this increase in scale stemmed from improved techniques, for example the use of bellows-blown shaft furnaces and larger developed bowl-type furnaces, instead of the induced-blast shaft furnace and small bowl furnaces of the pre-Roman and non-Roman Iron Age world. For our purpose this period starts with the birth of the Roman Empire in 29 BC.

The military and civil needs of Roman civilization created considerable demand for iron and non-ferrous metals. In the case of lead, where the minerals were always plentiful, the need in pre-Roman times had not been great, and little lead had been used. But the Roman standard of living brought about a great increase in the production of lead needed to provide the metal required for plumbing. The use of large amounts of copper-base alloys in coinage and general construction and adornment must have caused a similar increase in mineral output, but this is not so well documented and must have come from smaller, more widely-spread deposits.[1]

The copper deposits in the Palestinian Arabah (Negev) were worked again after a lapse of a thousand years.[2] The working of such deposits needed considerable organization and, after the cessation of the Egyptian expeditions of the 11th century BC, nothing was done until the Roman–Byzantine period, when copper smelting was resumed with slight technical changes. The need for copper would have levelled off at the end of the Late Bronze Age and the deposits then exploited would have been sufficient for the Early Iron Age and Roman needs.

The main requirement from both the civil and military point of view was for iron, and there must have been a considerable increase in the mining of ferruginous minerals. In Britain, the ore pits (bell pits) of the Weald and the 'Scowles' (open-cast workings) of the Forest of Dean are examples, although not the only ones, of this. Generally, the technique had not changed: the change is illustrated in the size and number of workings.[3]

Iron Production

Much of our information on iron production in this period relates to Eastern and Central Europe[4] and Britain. The European area covers 'Germania Magna' stretching from the Rhine to the Vistula, the major part of which is north and east of the Roman *limes*. There were many iron-producing sites in this area, the major ones being the Holy Cross Mountains of Central Poland, Northern Bohemia, Styria and Carinthia, Jutland, Schleswig-Holstein, and the Rhineland. These are the areas in which the most significant remains have been found, but in view of the ubiquity of iron ores one must assume that iron was smelted throughout the whole area. The position in the British Isles has been reviewed in detail by the present author,[5] and we have details for France,[6] but we know little about iron production in Spain at this time. Undoubtedly, North African deposits were exploited, and recent work at Sbeitla in Tunisia[7] has shown that local iron was being worked between the 2nd and 6th centuries AD.

The Roman occupation of Egypt appears to have done little to alter the picture in Iron Age Egypt itself, but it is possible that Roman penetration into Nubia was responsible for the improvement of technique in the Meroitic kingdom further south.[8] We know that nearer to the Roman heartland the Etruscan mineralized zone of Elba and Tuscany was exploited, and there is little doubt that full use was made of ore deposits in Macedonia and modern Turkey.

There is evidence that from the time of Diocletian (AD 245–313) all arms for the imperial forces were made in their own workshops. According to the *Notitia Dignitatum* there were about 32 such factories in the 4th–5th centuries.[9] Not all of these have been identified. Some of them specialized in certain types of arms, e.g. Lucca made swords, Irenopolitania spearheads, and Cremona made shields. Some did not specialize, such as that at Damascus, where both shields and general arms were made. The swordmaking tradition seems to have lasted a long time, as Damascus became a byword for swords after the Arab invasion (see the next chapter). The raw material, such as the iron blooms, was provided by the praetorian prefects who, in the role of purchasing officers, bought these from the local population either within or outside the *limes*.

Evidence of the efficiency of the Roman supply services was found in the enormous hoard of nails discovered at the legionary fort of Inchtuthil in Scotland.[10] This weighed over 5 t and contained nearly 900 000 nails of various sizes. As the fort was built in AD 83 and evacuated soon after AD 87, this shows that the supply services must have been operating efficiently before the period of the *Notitia*. From a metallurgical point of view there is nothing surprising about the composition or the structure of the nails. In many ways they resemble the construction of the iron cramps that key the stones of the Athenian Parthenon, i.e. they were made of forged heterogeneous bloomery iron. But they were well made to exact dimensional specifications, and the larger nails contain rather more carbon than the shorter nails (*see* Table 35). This was intentional, since the larger nails would have needed to be stronger to withstand the increased driving force and it was probably achieved by selection of suitable blooms or parts of blooms. There is no way of knowing where the iron was made: it was certainly made from low-phosphorus ores and the nickel content was also low. While such iron could have been obtained in Britain by this time, it would be shipped by sea. All one can say is that its composition is typical of the Roman period.

While the West was going through its Roman development, China was in the grip of the Han dynasty (206 BC to AD 220).[14] For this period we have definite evidence for ironmaking blast furnaces, showing that the EIA phase of wrought iron had rapidly given way to cast iron. It is even possible that the highly-sophisticated LBA had been transformed directly into a cast iron age.

At Chengchou, just south of the Yellow River, there are the remains of blast furnace working, in the form of 'bears' or 'salamanders' of waste cast iron, ore and blast furnace slag (*see* Table 36). This site was not water-powered and we have little evidence of suitable bellows for blowing such furnaces at this time. By the 14th century we have many examples of them.[15]

The remains show that the Chengchou furnaces were fuelled with charcoal, unlike the later medieval ones, but we have no convincing evidence of size and output. The surviving products are mainly agricultural imple-

ments, and moulds of cast iron for making them in either bronze or cast iron. There is evidence that the iron tools were made by a malleable process, which would mean heating them at 900–950°C for several days in an inert medium. This could have been done in clay boxes in a pottery kiln.

FURNACE TYPES AND TECHNIQUE

As in the EIA, furnaces can be divided into two basic types: shaft and bowl. The former, except for the addition of bellows, is little different from the natural or induced-blast furnace of the EIA (*see* Fig.40).

The only furnace to be found in the Mediterranean is that from the Etruscan site of Populonia. The remains of a non-slag tapping low-shaft furnace were found when an exposed section on the shore was cleaned up by Voss;[16] it was dated to the 2nd–1st century BC. But the simple bowl furnace grew in height and diameter until it had to have more than one tuyere[17] (*see* Fig.41). The indications are that slag tapping was almost universal in the Roman occupied regions, although this was by no

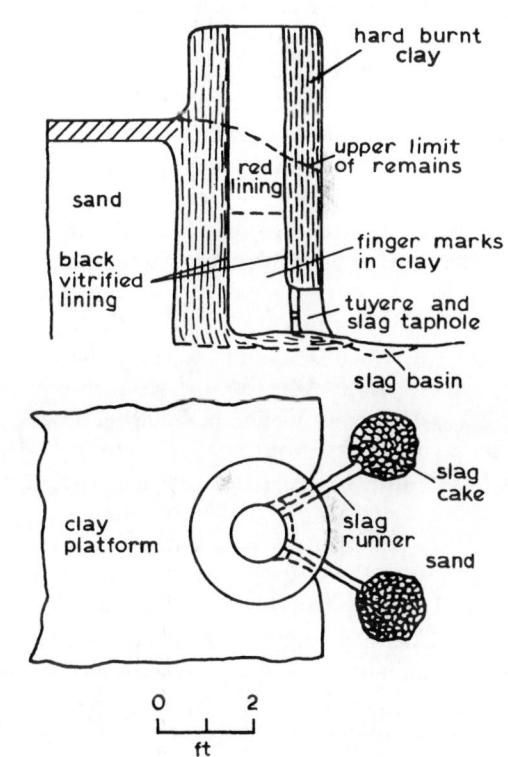

40 Plain shaft furnace of type found in Eastern Europe

Table 35 Composition of nails from Inchtuthil, Scotland (after Angus et al.[13])

| Group | Length, cm | Composition of heads, % | | | | |
		C	Si	S	P	Mn
A (i)	25–37	0·2–0·9	0·15	0·009	0·008	0·17
A (ii)	22–27	0·22–0·8	0·08	0·017	0·043	0·03
B	18–24	0·05–0·7	0·10	0·007	0·009	0·03
C	10–16	0·10–0·55	0·04	0·006	0·053	nil
D	7·2–10	0·05	0·08	0·01	0·035	0·03
E	3·8–7·0	0·06–0·35	0·05	0·003	0·16	nil

A (i) pyramidal heads; A (ii) flattened pyramidal heads

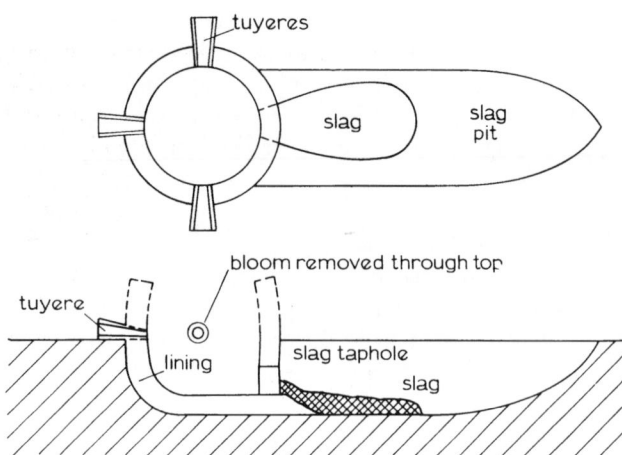

41 *Developed bowl furnace of the Roman period which was used well into the medieval period without change, scale same as Fig.40*

means the case outside. There is some doubt as to whether the induced-draught shaft furnace was used outside Africa and perhaps India, but there is a type of tall narrow shaft furnace which was in operation in Lower Saxony[18] in about the first century AD. The use of this type of furnace spread to North Jutland[19,20] (^{14}C; BC 210 ± AD 100), and then to East England in about the 4th-5th century AD.[21,22] This type was about 1.6 m high, 0.3 m dia., and had four tuyeres which may have been bellows-blown or induced (*see* Fig.42). These furnaces have similarities with Polish furnaces of the same period which seem to be somewhat shorter and fatter (1 m high × 0.5 m dia.), and which are definitely thought to have been bellows-blown.[23,24] One feature that was common to this was a pit beneath the furnace into which slag flowed at a certain moment in the smelting process. This slag flow may have been automatically controlled by the temperature–time regime in the furnace itself, or it may have been assisted by external stimulation.

The Saxo-Jutish furnaces had thin walls and, after one smelt, when the slag pit was full, the furnace shaft was moved in one piece and positioned over a new and empty pit. The slag lumps (Schlackenklötze) were allowed to remain *in situ* and whole fields of such furnace bottoms have been found in Poland.[25] The Polish furnaces seem to have had thicker walls than the Saxo-Jutish types and it is possible that they had to be rebuilt each time. On the whole, the idea of moving the furnace was not as good as that of allowing the slag to run into a pit in front or at one side, from which it could be removed when cold. This technique had long been normal practice in sophisticated copper-smelting sites, and the developed bowl furnace using the technique was carried forward into the Roman Iron Age (*see* Fig.41).

Several sites show the use of much larger furnaces spanning the Late Iron Age into the Roman period. Those at Unterpullendorf and Klostermarienburg in the Austrian Burgenland are as much as 1 m across.[26] Those in Britain on the Jurassic scarp in Northamptonshire can be as much as 1.4 m dia.[27] These are dated to the second century AD and it is difficult to see how they could have been worked, as the maximum range of primitive bellows in a fuel bed cannot be more than 30 cm. Clearly these are the largest bloomery furnaces known from any period.

Again there was no other change in technique. The product was a solid bloom of impure metal extracted through the side of the shaft furnace or the top of the bowl furnace. This situation persisted well into medieval times with little change. However, there was an improvement in what might be described as the 'division of labour'. Generally the ores were roasted before being charged with charcoal into the furnace: the working up of the bloom (5–10 kg) was done outside the smelting furnace in a second (forging) hearth.

We seem to meet more discarded pieces of cast iron in the Roman period, probably because of the greater output rather than a higher accident rate. Either the

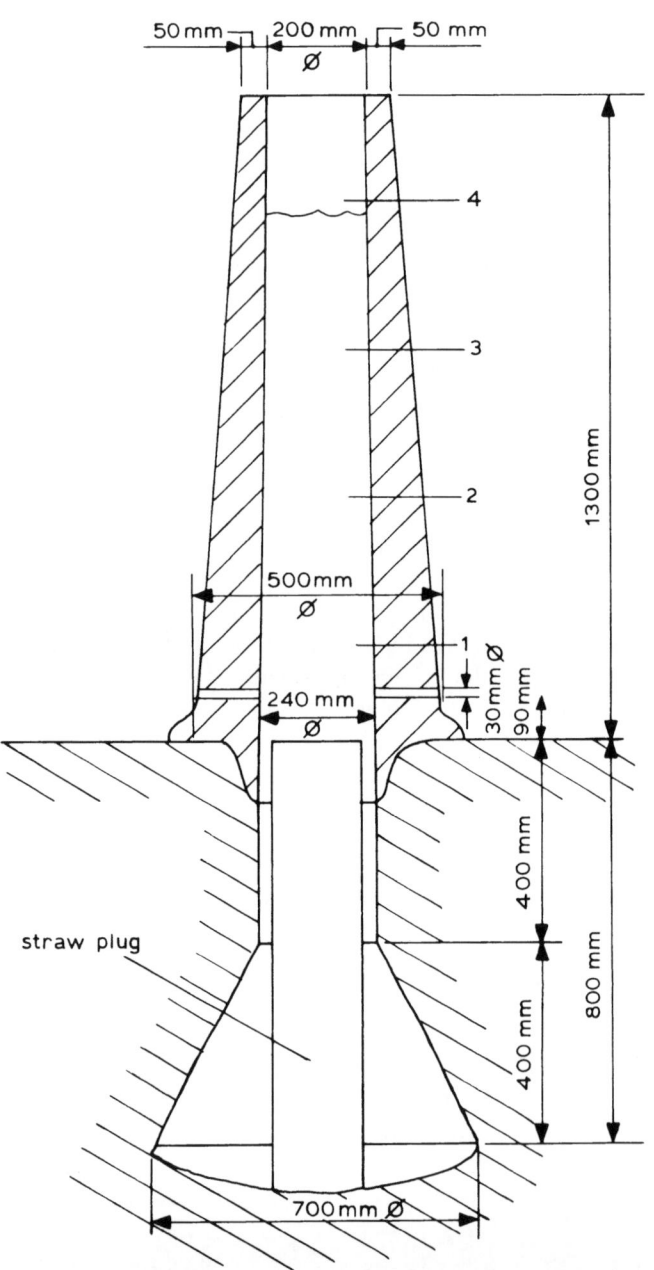

42 *Pit-type shaft furnace from Northern Europe and Jutland (reconstruction by R. Thomsen)*

significance of this accident was not appreciated or else the demand for wrought iron, rather than castings, put it out of mind, at least in the western world. There was little change in the composition of the slag (*see* Table 36). The addition of lime, which when it exceeds about 10% raises the free-running temperature, was clearly not practised. However, the quantity and the size of the slag blocks was far in advance of anything the EIA could produce. Tap-slag blocks from a British second century site now weighed 18.0 kg,[28] while the furnace bottoms from Jutland weighed as much as 170 kg. In France, in the forest of Aillant,[9] there were once 300 000 t of slag in mounds reaching a height of 15–20 m. These 300 000 t of ferruginous slag arose from a production of 75 000 t of iron, representing the manufacture of 15–20 million objects weighing 3–4 kg each. It is not certain whether the whole of this slag deposit is of Roman date but, even if only a part of it is Roman, it illustrates the scale of operation at this time. Needless to say, dumps of material of this size containing 30–50% Fe were in great demand for reworking in later periods in many European countries.

In Norway and Sweden, furnace bottoms have been found which date to the period 120 BC to AD 110. These seem to have come from non-slag tapping bowl furnaces, probably of the manipulative type which were quite common in this area.[30]

SMITHING TECHNIQUES AND THE ARTEFACTS PRODUCED

The size of the bloom in the Roman period varied from the 5 or 6 kg spindle-shaped currency bars still in use, to the 10 kg blooms of unknown shape incorporated into the bath-house beams from Corbridge[31] and Catterick[32] in the UK, and from Saalburg[33] in Germany (*see* Fig.43). There were probably smaller blooms supplied by local tribes to the Roman workshops. This iron would have been reworked and welded into structures weighing up to 500 kg: the four iron bars in the Lake Nemi anchors[34,35] totalled 414 kg and we have many anvils weighing up to

43 Fabricated wrought iron beams from Roman bath houses, Catterick Bridge, Yorkshire

50 kg[36-38] which were also made from smaller blooms (*see* Table 37).

Smithing furnaces for forging and welding were made in various shapes and sizes to suit the job (*see* Fig.44). As we can see from present-day primitive practice they do not need to be very sophisticate—merely a hole in the ground and a single clay tuyere to protect the bellows. The largest known is probably that erected for welding the bath-house beam found at Corbridge.[31] Charcoal can be piled around and a tuyere inserted wherever it is necessary to raise the temperature for forging or welding, and such a method was used in the manufacture of ships' stern frames even into the 20th century.

The blooms were, of course, no less inhomogeneous than those of the EIA,[39,40] and the variable carbon content has led many to talk about intentional carburization. But the fact is that reheating can remove or add carbon, according to the carbon level of the area of the bloom heated and the precise position of the metal in the hearth in relation to the tuyere.

Table 36 Analyses of iron slags from the Roman period

%	France Yonne[29]	Bohemia Prague[118]	Poland N. Slupia[62]	Austria Lölling[116]	England Ash-wicken[28]	England Forest of Dean[7]	Den-mark[117] Jutland AD 300 -500	Germany Pfakz[116]	Germany Aachen[116]	Scharm beck[18]	China Chung-chou (Han)[14]
FeO	46.9	23.57	52.08	47.7	62.1	40.5	41.2	39.38	65.42	54.3	3.74
Fe_2O_3*	4.8	39.29	7.38	3.36	7.7	13.2	3.6	0.44	5.18	16.9	-
SiO_2	31.8	29.02	25.21	27.3	21.2	27.5	22.7	34.93	17.19	18.9	53.74
Al_2O_3	9.9	2.38	5.32	6.6	3.2		1.0	9.40	4.95	2.1	12.14
CaO	2.1	2.30	1.05	2.2	0.4		1.4	2.26	2.73	1.1	22.7
MgO	0.75	-	tr.	1.08	1.4		1.13	1.89	1.68	1.3	2.52
MnO	2.2	-	1.84	12.1	0.5		16.8	7.08	2.17	0.39	0.63
P_2O_5	0.25	0.35	0.15	0.16	1.72	0.24	2.20	0.25	1.00	1.30	-
S	0.02	-	0.04	0.03	-	nil	-	-	0.22	0.10	0.114
TiO_2	0.35					-					
K_2O							} 0.05			} 1.18	-
Na_2O											
Loss			5.20								

* Dependent on oxidation after tapping

Table 37 Beams and anvils consisting of welded iron blooms

Object	Provenance	Weight, kg	Length, m	Section, * m
Anvil	Mainz[109]	–	0·20	0·20 x 0·20
	Sutton Walls[36, 37]	.50	0·27	0·23 x 0·23
	Stanton Low[90]	23·2	0·25	0·16 x 0·16
	Kreimbach, Pfalz[38]	–	0·21	0·18 x 0·18 (top)
Beam I	Chedworth, Glos.[108]	220	1·63	0·15 x 0·15
II	Chedworth, Glos.	162	0·96	0·15 x 0·15
III	Chedworth, Glos.	116	0·99	0·15 x 0·15
Beam I	Catterick, Yorks.[32]	250	1·68	0·15 x 0·15
II	Catterick, Yorks.	135	0·89	0·15 x 0·15
Beam	Corbridge, N'ld.[31, 7]	157	1·00	0·15 x 0·15
Beam	Leicester[113]	c. 130		0·15 x 0·15
Beam	Chesters, N'ld.	c. 14	0·25	0·11 x 0·14
Beam	Wroxeter, Shrop.[111]	Mass of iron found in stokehole of bath house (4th century)		
Beam I	Pompeii[7]	–	c. 1·0	0·12 x 0·09
II	Pompeii	–	c. 1·3	–
III	Pompeii	–	c. 1·3	–
Beam	St. Albans (Verulamium)	–	0·43	0·13 x 0·15
Beams (many pieces)	Saalburg[33]	220	1·40	0·12 x 0·11
	Saalburg	–	0·56	0·11 x 0·11

* This is to give a rough idea of the volume of metal represented; both the anvils and the beams taper somewhat

Most Roman material shows a substantial reduction in the amount of phosphorus by comparison with that of the EIA and this may point to better discrimination and the knowledge that low-phosphorus metal was better for carburizing (*see* Table 38). But there are plenty of examples of high-phosphorus iron in the Roman period and most of the metal used was of low-carbon content and therefore not carburized.[41]

However, the knowledge of carburizing and quench-hardening was patchy; although there is no lack of examples of properly executed carburizing followed by suitable heat treatment. Heat-treated cold chisels have

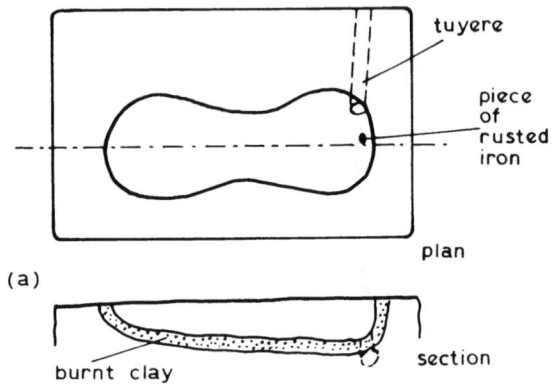

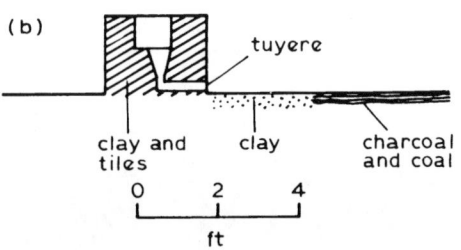

44 Typical Roman smithing furnaces

been found on two British sites, i.e. Wall[42] in Stafford-shire, and Chesterholm[43] on Hadrian's Wall in Northumberland. One of the chisels found on the latter site had been hardened on the edge but the lack of uniformity in carbon gave hardnesses varying from 579 to 464 HV, and a considerable range of metallurgical structures.

SWORDS

The principal weapon was still the sword and, as would be expected, this tends to show the best craftsmanship. While the Iron Age techniques of piling and surface carburizing were widely used, we begin to see more complex techniques which were to become common in the later migration period. One of these is the technique of pattern welding which used to be confused with damascening. Pattern welding is the welding together of strips or other small sections such as rods or wire to make a weapon of complex but easily recognizable structure: one might call it a 'hallmark', although the makers often added their own individual marks.

Pattern welding first appears in the Rhineland and at Nydam in Schleswig-Holstein early in the third century AD, and examples of swords made in this way are known from Poland and Britain[44-47] (*see* Fig.45). The rods and wires were twisted together and heated and forged: sometimes the process was repeated by folding over and starting again. Usually, a 'core' of unwelded material was sandwiched together between two pattern-welded plates, and sometimes the edges were welded on separately. There are, obviously, many possible combinations. The pattern is sometimes caused by imperfections in the lines of welding, by slag and oxide entrapment: in other cases, irons of different composition are the cause.

It is not clear that artefacts made in this way were appreciably stronger than most of the weapons which were made by simple piling but, if they were well polished, the structure would obviously be a delight in

Table 38 Iron and steel blooms and pieces of cast iron from the Roman period in Britain

| Metal | Weight, kg | Composition, % | | | | | |
		C	Si	P	S	Mn	N$_2$
Iron and steel							
Cranbrook, Kent,[106] 1st–2nd century AD	0·71	1·16–1·46	0·20	0·014–0·025	0·03	–	0·004
Lower Slaughter,[107] Glos., 3rd–4th century	11·0	0–0·8	tr.	0·0085	0·007	–	0·004
Nanny's Croft,[105] Sussex, 4–5th century	0·30	0–1·6					
Forewood, Sussex[105]	1·26	0–0·3					
Cast iron							
Wilderspool, Lancs.[121] 2nd century	–	3·23	1·05	0·76	0·49	0·403	–
Tiddington, Warw.[112]	0·57	3·52	1·92	0·77	0·049	0·63	–

itself and evidence that a good deal of work had been put into the weapon by a skilled craftsman. These artefacts were often adorned with non-ferrous inlay on the hilt end of the blades.

All of the three Nydam swords examined had some hardened parts.[44] The carbon content varies, as shown in Fig.46. In Fig.46(2), two parts had been effectively hardened; in Fig.46(4), only the central core on to which the patterned metal was welded had been successfully hardened owing to its higher carbon content. The maxi-mum hardness achieved in Fig.46(2) was on one of the edges, with a carbon content of 0.43%: this was 700 HV, which is what one would expect from such a steel today.

Some swords carry stamped inscriptions on the hilt end of the blade which closely resemble the potters' marks on Samian ware.[48] Such inscriptions include *CICOLLUS, RISSA-CUMA,* and *RICCIM(ANU),* the latter meaning 'made by Ricci'. Others carry non-ferrous inlaid designs.[46,47] Of course, there is more to a sword than its blade. It had to be mounted with a hilt, and these

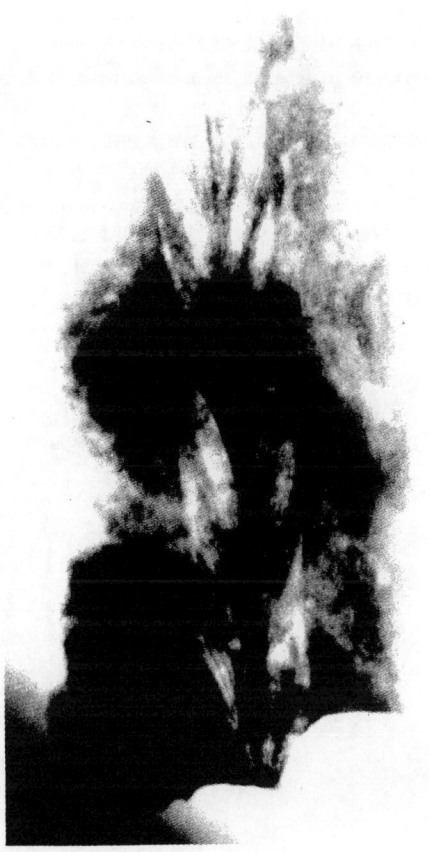
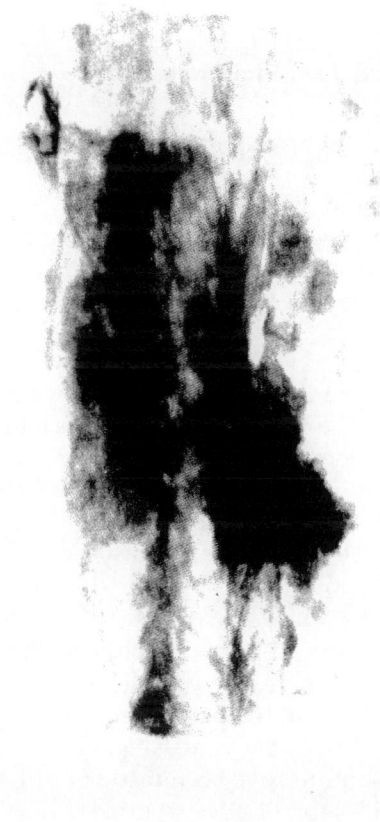

45 Radiographs of Roman Swords

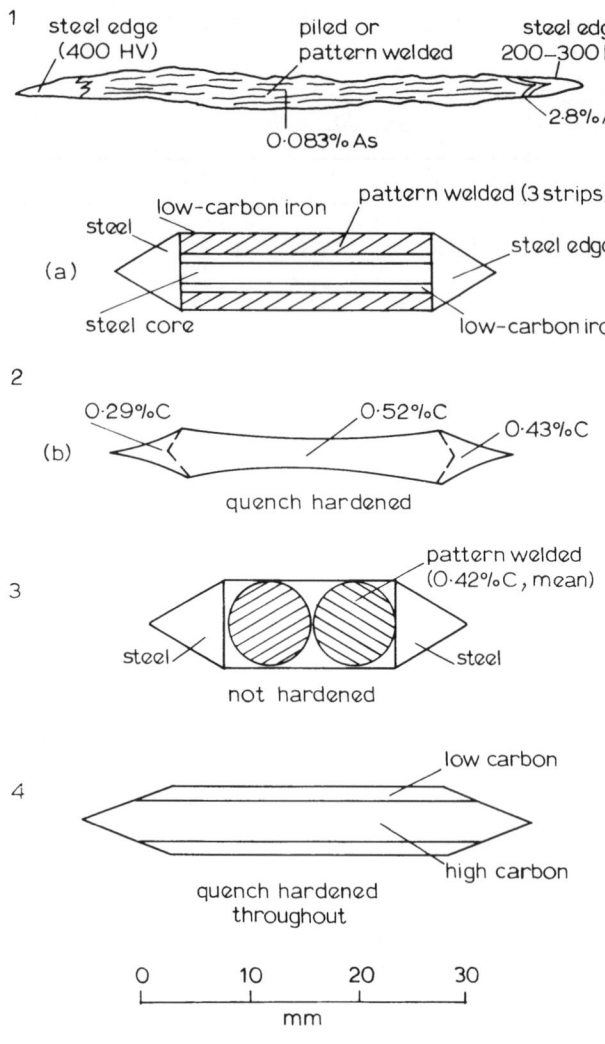

1 after Becker[96]; 2, 3 and 4 from Nydam after Schurmann[34]

46 Sections of pattern-welded swords of the Roman period

show less adornment in the Roman period than in the later migration and medieval periods. However, the scabbard often contained iron plates decorated with the sort of non-ferrous inlay found on some of the blades.[49]

OTHER TOOLS AND ARTEFACTS

The number of types of iron tools current in the Roman world was immense. The group of Assyrian tools found at Thebes gives some idea of the high standard that was being reached in the pre-Roman period.[50] One is, therefore, not surprised to find that the Roman world knew all the types of hand tools that are in existence today, with the possible exception of the wood screw.[51] We see, for example, iron smoothing planes[52] just like those available in modern hardware stores, spoon bits (but not twist drills), and the most amazingly complicated locks.[52] However, whereas one would expect modern scissors to be made of hardened steel, the Roman equivalent, i.e. the shears, do not seem to have been made of carbon steel nor to have been superficially carburized.[41]

The Roman *limes* stretched from the lower Rhine to the delta of the Danube and there is little doubt that the tribes to the north of it were in close contact with the

Roman world. Either they traded with the Romans, supplying iron blooms and other semi-finished products to be worked up in Roman military workshops, or they captured Roman material. Examples of this 'trade' are hoards of Roman coins and an inscribed sword, both from South East Poland: the latter is almost identical in inlay and construction with that from South Shields in Britain.[47] One good source of iron ore was near Rudki in the Holy Cross mountains of Poland, and there is no doubt that ironworking in this region was very active in the Roman period. Many of the tools found on Polish sites have been examined metallo-graphically and, like Roman material elsewhere, they show a varied technology. In a hoard found near Krakow,[53] six out of twenty-three knives and chisels had been quench-hardened to give hardnesses between 433 and 724 HV. On the other hand, a group found in Silesia[54] showed no sign of quenching. This tends to support the evidence from within the Roman world, and shows the continued existence of an EIA tradition with an increasing number of smiths capable of carburizing and hardening iron to obtain high quality tools. The poor average quality of Roman edge tools may have something to do with the lowly status of the smith.

Beyond the Roman frontiers, we know a good deal about Northern Europe and the East, but very little about other regions. There is no doubt that the Iron Age in China was developing steadily.[55] Large cast iron statues weighing up to 25 kg were produced by the sixth century AD. On the Indian Subcontinent, ferrous metallurgy must have reached the same stage of development as that of Europe. It has provided examples of the fabrication of large wrought iron pillars and beams, such as the pillars of Delhi[56,57] (*see* Fig.47) and Dhar,[58] and the temple beams of Konarak[58,59] (*see* Table 39). The techniques used were exactly the same as those in Western Europe so we can assume that this was a widely-known process at this time in spite of the lack of evidence in the area between. However, this lack may be due to the reworking of earlier material in the Middle Ages. We know that the iron cramps of Persepolis in Iran, and those in other buildings in the area, were removed for

47 Wrought iron pillar at Delhi, India, 7.2 m high. Dated to about AD 300

Table 39 Analyses and details of iron pillars, beams, and other objects from India, Ceylon, and Afghanistan (after Hadfield[57] and Graves[59])

Object	Provenance	Date	Length, m	Composition, %				Weight, kg
				C	P	S	Mn	
Wedge	India	125 BC	—	0·70	0·02	0·008	0·02	
Pillar	Delhi	AD 300	7·2	0·08	0·11	0·006	nil	6 000
Pillar	Dhar	—	12·5	0·02	0·28	—	—	7 000
Chisel	Ceylon	5th century AD	—	tr.	0·28	0·033	nil	
Nail	Ceylon	5th century AD	—	tr.	0·32	nil	nil	
Billhook	Ceylon	Medieval	—	tr.	0·34	0·02	tr.	2 900
2 Beams	Konarak (Orissa)	13th century AD	10·6, 7·7	0·11	0·02	0·02	tr.	4 300
Cast iron	Afghan	10–13th century AD	—	4·25	0·11	tr.	tr.	

reworking. The copper-clad statue known as the Colossus of Rhodes, which was erected in 292–280 BC and which undoubtedly had an iron framework, collapsed in 224 BC and was broken up and re-used in AD 672. In the Sudan, the Egyptian-orientated Meroitic civilization was overwhelmed in about AD 400 by the Kingdom of Axum, and this is the last evidence of Roman-type iron working in Africa.[11] Both East and West Africa were to continue with a basically EIA tradition until modern times.

Non-Ferrous Metals

COPPER-BASE ALLOYS

The really big change in this field was the large scale introduction of brass, principally for coinage. No doubt the relative scarcity of tin for bronze was showing itself in terms of high prices, and the opportunity of using brass rather than bronze would have been readily taken. For the first time use was made of true brasses, i.e. copper alloys in which the only intentional alloying element was 20–30%Zn. But this alloy seems to have been restricted to coinage. For most of the uses of copper-base alloys, bronzes, or the ternary Cu–Zn–Sn alloy, now known as gunmetal, were preferred. This may have been because these were better casting alloys and castings were more widely used except for coinage.

The brass was made by the addition of 'calamine' ($ZnCO_3$) to copper, under reducing conditions, in a crucible.[60] In this process some of the zinc was reduced by the charcoal before the copper was molten. The zinc vapour entered the copper and lowered its melting point so that some of the charge gradually melted at about 900°C. Zinc ores came from the Mendips of Britain, the Ardennes of Belgium, and doubtless from many other deposits in the East.

Mining of lead-zinc ores goes back to before the Roman period in Rajastan in India.[61] We know that vertical distillation furnaces were used for zinc in medieval times (Chapter 6), and this process could have been used earlier. But it now seems likely that the Western cementation process was used and zinc oxide, made by roasting zinc blends (ZnS), was added to copper.

It might be expected that, once the calamine process for making brass had been developed, and experience had been gained in its use—as must have been the case

in the Roman period—this alloy would have tended to oust bronze from its role of principal non-ferrous metal. However, this was not to be the case: while brass was used in the coinage 'brasses' as their name correctly implies, we find that in Britain the average zinc content of casting alloys was only 2.7% with 13.3%Sn and 7.1%Pb,[7] and thus they were gunmetals. The situation was a little different in the case of wrought alloys which, on average, contained 6.1%Zn and 5.7%Sn, the lead content being low or absent.[62] The same reluctance to use true brasses existed elsewhere.[63] In fact, in a sample of copper-base alloys used for statues in Roman Italy[64] the maximum zinc content was only 0.14%, and a plug cock from Pompeii was free of zinc. However, if we look through a list of material from bath houses[65] we find a preference for the ternary alloys of copper, tin, and zinc, with the zinc content varying from 1 to 9%.

In the early part of the Roman period it would seem that the average zinc content of the copper-base artefacts was little different from that of the EIA. The introduction of brass as a coinage metal starts about 45 BC[66] and brass coins seem to have had a higher value than those of bronze. Whatever the reasons for making brass coins, the brass content steadily decreases until AD 162,[67] when the alloy is more of a gunmetal, with 7.87%Zn and 2.42%Sn.

It would appear, therefore, that a large proportion of the zinc in the later gunmetals could have come from the brass used in the early coinage. It is clear that, for one reason or another, a simple brass was not generally preferred in the Roman period, although the method of its production was well understood, in at least some parts of the Empire.

We have very little information on the making of brass or gunmetal in the EIA or the Roman period. It is possible that calamine was added to melts of bronze or copper under reducing conditions and it would be noticed that there was a definite increase in weight. For Britain, we have some closed crucibles which could have been used for this purpose[60] and an inscribed plate of standard brass from Colchester.[68] Many crucibles of this type could have been heated in a simple wood-fired reverberatory furnace such as that found by May at Wilderspool.[69]

Plano-convex copper ingots dateable to the Roman

period have been found on many sites in Western Europe.[70] They are to be found mainly in North Wales,[71] through Brittany[72] to the South of France. Many of the latter come from shipwrecks off the coast.[73] They are much larger than those of the LBA, mostly in the range 10–100 kg and, like that shown in Fig.48 have diameters of the order of 25 cm. Some are inscribed around the rim of the ingot. The composition of the British ones, and some of the French, are given in Table 40.

Two hoards of purse-shaped tin ingots have been found off Port Vendres on the Franco-Spanish border and they almost certainly come from the Galician mines.[74] They are of variable form and weigh about 11 kg each. Other hoards such as that from the Sardinian coast near Cap Bellavista[75] are of similar shape, and there is little doubt that this was a common form in the Roman period. Clearly, their shape was convenient for transport.

ROMAN COINAGE: ALLOYS AND TECHNIQUES

In this period, as in the last, coinage was one of the main uses of non-ferrous metals. The widespread trade within the Roman world required an ever-increasing amount of coinage and this required the establishment of a series of well-known mints. After the fall of the Roman Empire this coinage continued to circulate for the next hundred years or so until replaced by local coinages.

48 *Copper ingot from north Wales; weight = 19 kg (after Boon[115])*

The high value Roman coinage, such as the *denarius*, was made of silver and, in the days of the Republic (*c.* 150 BC), contained as much as 94%Ag. By the time of the Empire this had been debased somewhat to 80% and, by AD 220 to 44%, following the pattern of the Athenian tetradrachm (*see* Table 40). The *antoniniani* of the third century AD contained 46–60%Ag, whereas the *follis* of the fourth century contained as little as 1.76%, and this amount was no accident.[76] There were various official ways of increasing the surface silver content of the coins with lower silver content. One method was oxidation (heating in air) and the dissolution of the oxidized copper by acids, leaving a bright silvery surface. Probably, this was the reason for the presence of such small amounts as 1.76%Ag. Others were plated by dipping in molten silver,[77] or wrapping in silver sheet.[78]

We have already mentioned the use of brass for Roman coinage. This seems to date from about 45 BC when we see the introduction of a true alpha brass containing as much as 27.60%Zn. By AD 79 this was down to 15.9%, and, by AD 161, to 7.87%. We should like to know more about the reasons for this extraordinary decrease, but there is little doubt that the metal was made by the calamine process, and not by the addition of metallic zinc.

Low-tin bronzes were in use in Ptolemaic Egypt (169 BC) and continued in use, often with low silver contents, into the fourth century AD. Impure coppers (1%Zn–1%Sn) were also used over a long period and these resemble the composition of the recent British low-value coins. The bronze coinage of the fourth century consisted almost entirely of leaded argentiferous tin bronzes. The lead content tended to increase with coin size. Before striking, the blanks were homogenized to dissolve the brittle delta (tin-rich) constituent. The larger coins were struck hot, in some cases from cast blanks, in others from previously worked or preformed blanks.[79]

Most coinages have undergone devaluation and revaluation. Devaluation is a simple process for a coiner as all he has to do is to add more base metal, but revaluation would require a process for the removal of the base metal unless adequate supplies of fresh noble metal were available. In Britain, both Silchester and Hengistbury,[66] have yielded cupellation furnaces which look as though they had been used for the recovery of noble metal from debased coins. Neither of these sites was an official mint and it is possible that the process had been used illegally. Globules from the hearth at Silchester contain 2.98%Ag, 78.13%Cu, and 16.14%Pb, and represent typical coinage alloys of the period AD 250–275. The process of cupellation is discussed in the next section.

Coining necessitates the use of dies and in the Roman period these were either made of tin bronze, containing 15%Sn, or of iron, but it is probable that the great majority of dies were bronze. The use of hot striking would have given a reasonable life to a bronze die provided that the bronze tool itself was not allowed to get overheated.[81] It has been estimated that the lower (pile) die would give about 16 000 impressions, while the top (reverse) would give about 8 000. But at least one

Table 40 Coinage alloys of the Roman period (after Gowland,[66] Caley,[67] and Cope[76,79])

Type	Date	Composition, %					
		Au	Ag	Cu	Sn	Zn	Pb
Roman Denarius	150 BC	0·53	94·34	4·40	0·23	–	0·39
Ptolemaic tetradrachm, Egypt	1st century BC	0·24	52·51	40·45	1·74	0·11	1·36
Ptolemaic copper coin, Egypt	169–146 BC	–		65·11	5·12	0·10	28·78
Brass	45 BC			71·10	–	27·60	–
Brass	AD 79			81·13	–	15·90	–
Brass	AD 161–162			88·96	2·43	7·87	0·18
As	AD 14–37			99·65	0·01	tr.	tr.
Antoninianus	AD 238–244	0·13	58·90	40·65	0·10	–	0·22
Denarius	AD 244–249			98·36	1·03	–	0·51
Antoninianus	AD 254–255		16·25	80·79	2·52	0·03	0·70
Folles; Carthage	AD 307		1·20	81·25	5·45	0·01	11·90
Folles; London	AD 310		1·76	86·78	5·54	nil	6·01
Folles; London	AD 318		2·11	87·89	4·33	0·01	5·57

top die or trussel, that from Trier,[82] has been found to be steel faced. The shaft contained about 0.038%C but the working face had contained 1.03%C, showing it to have been carburized. The design was cut into the soft iron and then the surface was carburized and quench-hardened. This shows a most remarkable level of technique.

We know very little of the technique used in the preparation of the blanks, but it is possible that the cast blanks were made in tier moulds rather like those used by counterfeiters in Roman times in Britain,[83] and by legitimate coiners in some other countries.[84] But the design would have been struck on to the blank and not cast in the mould as with the counterfeit coins. Some late coins seem to have been made by the cutting of blanks from a circular rod. There seems to be no evidence for the use of 'blanking', i.e. stamping out discs from sheet metal, a process which would be very difficult to carry out at such an early time. It is more probable that the blanks were cut with shears, as in later times, or cast in 'strings' with narrow 'runners' between.

Relative alignment of the upper and lower dies could have been obtained by fixing them in a hinged frame or by inserting the two dies into a square socket. Simpler methods are possible. External reference marks can be made, or pins can be arranged to project from both the trussel and the pile, but neither of these methods would have permitted accurate double-striking of a hard flan.

LEAD AND SILVER

The production of lead in Greece was largely a byproduct of silver production. The mines of Laurion were worked from very early times, perhaps as early as Mycenean times. The use of lead in the Roman period increased very considerably. In order to sustain the higher standard of Roman civilization over a large part of Europe, new lead deposits were opened up as soon as the conquest of new territory was complete. In Britain, the lead mines of the Mendips were worked by AD 49, only six years after the conquest, and the low-silver lead deposits of Flintshire were worked by the late first century AD, while the virtually silver-free deposits of Derbyshire were worked by AD 117–138. This lead not only supplied

local requirements, but some at least was sent through France on the road to Rome.[85]

Similar activities were organized in Spain. Inscribed lead pigs have been found at Orihuela near Cartagena,[85] and other sites.[86,87] The Rio Tinto ore deposits were worked for silver as well as copper in this period.[88]

Examination of the debris at Rio Tinto shows examples of slags which, like those from Wales, show low lead-silver levels but appear to have been associated with silver recovery.[89] We have little evidence of the actual process used but it is possible that it was like that used at Laurion (*see* Chapter 5). In this case lead would have to be brought from elsewhere as it would be needed to dissolve the silver from the jarositic ore (*see* BR, Tite and PTC).

We have other examples of lead production, such as lead slags, speiss and litharge (*see* Table 41) and the remains of smelting or melting furnaces near Cologne.[89,90,91] It is clear however, that the exploitation of lead minerals in this period was very much more extensive than the known remains would suggest.

At Rio Tinto, silver was the main product of the Republican period. Imperial Rome continued this tradition but increasingly exploited the copper ores.[89] It is likely that the jarositic silver ore was exhausted by this time and more use was made of the British and Welsh lead-silver ores.

Numerous lead ingots of plano-convex and the normal rectangular shape have been found off the coast of Brittany near Ploumanac'h.[92] The total weight was of the order of 22 tonnes. The rectangular ingots with truncated cone section are inscribed *CIV BR* and *ICENI* on the sides, suggesting a British origin. The inscriptions are more primitive than the usual ones of the first to the early second centuries and are believed to date from the second to the fourth centuries. The inscriptions have been written on the sides and not on the bottom (as cast), and appear to have been scrawled on the inside of the mould with a finger. While the letters *BR* can be interpreted as *BRIGANTES*, and therefore connected with the lead producing areas of Derbyshire and north Yorkshire, the word *ICENI* comes as a bit of a surprise, as this

Table 41 Examples of Roman lead and silver slags and litharge

| % | Lead smelting slag | | | Silver | | | |
	P.Ffwrdnan Wales	Laurion Greece	Scarcliffe Britain	Slag Rio Tinto (Spain)	Speiss	Litharge Nordeifel Germany
PbO	32.3	10.7	0.43	1.50	Pb 2.43	77.34
SiO$_2$	58.2	33.8	28.16	26.20	0.35	-
CaO	8.0	13.8	17.56	-	0.50	-
FeO	0.8	15.2	3.01	58.5	Fe66.27	0.58
Al$_2$O$_3$	-	3.9	14.46	-	-	-
K$_2$O	-	-	2.07	-	-	-
BaO	-	-	26.10	-	-	-
Ag	4ppm	0.06	-	0.00632	111ppm	-
Zn	-	5.4	-	-	-	-
CuO	-	-	-	-	Cu 1.32	4.71
CO$_2$						pres.
As					18.75	-
Sb					6.57	-
S					2.78	-
Au					8.5ppm	-

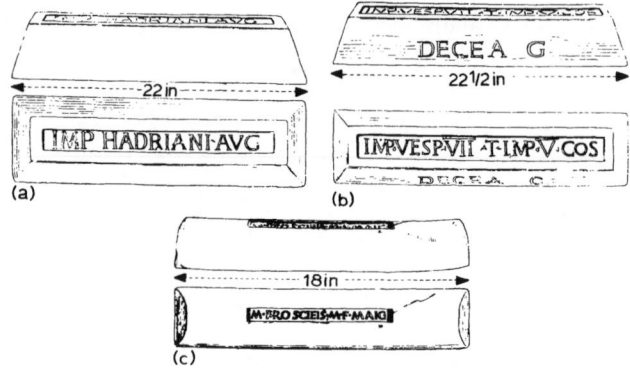

a Near Westbury, Shropshire; *b* Hints Common, Staffordshire; *c* Near Orihuela, Valencia, Spain

49 Typical lead pigs of the Roman period (after Gowland[98])

tribe is not thought to have opreated as far north as the lead-producing area on the east coast. Perhaps this shows a new role for the powerful *ICENI*, that of middleman. It should be noted that in the 16th century AD some Derbyshire lead was exported through ports south of the Humber.[93]

SMELTING FURNACES AND SILVER RECOVERY

Smelting installations have been found at Horath[90] in the Rhineland, and at Pentre Ffwrndan in Flintshire,[94,95] North Wales. These are fragmentary and do not give enough detail to enable us to reconstruct the actual furnaces used. But we know from Roman and Medieval records and the slags found on Roman sites that the process was of a primitive type, and that induced-draught shaft furnaces a metre or so high were probably used. Strabo[96] (63 BC–AD 25) says that they had 'tall chimneys', and therefore they appear to have been tall shaft furnaces. With this type, the lead recovery is poor and a good deal is lost in fume and in the slags, but it has the advantage that most of the silver in the ore is recovered in the lead produced. In the best-organized sites the lead was allowed to flow into hard clay moulds, at the base of which inscriptions were impressed so as to be cast on the pigs produced (*see* Fig.49). We know that these moulds were strong enough to cast several pigs, although none has yet been found. It is probable that, when the lead was destined for silver recovery by cupellation, simpler forms of ingot would have been made at first and only in the last stage of the process would an inscribed pig have been made.

It seems that in the Roman period it was only economic to recover silver from lead when it exceeded about 0.01% or 100 g/t. This would have entailed melting down the smelted lead in shallow hearths of bone ash, examples of which have been found in Britain at Hengistbury Head[97,80] and Silchester.[98] The lead would have been heated to 1 000°C and oxidized to litharge (PbO) with a blast of air from bellows directed over the surface of the molten lead. Most of the litharge would have been skimmed off, and much has been found on Roman sites in the Mendips,[99] but some would have been absorbed by the bone ash of the cupels, leaving behind a button containing the silver and any gold present in the ore.

The silver was extracted when its recovery became economic, but it is clear that lead was produced for its own sake and not merely as a byproduct of silver production, as in the Classical Greek period at Laurion. Much of the lead went into sanitary engineering equipment such as pipes[100,101] and cisterns.[102] But some went into statues, pewterware,[103] and into anchors which, when they were not of iron, were mixtures of wood and lead.

THE PRODUCTS OF THE ROMAN LEAD–SILVER INDUSTRY

The cisterns, like the pipes, were made of cast sheet. To make them, lead was cast into a shallow sand or dried clay–sand mould with various forms of embossed ornamentation. The sheet so formed was cut, bent and welded at the edges by the formation of a temporary mould and by the running-on of pure lead or lead-based alloy, superheated sufficiently to cause the sheet to melt at the edges to be joined. Some thin sheets would be soldered, but this was rare. Coffin sheets were usually left unwelded at the corners.

Pipe was made from long narrow sheets of lead bent into the form of a tube and welded along the top edge with the aid of temporary mould strips. The lengths of pipe often measured 7 m or more and these were joined when necessary by a process of inserting one end into the other and sealing with a moulded 'box' of cast lead. These joints were made of a quality sufficient to withstand considerable pressure, but could be reinforced with 'concrete' poured round them.

The metal used in pewter tableware averages 50%Pb, the rest being tin. This was, of course, the poorer man's metal for tableware. Those who could afford it would use silver, as we know from the finds from Pompeii and other famous sites. The pewter had been cast in stone moulds, thus showing that it had been intended for mass production, while the items of silver plate had been hammered individually from sheet. This was joined either by lead–tin solders or by solid phase welding, which in the case of silver would have involved hot hammering. Examples of the latter technique can be seen in vessels from Traprain Law in Scotland.

References

1   O. DAVES: 'Roman mines in Europe', 1935, Oxford, Oxford University Press

2   B. ROTHENBERG: 'Timna; valley of the biblical copper mines', 1972, London, Thames and Hudson

3   H. R. SCHUBERT: 'History of the British iron and steel industry', 1957, London, Routledge and Keegan Paul

4   H. F. CLEERE and D. W. CROSSLEY: 'The Iron Industry of the Weald', Leicester, 1985

5   D. E. BICK: 'Early iron ore production from the Forest of Dean and district', *JHMS*, 1990, **24**, (1)

6   R. PLEINER: *BRGK*, 1964, No.45, 11

7   R. F. TYLECOTE: 'The Prehistory of Metallurgy in the British Isles', *Inst. Metals*, London, 1986

8   C. DOMERGUE, A. REBISCOUL and F. TOLLON: 'Les fours antiques dur fer dans la Montagne Noire (Aude)', In: *Mines et Fonderies antiques de la Gaulle,* (ed. C. Domergue), *CNRS*, Paris, 1982, 215–236

9   A. BOUTHIER: 'Données nouvelles sur l'utilisation du mineral de fer dans le Nord-Ouest de la Nièvre a l'époque Gallo-Romaine', In: *Mines et Fonderies antique de la Gaulle,* (ed C. Domergue), *CNRS*, Paris, 1982, 137–156

10   G. SIMPSON: Ann. Report of the Brathay Exploration Group, 1964, 25

11   R. F. TYLECOTE: 'Iron making at Meroe, Sudan', Meroitica, 6, *Meroitic Studies,* (ed. N. B. Millet *et al.*), Berlin, 1982, 29–42

12   A. H. M. JONES: 'The Later Roman Empire', 1964, Oxford, Oxford University Press

13   N. S. ANGUS *et al.*: *J. Iron Steel Inst.*, 1962, **200**, 956

14   HUA JUEMING and LI JINGHUA: 'Preliminary researches on iron smelting in Henan during the Han Dynasty', Koagu Xuebo, 1978, **1**, 1–23

15   J. NEEDHAM: 'The development of iron and steel technology in China', Newcomen Society, 1956, Plate 15, Fig.25

16   O. VOSS: Personal Communication

17   J. H. MONEY: *JHMS*, 1974, **8**, (1), 1

18   W. WEGEWITZ: Nachr. Niedersachsens Urgeschichte, Special Report no.26, 1957, 3

19   O. VOSS: *Kuml*, 1962, 7

20   R. THOMSEN: *ibid.*, 1963, 60

21   M. U. JONES: *Ant. J.*, 1968, **48**, 210

22   Unstratified slag block in museum in Norwich, Norfolk

23   K. BIELENIN: *PZ*, 1964, **42**, 77

24   K. BIELENIN: Starozytne Gornictwo i Hütnictwo Zelaza w Gorach Swietokxrzyskich, Warsaw-Krakow, 1974

25   M. RADWAN and K. BIELENIN: *RHS*, 1962-3, **3**, 163

26   K. BIELENIN: 'Einige Bemerkungen uber das altertumliche Eisenhuttenwesen im Burgenland', Wiss. arbeit Bgld (Eisenstadt), 1977, **59**, 49-62

27   D. JACKSON and R. F. TYLECOTE: 'Two new Romano-British ironworking sites in Northamptonshire', *Britannia*, 1988, *19*, 275–298

28   R. F. TYLECOTE and E. OWLES: *NA*, 1960, **32**, 142

29   J. MONOT: *RHS*, 1964, **5**, (4), 273

30   HELEN CLARKE (ed.), 'Iron and Man in Prehistoric Sweden', Stockholm, 1979

31   SIR H. BELL: *J. Iron Steel Inst.*, 1912, **85**, 118

32   J. H. WRIGHT: *Bull. HMG*, 1972, **67**, 24

33   L. JACOBI: Das Römerkastell Saalburg, 1897, Hamburg v.d. Höhe

34   G. CALBIANI: *Met. Ital*, 1939, 359

35   G. C. SPECIALE: *Mariner's Mirror*, 1931, **17**, 304

36   K. KENYON: *Arch. J.*, 1953, **110**, 1

37   R. F. TYLECOTE: *TWS*, 1961, **37**, 56

38   L. LINDENSCHMIT: *RGZM*, Mainz, 1911, **5**, 256

39   G. BECKER and W. DICK: *Arch. Eisenh.*, 1965, **36**, 537

40   A. KRUPKOWSKI and T. REYMAN: *Sprawozdania Panstwowego Musea Arch.*, Warsaw, 1953, **5**, 48

41   R. F. TYLECOTE and B. J. J. GILMOUR: 'The metallurgy of early ferrous edge tools and edged weapons', *BAR Brit. Ser.*, No. 155, Oxford, 1986

42   J. GOULD: *Trans. Lichfield and S. Staffs. Arch. Soc.*, 196–4, **5**, 1

43   C. E. PEARSON and J. A. SMYTHE: *PUDPS*, 1938, **9**, 141

44   E. SCHURMANN and H. SCHROER: *Arch. Eisenh*, 1959, **30**, 127

45   E. SCHURMANN: *ibid*, 1959, **30**, 121

46   A. M. ROSENQVIST: 'Pattern welded swords from the Roman period with figured inlays', 1967–8, Oslo, 1971, Universitets Oldsaksamlungs Arbok

47   J. PIASKOWSKI: *Z. otchlani wiekow*, 1965, **31**, (Zesz.l), 36

48   C. BOHNE: *Arch. Eisenh.*, 1963, **34**, 227

49   K. J. BARTON: *Discovery*, 1960, **21**, 252

50   A. R. WILLIAMS and K. R. MAXWELL-HYSLOP: 'Ancient steel from Egypt', *J. Arch. Sci.* 1976, **3**, 283–305

51   W. H. MANNING: 'Catalogue of the Romano-British iron tools, fittings and weapons in the British Museum', BM Publ., 1985

52   G. BOON: 'Roman Silchester', 1957, London

53   SIR J. EVANS: *Arch.*, 1894, **54**, 139

54   J. PIASKOWSKI: *Materialy Starczytne*, 1965, **10**, 169

55   J. PIASKOWSKI: *Przeglad Arch.*, 1962, **15**, 134

56   J. NEEDHAM: 'The development of iron and steel technology in China', 1958, London, Newcomen Society

57   SIR R. HADFIELD: *J. Iron Steel Inst.*, 1912, **85**, 134 (see also, S. V. BRITTON: *Nature*, 1934, **134**, 239, 277)

58   W. E. BARDGETT and J. F. STANNERS: *ibid*, 1963, **201**, 3

59   H. G. GRAVES: *J. Iron Steel Inst.*, 1912, **85**, 187

60   J. BAYLEY: 'Non-metallic evidence for metal working'. In: *Archaeometry*, Proc. 25th Int. Symp. (ed. Y. Maniatis), Athens, 1986, Amsterdam, 1989, 291–303

61   L. WILLIES: 'Early metal mining in India', In: 'Aspects of ancient mining and metallurgy', (ed. J. Ellis Jones), Bangor, 1986, 129–135

62   J. A. SMYTHE: *PUDPS*, 1938, **9**, 382

63   M. PICON *et al.*: *Gallia*, 1967, **25**, 153

64   M. LORIA: Actes XI Congrès Int. Hist. Sci., 261, 1965, Krakow and Warsaw

65   A. MUTZ and L. BERGER: Studien zu unserer Fachgeschichte', 1959, Baden, A. G. Oederlin & Cie

66   W. GOWLAND: *Arch.*, 1899, **56**, 267

67   E. R. CALEY: 'Analysis of ancient metals', 1964, Oxford, Pergamon Press

68   J. MUSTY: A brass sheet of 1st Cent. AD date from Colchester, Ant. J. 1975, 55(2), 400–410

69   T. MAY: *Iron and Coal Trades Review*, 1905, 71, 427, *et seq.*

70   F. LAUBENHEIMER-LEENHARDT: 'Recherches sur les lingots de cuivre et de plomb d'époque Romaine dans les regions de Languedoc-Rousillon et de Provence et Corse', *Rev. Arch. Narbon*, Suppl. 3, 1973

71   G. C. WHITTICK and J. A. SMYTHE: 'An examination

of Roman copper from Wigtownshire and N. Wales', *PUDPS*, 1937, **9**(2), 99–104

72   P. ANDRE: ' A copper ingot from Brittany', *Bull. Board of Celtic Studies*, 1976, **27**(1), 148–153

73   LUC LONG: 'L'épave antique Bagaud 2', In: *VI Congr. Int. Arch. Submarina*, Cartagena, 1982, 93-98

74   D. COLLIS, C. DOMERGU, F. LAUBENHEIMER and B. LIOU: 'Les lingots d'etain de l'épave Port Vendres II', *Gallia*, 1975, **33**(1), 61–94

75   R. D. PENHALLURICK: 'Tin in Antiquity', *Inst. Met.*, London, 1986, 108

76   L. H. COPE: *Num. Chron.*, 1968, **8**, 115

77   ST. G. WILLMOTT: *J. Inst. Metals*, 1934, **55**, 291

78   E. KALSCH and U. ZWICKER: *Mikrochim. Acta.*, 1968, Suppl.3, 210

79   L. H. COPE and H. N. BILLINGHAM: *Bull HMG*, 1968, **2**, 51

80   B. W. CUNLIFFE (appendix by J. P. NORTHOVER): 'Hengistbury Head, Dorset, Vol. I, the prehistoric and Roman settlement (3500 BC– AD 500)', OUCP Monog. No. 11, Oxford, 1987

81   D. G. SELLWOOD: *Num. Chron.*, 1963, **3**, 217

82   G. BECKER and W. DICK: *Arch. Eisenh*, 1967, **38**, 351

83   G. C. BOON and R. A. RAHTZ: *Arch J.*, 1966, **122**, 13

84   F. C. THOMPSON: *Nature*, 1948, **162**, 266

85   M. BESNIER: *RA*, 1920, **12**, 211; 1921, **13**, 36; 1923, 14, 98

86   G. C. WHITTICK: *Ur-Schweiz*, 1965, **29**, 17

87   H. D. H. ELKINGTON: 'The development of the mines of lead in the Iberian Penninsula and Britain under the Roman Empire until the end of the 2nd century AD', thesis, 1968, Durham

88   L. U. SALKIELD: 'A Technical History of the Rio Tinto Mines', (ed. M. J. Cahalan), *Inst. Mining and Metallurgy*, London, 1987

89   B. ROTHEBERG and A. BLANCO-FREIJEIRO: 'Ancient mining and metallurgy in S.W. Spain', *IAMS*, London, 1981

90   H. von PETRIKOVITS: *Germania*, 1956, **34**, 99

91   H. G. BACHMANN: Vl Congreso Internacional de Minera, 15, 1970, Leon

92   M. L'HOUR: 'Un site sous-marin sur la côte d' l'Armorique; l'épave de Ploumanac'h', *Rev. Arch. de l'Ouest*, 1985(2), 1–19

93   D. KIERNAN: 'The Derbyshire Lead Industry in the 16th Century', Chesterfield, 1989

94   D. ATKINSON and M. V. TAYLOR: *FHS*, 1924, **10**, 5 (appendix by F. C. THOMPSON)

95   J. A. PETCH: *Arch. Cambr.*, 1936, **91**, 74

96   STRABO: Geography Book III, 2, 8

97   J. P. BUSH-FOX: 'Excavations at Hengistbury Head, Hampshire in 1911–12', 1915, Rept No.3, Society of Antiquities, London (appendix by W. GOWLAND)

98   W. GOWLAND: *Arch.*, 1900, **57**, (1), 113; 1901, **57**, (2), 359

99   F. W. W. ASHWORTH: 'Romano-British settlement and metallurgical site at Vespasian Farm, Green Ore, Somerset', Mendip Nature Reserve Committee J., Mar. 1970

100   W. A. COWAN: *J. Inst. Metals*, 1928, **39**, 59

101   J. A. SMYTHE: *Nature*, 1939, **143**, 119

102   M. CHEHAB: *Syria*, 1935, **16**, 51

103   W. J. WEDLAKE: 'Excavations at Camerton, Somerset', 1958, Camerton

104   H. C. LANE: 'Field Surveys and excavation of a Romano-British native settlement at Scarcliffe Park, E. Derbys', *Derwent Arch. Soc. Res.*, Report. No. 1, 1973, (The slag analysis was done at a later date by E. Photos)

105   J. A. SMYTHE: *TNS*, 1936–7, **37**, 197

106   G. T. BROWN: *J. Iron Steel Inst.*, 1964, **202**, 502

107   H. E. O'NEIL and G. T. BROWN: *Bull. HMG*, 1966, **1**, 30

108   C. BUCKMAN and R. W. HALL: 'Notes on the Roman Villa at Chedworth', 1959, Cirencester

109   G. BEHRENS and E. BRENNER: *Mainzer Zeit.*, 1911, **6**, 114

110   A. MAU and F. W. KELSEY: 'Pompeii: its life and art', 1904, London

111   VCH Salop., Vol.l, 232

112   W. J. FIELDHOUSE *et al.*: 'A Romano-British industrial settlement near Tiddington, Stratford upon Avon', 1931, Birmingham

113   Personal communications from C. M. Daniels and J. Wacher: Beam now in museum at Leicester

114   M. U. JONES: *Records of Bucks.*, 1957–8, **16**, 198

115   G. C. BOON: *Apulum*, 1971, **9**, 455

116   B. NEUMANN: 'Die Altester Verfahren der Erzeugen Technischen Eisens', 1954, Berlin, Akademie Verlag

117   R. THOMSEN: Personal communication

118   R. PLEINER: 'Zaklady Slovanskeho Zelezarskeho Hutnictvi v Ceskych Zemich', 1958, Prague, Czechoslovak Academy of Sciences

119   W. GOWLAND: *Arch.*, 1900, **57**, 393

120   G. BECKER: *Arch. Eisenh.*, 1961, **32**, (10), 661

121   T. MAY: 'Iron and coal trades review', 1905, **71**, 427, et seq.

Chapter 7
The Migration and Medieval Period

The decline and break-up of the Roman Empire was assisted by the constant aggression of the Iron Age tribes. Their success was not due to any technical superiority, although clearly there had been a considerable interchange of techniques during the Roman period. The decline was due to the constant pressure of people with a developed Iron Age tradition against a long established and highly civilized power.

It is not surprising, therefore, to find that in the Migration period there was a fairly widespread knowledge of the art of pattern welding in swordmaking throughout Western Europe. Evidence lies in the Anglo-Saxon weapons of England and those of Scandinavia,[1] the Baltic States,[2] and the Merovingian Empire.[3] The techniques used were those known in the Roman period which have been described in the last chapter. They were to be brought to full fruition in the early phases of this period. Knives and other tools were often decorated with an inlay of non-ferrous metal.[4,5] Many of the examples of this technique come from burials in which both sexes were accompanied by weapons and tools of office—the men by swords and spearheads, and the women by 'weaving swords', i.e. blades of pattern-welded iron which could have been used for beating the weft but which were possibly no more than ceremonial badges of office.[6]

As far as the non-ferrous metals are concerned the most common alloy was bronze, although its contamination with brass was fairly widespread owing to the use of scrap metal of Roman origin. It is only towards the end of the period when church 'brass' was being introduced that we see the use of a true brass. In contrast, the three-dimensional effigies seem to be gunmetals, i.e. ternary alloys of copper, tin and zinc. Their use was probably intentional as they make very satisfactory casting alloys.

Iron
THE PRODUCTION OF WROUGHT IRON
In the early stages of the period one can detect little change in the methods of iron production from the Roman, or indeed the EIA. This is to be expected, since the Migration people were essentially in an EIA phase. There was certainly no immediate increase in the size of the bloom: on the contrary, most of those found have proved to be somewhat smaller than the 8 kg Roman maximum that has been surmised. Some Irish blooms weigh 5.2, 5.2, and 5.4 kg, respectively.[7,8] In Britain, historical evidence suggests that 14 kg was reached by AD 1350, and without the use of water power.[9]

It is possible that the limit of bloom size was dictated more by the problems of smithing and working than by smelting. Some peoples, such as the Swedes[10] and the Slavs,[11] went on producing slender axe-shaped 'currency' bars which were up to 32 cm in length. In most countries, the larger blooms were smithed into small pieces which were usually bar-shaped. The furnaces themselves appear to be of two basic types:

(i) the horizontally developed bowl furnace of the Roman period with slag-tapping facilities: we have many examples of this type from Britain[12] (see Fig.41)

(ii) the vertically developed bowl furnace.

The latter is perhaps shown by the furnace found at Bargen Hofweisen[13] near Schaffhausen (see Fig.50a). Unfortunately, like many furnaces, it clearly did not survive to its full height and it is difficult to be certain as to what this may have been. It seems likely, however, that there was a type of early medieval furnace from which the bloom was extracted through the top rather than at the side, as in the developed bowl furnace. This type eventually became the Catalan hearth of the Pyrenees, and appears to have been used in Northern England into the post-medieval period.[14]

The more primitive migration peoples continued to use simple bowl furnaces and there were other types such as the Slav furnaces[15] found at Zelechovice (see Fig.50b). The most interesting problem is that of the fate of the shaft furnace found in several Roman locations. It seems that the tall slender Saxo-Jutish shaft furnace made its way across the North Sea to Eastern England and then died out. However, it is probable that it lingered on in Central Europe, to blossom finally as the Styrian high bloomery furnace (Stückofen) and the blast furnace, but we lack examples demonstrating the continuity of this type. Unfortunately, the shaft furnace at Gyalar in South East Hungary is not well dated.[16]

The next aspect of interest in ironworking in this period is the application of water power. Many of the smaller production units lay near streams, but it is clear that this water was not used as a source of power.[12] For one reason, the power necessary to produce the required airflow of 300 l/min is not great and could easily be produced by manual means. Secondly, these units were

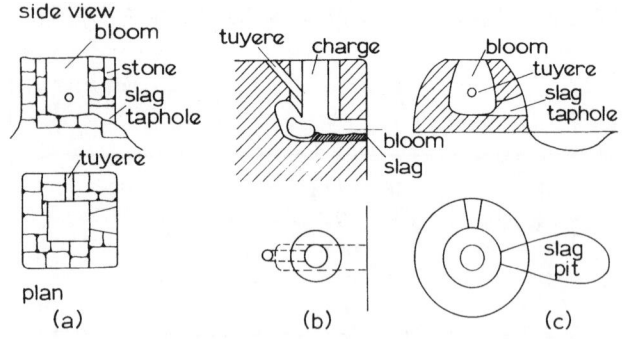

a furnace found at Bargen-Hofweisen, near Schaffhausen (after Guyan[13]); b Slav furnace from Zelechovice, Czechoslovakia (after Pleiner[15]); c domed
furnace from Co. Durham; essentially the same type as
that shown in Fig.37

50 *Medieval bloomery furnaces; inside of (a) is 30 cm square*

obviously based on very small capital resources and the
exploitation of water power required considerable capital. Its use in metallurgy, therefore, had to compete with
its use in non-metallurgical processes such as corn
grinding and fulling, and since these were greater con-
sumers of power than blowing, they were given priority
in the West. But in the East, i.e. in China, the development of metallurgical blowing engines was one of the
reasons why the Chinese gained superiority, and probably accounts for the early adoption of a cast iron tradition.

Vitruvius[17] describes the use of water mills in the
Roman world, and Gregory of Tours[18] (AD 600) can be
quoted as evidence for their medieval use. In Britain, we
have references to more than 5000 water mills,[19] including
some 'molindini ferri'; in the Domesday accounts (1086)
some of the rents were occasionally paid in iron.[20] Unfortunately, there is no reliable documentary or archaeological evidence for the use of water power in Europe for
metallurgical purposes before AD 1408 in Britain,[21] and
1440 in Italy,[22] both for bloomery furnaces, and in Italy
for the blast furnace of 1463 described by Filarete.[23] It is
thought that water power was used for bellows in
Hungary in the 13th century.[24]

It seems that in Europe only monastic or episcopal
bodies had sufficient capital to invest in the iron industry, and that the large scale development of this industry
depended on the rise of the religious institutions. Thus,
in 1408 the Bishop of Durham[21] established the first
documented water powered bloomery in Britain and, in
view of the absence of references to hammers, we can
presume that the power was made available for bellows.
Once the principle was established, the way was open
for the production of larger blooms, since power hammers based on the principle of the fulling mill would
have been available to split them. Furthermore, the
continuous operation made possible by the application
of water power to bellows made the blast furnace possible. Outside China there is certainly no evidence for a
manually blown blast furnace and, on the basis of the
continuity of operation required of such a piece of
equipment, one would not expect it, apart from tempo-

rary stoppages when the treadmilling of a waterwheel
might just have been a possibility.

Before considering the early stages of the blast furnace, it is necessary to complete our discussion of the
medieval bloomery. Once the application of water power
had been introduced, by the early 15th century at the
latest, the bloom increased in size to over 100 kg. It was
cut up and the pieces were reworked in a second hearth,
known as a string hearth in England, which appears to
have been very like the low bloomery hearth but worked
intermittently. Unfortunately, we lack material evidence
of this period and have to extrapolate earlier and post-
medieval data.

In Europe, the ultimate design in bloomery hearths
was the Catalan hearth of the 19th century.[25] The low
bloomery hearth in which the bloom was removed
through the top was almost universal, as shown by the
hearths near Sheffield,[14] and in the Lake District;[26] in
Norway they were 1 m high.[27] The exceptions appear to
be the Stückofen in Styria which, being 3–5 m high, was
designed for the bloom to be removed through the
side,[28] and the Osmund furnace of Sweden which seems
just too tall for top removal[29] (about 2.0 m). Tall shaft
bloomery furnaces were almost certainly being used in
Africa and parts of the East at this time.

THE BLAST FURNACE

The development and introduction of the blast furnace
in Europe is one of the most interesting subjects in the
history of ferrous metallurgy. We know that the blast
furnace was being used in China long before it was used
in Europe, but it cannot be assumed that it did not have
an independent origin in Europe. On the other hand, its
introduction in Europe came at a time when contact
between East and West was well established, and all that
was needed in Europe was an appreciation of the useful-
ness of cast iron. It had been produced accidentally in
the Roman period, and experimental work has shown
that it is possible to produce cast iron in a 2 m high
bloomery furnace with a sufficiently large charcoal/ore
ratio.[30]

Present evidence suggests that the Swedes were the
first to use blast furnaces for the smelting of iron. This is
supported by two sites Lapphyttan[31] and Vinarhyttan,[32]
which give ^{14}C dates in the range AD 1150–1350. So far no
dates as early as this have come from Western Europe,
previously thought to have been the centre of blast
furnace development. The close contacts which existed
between Sweden and the East via the Volga suggests
that the impetus might have come directly from China
via the Mongols.[33] So far there is no proof of this in the
design or working of the blast furnaces apart from the
conversion hearths (fineries), which could have been
modelled on the Chinese hole-in-the-ground pattern. It
is clear that most of the cast iron was converted into
wrought iron, which suggests that the intention had
nothing to do with cast iron, gun-making and the metal
was destined for the more common wrought iron objects
as noted by Filarete[23] below.

The analyses of slag from the two Swedish sites

mentioned above are given in Table 42. It is probable that the small amount of lime came from the wood ash and the ore.

There was, of course, a need for cast iron in the military sense since wrought iron was not efficient for guns. The first guns were made from strips of wrought iron held together by hoops and for several reasons they were not very successful. While bronze was a much better metal for the purpose it was comparatively rare and was more expensive than iron. Wherever the idea of cast iron originated, it was clear to the medieval armament manufacturer that cast iron was more suitable and could produce a one-piece weapon with good pressure-retaining properties.

At some stage during the 15th century the incentive was sufficiently strong to cause the development of either the tall shaft or the low shaft bloomery furnaces for the making of cast iron guns. However, we find that the first well-documented blast furnace, the furnace at Ferriere in Italy described by Filarete, was designed, apparently, for the making of granulated cast iron by tapping the molten metal into a water bath or channel.[23,35] On no account could the material have been used for guns: it was most probably used for mixing with wrought iron for producing steel, as in the Japanese process. The action of the bellows seems to have been on a horizontal plane, rather than the vertical one of the 16th century and later blast furnaces. There are similarities with later Persian blast furnaces,[36] and we know that Chinese furnaces were powered with bellows operated in this way,[37] at least as early as AD 1313. Here then, there are two traits with a decidedly Eastern flavour, and one is forced to agree that not only the idea of cast iron but some of the details of technique came to Europe from the East.

Technically, the main difference in operation between a bloomery and a blast furnace lies in the withdrawal of liquid iron in addition to slag from the bottom of the furnace. For a furnace of a given height (about 2 m), the main change required was that of an increase in fuel/ore ratio to make the atmosphere more reducing and carburize the iron lowering its melting point from 1 540°C to about 1 200°C. Under these conditions, how-ever, the slag is very viscous as the iron content is reduced, and it does not run easily. The early smelter would have found that certain ores were more self-fluxing than others, that is, that they produced more free-flowing slags. These would have contained an appreciable amount of lime in a combined form. In these early furnaces it would not have been easy to add additional lime as limestone in lump form, as it is not easily dissolved, but it could have been powdered. The result of these changes would have been the need for a working temperature of at least 1 300°C to get the limy slag to run: this temperature was certainly obtained in some bloomery hearths, so there would have been no problem in this case.

As time passed it would have been found that fuel could be saved by using a taller furnace to give a larger residence time in the strongly reducing conditions. Following the construction of larger furnaces, it would have been found that working needed to be continuous. Intermittent working, as used with bloomery hearths, would have been out of the question as there would have been too much heat and material bound up in the process, so that it would be uneconomical to stop it after each batch of iron had been tapped. It would also have been clear that a constant supply of water was more important than quantity, and small streams with a constant flow would have been more sought after than large irregular rivers. Ponds would have been built to regularize the flow in areas where rainfall might have been interrupted by summer drought. So, when once a blast furnace had been blown-in, its 'campaign' would not have been terminated until either it had worn out, i.e. the lining needed replacement, or a failure of water supply, wood or ore had occurred.

The fact that cast iron was not used in the early bombards, and that wrought iron continued to be used for these massive pieces of ordance, shows that the production of cast iron in large amounts was fraught with difficulty well into the 15th century.

THE PROCESSES OF DAMASCENING AND PATTERN WELDING

The seventh century saw the beginning of the Arab conquests and Damascus was becoming better known to the Western world. It is at this point that we should try to relate the apparently new process of 'damascening' and the older process of pattern welding, as applied to sword and steel production. It is believed that, at some time in medieval chronology, Damascus was the point at which Oriental or Indian steel entered the West. This steel was probably that which later went under the name of 'wootz', which was often a high-carbon cast steel in which the carbon content averaged about 1.6%. If the small ingots of this steel are hammered out into thin strips some surface decarburization will occur and, when these are welded together at a very low temperature, at which carbon does not readily diffuse (i.e. 700°C), the result will be a heterogeneous structure not unlike that obtained by pattern welding. An alternative method, by which much the same result could have been obtained, is the segregation of phosphorus in the original steel

Table 42 Analyses of slags from early Swedish blast furnaces

%	Vinarhyttan Serning[32]	Lapphyttan Bjorkenstamm[34]
SiO	69.27	53.90
Al$_2$O$_3$	10.10	6.20
Fe$_2$O$_3$	1.34	-
FeO	4.74	3.50
MnO	1.44	11.80
MgO	4.12	10.00
CaO	4.23	12.10
Na$_2$O	2.63	0.91
K$_2$O	1.86	0.90
TiO	0.27	-

ingot. This would control the distribution of carbon during later working at low temperatures. Pattern welding, as we have seen from the Roman examples, differs from damascening in that the starting material is low-carbon iron and the strips are twisted together in a more elaborate way.[39] Like some of the Roman blades, the later Anglo-Saxon swords were often edged with steel and quench-hardened[40] (*see* Fig. 46).

Both damascening and pattern welding were designed to solve the same problem, that of the brittleness of the stronger higher-carbon steel which tended to show itself in swords which either broke if strong and hard, or bent if soft and ductile. In Japan, the swordmaker was faced with the same problem and, by the end of the 10th century AD, had evolved a slightly different technique from that of pattern welding in which comparatively few pieces of steel of different carbon content were welded together to make a composite single-edged blade, which was finally heat treated.[41,42] The low-carbon (iron) core of the blade was enclosed by three pieces of relatively homogeneous high-carbon steel, and all four pieces were hammer-welded together at a red heat. This was the forerunner of the later 'shear' steel that was to prove so popular in Western Europe after the Industrial Revolution. The composite blade was then wrapped in clay and heated to about 800°C: the clay envelope in the vicinity of the cutting edge was quickly removed and the edge quenched or allowed to cool in air. The clay envelope would have slowed down the rate of cooling of the core and the back of this one-edged blade so that a 'gradient' hardening would have been achieved, with the maximum hardness on the cutting edge and a low enough hardness in the core to give the necessary toughness[43](*see* Fig.51).

It is possible that there is no real difference in principle between pattern welding and damascening. Some authorities believe that the latter process is based upon the structure obtained during the solidification of the wootz ingot.[44] However, it is difficult to be sure as to exactly what the structure of the medieval wootz ingot was. Much of our knowledge of so-called 'wootz' is based upon 19th century ingots brought back to Europe by colonial administrators. In this period, there seem to have been at least two varieties of wootz. The first was made from wrought iron and wood placed in a crucible and heated until they produced a melted steel ingot the shape of the inside of the crucible: this is the Salem process.[45] The total carbon content of this material seems

to have been about 1.6%, and the ingots seem to have had a relatively coarse but chemically homogeneous structure of cementite and pearlite.

In the second process, which was used in the Hyderabad region, the crucibles were made from refractory clay, derived from the weathering of granite, to which was added rice husk.[46] Neither charcoal nor wood was put into the crucibles, which were heated in a circular charcoal furnace, the air being supplied by four bellows placed near the top and facing downwards: this heating lasted for 24 hours (*see* Fig.52).

Ingots from this second process were recently examined[47] and it was found that the carburization had come from and through the wall of the crucible and that, in one of the two cases examined, the carbon content varied through the section, being a maximum at the outside, where it was 0.8%, while the top centre was pure ferrite. The crucibles themselves had contained about 10% of carbonaceous material and had been heated to at least 1 300°C and in places to over 1 470°C. It is clear that the carburization had taken place in the solid state and that, towards the end of what must have been a very lengthy process, the outside had reached the melting point of a 0.8% carbon steel, permeating the solid pieces of lower carbon material in the centre with liquid metal, but not sufficiently to produce a homogeneous steel.

This process was not confined to India. It was used for making cast iron in China[48] (see below) and we have a reference to its use in the Sudan in more recent times, also for the making of cast iron. Pieces of rich ore were charged into lidded pots; these were heated in a charcoal fire and the reducing gases penetrated the pots as in the Hyderabad process and converted the ore into cast iron. The lids were removed and the iron was poured out or allowed to solidify as an ingot.[49]

So it can be concluded that Eastern weapons could have been made from homogeneous cast steel ingots called wootz, or some mixture of ferrous materials. It is certain the pattern could not have been derived from a distribution of carbon obtained by solidifying a wootz ingot as this would, on a microscopic scale, have readily diffused while at high temperature as carbon is so highly

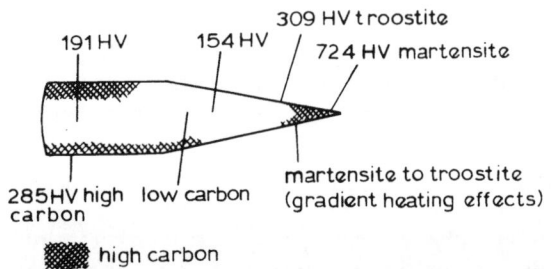

51 *Section through a Japanese sword of the 17th century (after O'Neill[43])*

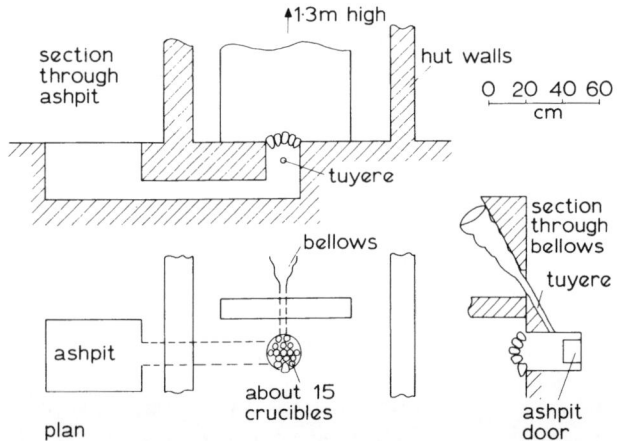

52 *A crucible furnace from Southern India for the making of wootz (Buchanan[46])*

diffusible. However, it is just possible that the final distribution of carbon, that is responsible for the pattern on some damascened swords, could be obtained by the segregation upon solidification of some other element, such as phosphorus. This would control the final distribution of carbon as the specimen cooled below 700°C after forging. The only analysis for phosphorus gave 0.27%, which is quite sufficient for this purpose.

There is ample evidence that the Western swords were made by pattern welding and it is extremely likely that some Eastern weapons were made by a very similar process. The Malayan or Indonesian 'Kris' shows a pattern known locally as *pamor* (mixed) and it also was made by welding several kinds of metal.[50]

In fact, some blades contain laminations of nickel which can arise in two ways: by the incorporation of some high nickel iron, possibly of meteoric origin, or surface enrichment of nickel during heating for forge welding.

One blade has clearly contained a lamination of meteoric iron to give an overall composition of 5.5% Ni + 0.6%C.[40,51] This technique is said to date from the 14th century in this area, and the 400 year gap between the disappearance of pattern welding in Anglo-Saxon Europe and the appearance of pamor in Indonesia is merely a measure of the rate of diffusion of the process.[52]

A recent examination was made of a sword of North Indian origin which has a pattern on it. Like the Malayan *kris* this also was found to be made of a number (in this case 100–150) of layers of wrought iron of varying carbon content.[53] The carbon content varied from 0.04 to 0.3% and the blade had been finally hardened by quenching. It was found possible to make such material either by forge welding lightly carburized wrought iron strips or by interposing a layer of finely powdered white cast iron between the wrought iron strips and heating in the range 950–1 000°C. By suitable forging and final heat treatment, the hard 'water marked' structure of the original weapon was obtained. It was also found that by heating at 900°C for 8 hours, full diffusion of the heterogeneous structure could be obtained, giving an even carbon content: this showed that the forging temperature must have been lower than this, or the time taken must have been much shorter .

From all the work that has been done, one can conclude that damascening and pattern welding are, in principle, the same thing. The possible methods of obtaining the desired structures range from welding wrought iron strips, welding steel strips with surface decarburization, welding wrought iron with carburization, mixing high and low-carbon material, or even different steels and meteoric iron, and incorporating layers of powdered cast iron or high-carbon steel. It is important that the working temperatures be kept to a minimum or else complete diffusion will occur. The use of wootz is an example of the welding of decarburized steel strips or inhomogeneous ingot material. As yet there is no convincing evidence for a structure based on segregation during solidification but this would seem to be possible, especially in steels with a high phosphorus

content. Much of the 'watered silk' or water-marked pattern was a by-product of the smithing technique, but it could be accentuated by disturbing the normal pattern using localized working with suitably shaped tools, and by forging the disturbed material flat again.

THE MEDIEVAL SWORD

The later swords, such as the heavier swords of the Viking period in Europe, seem to have been relatively homogeneous and of much higher carbon content than the pattern-welded types of the Migration period. Norwegian examples have shown carbon contents in the range 0.4–0.75%. They were made of piled layers of carburized iron and, if the phosphorus and arsenic contents were low enough, diffusion during forging at high temperatures was capable of giving a homogeneous carbon steel. The swords often had an inscribed blade inlaid with iron wire, and the guards and pommels were sometimes decorated with silver inlay and niello.[55] Niello is an alloy of copper, lead, and silver sulphide that was used for decorative purposes as a black 'ground'. It has had a long tradition but seems to have been first mentioned by Theophilus.[56]

Some 12th century swords show evidence of the piling of as many as 16 layers of comparatively low-carbon material in the centre with a high-carbon layer on the surface, containing as much as 0.54%C in the case of an Italian sword.[57] This suggests that a plate of high-carbon steel had been wrapped around and welded to the low-carbon core in much the same way as in the Japanese technique. However, some Frankish swords show that the high-carbon 'case' has been obtained by surface carburization.[58]

OTHER ITEMS OF MILITARY IRONWORK

Of course, the sword is not the only piece of military equipment to be considered. The Bayeux Tapestry, which was woven between AD 1066 and 1080, shows a typical assemblage of military weapons of the Norman period[59] (*see* Fig.53). First, the helmet was made of sheet iron, usually in one piece except for the riveted nasal. The shield bosses, which are quite common in Viking burials, are made, like the helmets, of one piece of iron carefully hammered out on a stake anvil.[60] The battleaxes are quite different in design from the carpenter's axe and are often called 'bearded' axes. These are usually made of pieces of piled low-carbon iron. One from Stratford in Essex is typical of the period AD 850-950. The grain flow is parallel with the cutting edge, and the socket was made by lapping over a tongue and welding it to one side of the blade.[42] The hardness of the ferritic socket was 165 HV, owing to the high phosphorus content. The cutting edge was locally carburized and quenched to give a hardness of 350–450 HV, the martensite having a rather low-carbon content.

The arrowheads were usually of iron but, by the 14th century, they were being tipped with steel. The spurs of goad type were usually tinned as it was not otherwise easy to protect them from rusting. The method has been described by Theophilus.[56] The stirrups, when of wrought

53 *A scene from the Bayeux Tapestry showing the death of Harold; this shows the main items of ferrous military equipment in use at the time of the Normal conquest of England*

iron, were not tinned but they were very often made of bronze. The spearheads were socketed and fastened to the haft with bronze rivets. In these, the pattern welding process lasted longer than in the swords. An example from Kentmere in Westmorland, dated to the 11th century, was 34 cm long and 3.6 cm wide across the blade and had a herringbone-type pattern: another, from Reading, had edges of spheroidized 0.45%C steel with a hardness of 219 HV.[63]

Finally, we come to the mail shirt which, as can be seen from the small figures in the bottom margin of the tapestry (*see* Fig.53), was very valuable and the wounded were quickly stripped of them in the heat of battle.[63] Body armour had been in use since early Greek times when it appeared as the corslet or cuirass.[66] This proved rather inflexible, and during Roman times it became segmented finally to become scale armour or *lorica squamata*, in which small plates of brass or bronze were fastened together and stitched to a cloth or leather backing. Mail, or *lorica segmentata*, appeared about the second half of the first century AD, when it started as brass or bronze rings with a fine equiaxed worked structure, possibly of drawn wire, with a hardness of 175 HV. Rows of punched rings alternated with rows of rings of riveted wire. In the same period iron mail, in which solid welded rings alternated with riveted rings was introduced, and this is probably what we see in the tapestry. Medieval mail was sometimes made of steel but wrought iron mail was strong enough. By the time the mail itself was damaged the blow would in any case have proved fatal. The riveted rings were made of drawn iron or steel wire, the holes drifted, and the rivets made of wrought iron wire.[67]

Plate armour began to replace mail in about 1250, but mail continued to be used into Egyptian (Mamluk) and Turkish times, i.e. into the 19th century. Most European mail of the second half of the 15th and the 16th centuries has all the rings riveted. A number of specimens of plate armour have been metallurgically examined. Italian body armour of about AD 1400 was found to consist of wrought iron with a lightly carburized surface consisting of ferrite, pearlite, and grain-boundary cementite. Late 15th century armour was of steel rather than wrought iron. A full cross-section of one piece showed that it had been slow quenched on the surface, leaving ferrite and angular pearlite in the centre. German armour, dated to about 1550, varied between carburized wrought iron containing spheroidized pearlite, and decarburized high-carbon steel.[68] It seems that it was difficult to shape this material satisfactorily while maintaining the optimum structure.

STEEL

Much of the iron produced in the medieval period seems to have contained considerable amounts of phosphorus and was therefore not suitable for making steel by cementation, as this element reduces markedly the rate of penetration of carbon. For this reason steel was an article of trade, being made from iron imported from areas with suitable ores; in some cases the finished steel was imported. We now know that it is possible to make iron-carbon alloys directly by the bloomery process, given suitable ores, but we do not know whether this was done intentionally or whether all medieval steel was made by cementation and subsequent piling. It was certainly expensive and used sparingly (and only where necessary) so that a good many artefacts were made by 'steeling' or welding pieces of steel to wrought iron.

Early metalworkers had to make their own tools, and Theophilus[56] tells us how to make steel cutting tools. Engraving tools were made of solid steel filed to a suitable shape and quenched in water. Some files were also made of solid steel but others of iron 'covered' with steel. They were forged to shape, smoothed on a grindstone, and 'cut' with a hammer which had a cutting edge at each end of the head, or with a chisel. To harden them they were sprinkled with a hardening medium consisting of a mixture of two-thirds burnt ox-horn and one-third salt. They were then heated in a fire to a red heat and quenched evenly in water.

Another method was to smear them with pig fat, wrap them in strips of goat leather, cover them in clay, dry them, and then heat them in the fire. This burned the skin to carbon and nitrogen compounds and, when carburized, the tools were quickly removed from the clay envelope and quenched.

In view of the different conditions needed in the smith's hearth for heating to 900°C and carburizing, the wrapping in clay or leather was a very clever way of achieving this.[69] Today we would use an iron or steel 'box' to separate the carburizing conditions from the external heating conditions that would be too oxidizing. In some cases Theophilus mentions quenching in goat's urine, but nowhere is there any mention of tempering so one presumes that the carbon content was not high enough to require this.[56]

One of the striking things about early steel compared with iron is its comparative cleanness. Wrought iron often contains as much as 10% of slag, while steel is much cleaner. This may point to its method of manufacture. Steel derived from the bloomery process would contain much slag, while that made by carburizing wrought iron would be expected to contain less, as reduction of the fayalitic slag inclusions to silica would be expected to reduce their size by evolving CO and CO_2; this would leave blisters which would be closed by forging.

The value of steel compared with that of wrought iron was very high, perhaps of the order of 4 to 5 times. We have no definite idea as to how it was made apart from the wootz process. The cleanness of steel used in Anglo-Saxon times was very high and this suggests a wootz-type process in which the slag could float out of the molten steel. The selection of high carbon areas of bloomery iron was unlikely to give clean steel. The process mentioned by Biringuccio, in which wrought iron was put into molten cast iron, would also leave the original slag inclusions in situ.

IRON ARTEFACTS IN EVERYDAY LIFE

Much of the agricultural metalwork would be of wrought iron, sometimes as an edge to wooden tools. One would expect the ploughshare to be steeled, i.e. the life of the wearing surface would have been extended by the welding on of steel to the most worn places, but this has not yet been confirmed. The smith would have made knives and other edge tools such as scythes: by the 13th century there are signs of specialization, and the craft of cutler came into being. Knives were rarely made of solid steel as it was much too expensive, and the welding of the steel cutting edge to the iron back was a process offering an immense number of variations, so much so that it is possible that each smith had his own personal technique.

Some Migration and medieval artefacts show very thin layers of high arsenic concentration (more than 1%As), and layers of this type have been identified in the blades of a Scandinavian sword and a Viking axehead where two pieces of metal, either iron or steel, have been joined.[71] When iron is heated in the smith's fire the arsenic that it contains accumulates in the outer layers of the iron under the scale, so that when the smith comes to weld, for example, a piece of steel to a piece of iron to make a hard cutting edge he incorporates a layer of high arsenic metal.[72] 'White lines' arising from this effect are often visible in the structure of the finished artefacts and show the joins in the various pieces that make up the tool (*see* Fig. 46*i*).

Three examples of the sort of joint used in the making of knives are shown in Fig.54. Figure 54*a* shows the steel used as a core with a plate of wrought iron folded over the back and forming the sides; Fig.54*b* shows a steel cutting edge welded to a piece of piled iron consisting of layers of high- and low-carbon iron carefully welded together. The white lines are probably areas of high arsenic content, owing to oxidation during heating in the smith's hearth. Scissors were invariably steeled in the same manner as that shown for the knife blade (*see* Fig.54*c*). The heat treatment of the cutting edges was very efficient, the steeled areas in Figs. 54(*a*), (*b*), and (*c*)

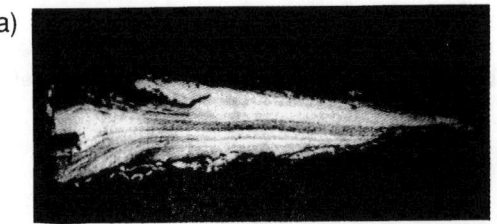
(a)

(b)

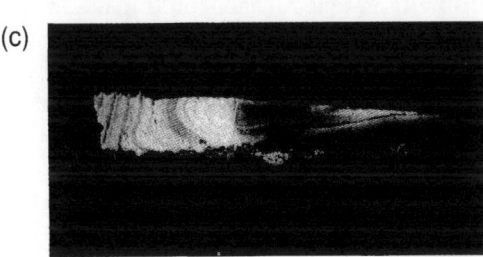
(c)

54 Photomicrographs of sections of medieval cutlery from Winchester showing typical examples

having hardnesses of 557, 575, and 857 HV, respectively, which is as good as any butcher's knife today.

WELDING AND BRAZING

Hammer welding comes naturally to a smith working wrought iron and we see that he often had to weld steel to iron to make effective cutting tools and wearing surfaces. On the whole he preferred this kind of joining to 'brazing', i.e. the use of liquid alloy interlayers of ferrous or non-ferrous metals. It is possible that the diffusion of phosphorus to the surface, or the concentration of arsenic in the surface layers that occurs during heating before joining, could give layers with low melting points which would assist joining because of their liquid character.

There is evidence of the use of arsenical cast irons[73] in the manufacture of cannon balls (or shrapnel) as soon as the gun came into general use (1454). We do not know the source of the arsenical iron but it may have come from the iron-arsenic minerals that form the gossan or cap of a non-ferrous mineral deposit, such as that at Rio Tinto[74] which contains 55%Fe and 1.0%As. Another possibility is speiss, the complex arseno-sulphide of iron, nickel, and cobalt arising from non-ferrous smelting.[79] All these alloys have low melting points and could have been used for joining. In some ways they resemble niello, an alloy of copper, lead and silver sulphide that was used for decorative purposes, as mentioned by Theophilus, and which has had a long tradition.

IRON IN CHINA AND JAPAN

In China, the iron industry is not well documented until the 11th century AD. While China was clearly self-sufficient in cast iron made from coal, there are signs that steel was being imported from India in the 5th century AD, although some good steel was being made locally. It is possible that this is because of a charcoal shortage in China — the steel made from charcoal being superior to that from decarburized cast iron made from coal.

By the beginning of the 11th century, 125 000 t of iron were being made each year with ore from the Northern Sung mines.[76] This compares with the total estimated European output for the year 1700 of 150–185 000 t. Naturally, the distribution was not uniform although there were considerable iron deposits in most provinces. It seems that the industry was organized on the Roman pattern, with major workshops employing many workers. It is estimated that there were as many as 36 smelting plants, each employing about 100 men mining ore, getting fuel and smelting. One furnace could make 15 t/a. One tonne of iron would require about 2.5 t of ore and 3 t of fuel, and 100 workers could produce about 500 t/a. Table 43 gives the composition of some of the objects produced.[77]

Clearly, at this rate there would soon have been a shortage of charcoal and, after 1078, coal was almost certainly the primary fuel. The crucible process for making cast iron and steel would have used anthracite, as bituminous coal was unsuitable for this purpose. The crucibles were rather larger than those used in India and held 12 kg of ore, a similar quantity of anthracite dust, and some flux. A stall furnace could hold 64 of these crucibles surrounded by 1.2 t of anthracite. The ingots of cast iron would have weighed 4.8–6.6 kg and could have been converted to iron and steel by fining.

In Japan, the industry diverged somewhat from the Chinese model and the 'Tatara' process, with its single furnace, could make cast iron, steel, and wrought iron all in one operation.[78] The furnace was consumed in the process. After the cast iron had been tapped, the remains of the furnace were broken apart and the 'bloom' of mixed steel and wrought iron was extracted. As much as 4 t of metal were made in a single smelt.[78,79]

Copper and copper-base alloys

We know very little of the immediate post-Roman period of non-ferrous metallurgy. However, it seems that Frankish miners started to work the *fahlerz* ores near Frankenberg and Mittweida, on the northern slopes of the Saxon Ore mountains in Germany, before the end of the 10th century. They also worked the argentiferous ores on the Rammelsberg near Goslar, and the Mansfeld copper schist was worked at Herrstadt from 1199.[80] By this time, German miners were gaining their international reputation, and miners of the Harz mountains went to Tuscany in 1115 to resume the mining of the Tuscan copper ores at Massa Maritima. The famous Falun copper mine in Sweden was started in 1220.[81]

The extraction process, used in the 14th and 15th centuries, is often known as the 'German process' to distinguish it from the later 'English process'. Most ores were by now sulphides and, in the German process, these were roasted to completion, i.e. the Cu_2S was converted entirely to CuO and the oxide reduced to metal in a reducing furnace. The roasting was done in the open over burning wood in shallow cavities in the ground, and would take more than 30 days to complete. The oxide was then reduced in low-shaft furnaces with forced blast, and with charcoal and fluxes. The main difficulty in this process is how to avoid reducing the iron oxide as well, although the low solubility of iron in copper would have tended to restrict this to an acceptable level for the times. However, the iron contents of 0.5–2.6%, which are present in medieval copper, illustrate this difficulty.

Theophilus tells us a little about the smelting of copper in the early years of the 12th century. The fact that he was familiar with a green ore indicates that it was of the oxide type. It contained lead as an impurity and was well roasted, presumably to oxidize any residual sulphide, and also to make it easier to break up. (The latter is a very good reason for roasting any ore.) It was then charged with alternate layers of charcoal into a furnace and blown with bellows.[56]

Theophilus then goes on to discuss the refining of copper in a clay-lined crucible in a smith's forge, the blast from the bellows passing partly into and partly

Table 43 Cast iron objects from China

| Date | Object | Carbon, % | | Composition, % | | | | |
		Total	Graphitic	Si	P	S	Mn	Reference
430 ± 80 BC	(From Loyang)	4·19	nil	0·055	0·08	0·014	0·011	55
110 ± 80 BC	Stove	4·32	nil	0·11	0·38	0·027	0·07	55
AD 502	—	3·35	2·30	2·42	0·21	0·07	0·13	61
AD 508	—	3·22	2·26	2·39	0·17	0·08	0·23	61
AD 550	—	3·35	3·02	1·98	0·31	0·06	0·78	61
AD 558	—	3·33	3·17	2·12	0·19	0·06	0·64	61
AD 923	—	3·96	0·61	0·61	0·23	0·02	nil	61
AD 1093	—	3·58	0·04	0·04	0·13	0·02	0·25	61
AD 1550 ± 10	Statue of cow	2·97	—	0·06	0·29	0·067	0·09	55

above it but not below it. The copper is melted under oxidizing conditions and the oxides are slagged with wood ash: finally, the copper is 'poled' with a slender dry stick to reduce it. This is, of course, a perfectly normal refining or 'flapping' operation followed by 'poling back'. He probably did this latter process himself: from his description it seems that he had not taken part in the smelting operation which was probably done elsewhere, near the source of ore.[82,83]

The description of the process used at Masa Marittima seems to be the first involving the matting process which we know was used in Byzantine times and earlier.[82] At Massa Marittima the ore was first calcined in the open, smelted in a shaft furnace several times, as in the later Agricolan process, and finally the matte was converted to metal and refined in wood-fuelled reverberatory furnaces. The metal was probably run into circular hearths for final treatment and casting.[83] The furnaces seem not unlike those used at Rio Tinto in Spain in the 19th century.

To Theophilus we owe the first description of an induced-draught crucible furnace. This consisted of an iron grid raised above ground level on four stones. The height of the clay-lined shaft erected on this grid was to be less than its diameter (*see* Fig.55). For the most efficient heating of crucibles there should be a layer of fuel between the base and the crucibles. A side entry hole would be advantageous as it would avoid the need for emptying the furnace before removing the crucibles. Such a melting furnace was found in the Outer Hebrides, dated to about AD 200.[84]

From China we have contemporary illustrations showing bronze casting in the Ming period (AD 1368–1644). We see as many as seven cupola (melting) furnaces blown manually with piston bellows, supplying metal to four launders leading to the mould assembly which is placed below launder level.[85] From the caption, a *Ting*-cauldron was being cast with separate legs which were later joined to the body. Such vessels weighed as much as 1 400 kg. The furnaces are sometimes fixed and at other times portable. They seem to have been about 130 cm high and about 60–100 cm in dia. The air entered the furnace about half way up at the back, and the furnace seems to have been tapped from about the same level at the front. In the case of the portable furnaces, the joint between the furnace body and the bellows could easily be broken and the furnace carried by two men by means of two poles pushed through horizontal holes in the bottom of the body.[37] Thus, the furnaces could not have weighed more than about 110 kg when full, and could not have held more than about 45 kg of metal. Since the metal could not have come above the tuyere level, they must have been something like the reconstructions shown in Fig.56. In normal use pieces of metal would have been charged with the fuel and the whole lot melted down, as is cast iron in the modern cupola. In some cases metal was transferred by means of a ladle into permanent moulds, and water power was used for blowing.

The crucible furnace, or cupola, was also in use in

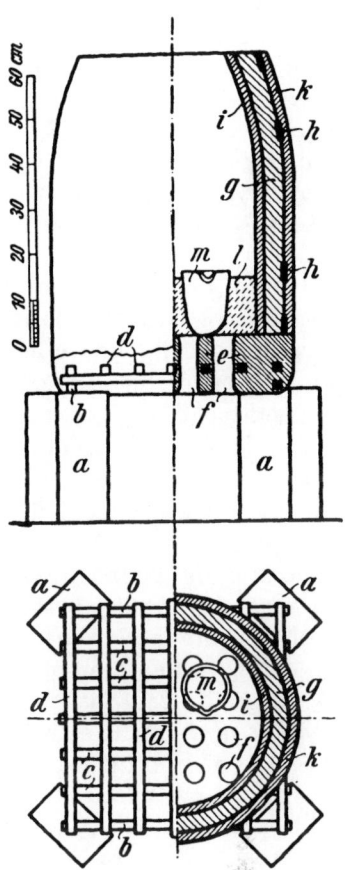

55 *Theophilus' crucible furnace as shown by Hawthorne and Smith* [56])

Europe for heating metal for bell and gun casting. According to Theophilus, round-bottomed iron pots were coated and lined with clay about 3 cm thick: these had two handles on each side and two to three pots were required according to the size of the bell. The pots were

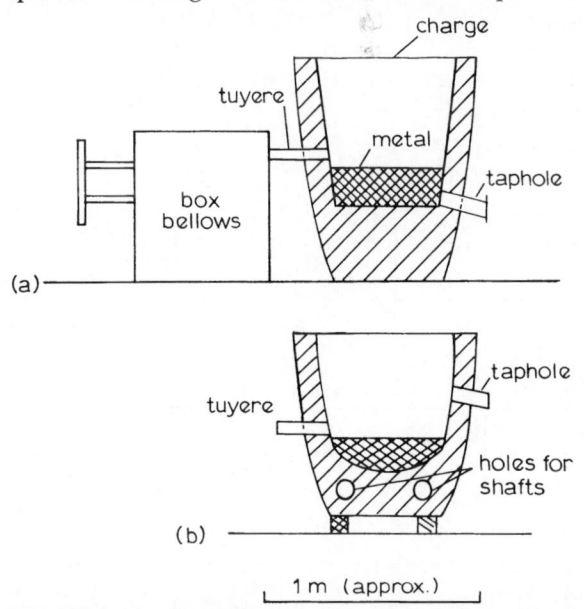

a fixed furnace; *b* portable furnace

56 *Chinese bronze or iron melting cupola furnaces of the Ming period (after Barnard* [85])

placed on the ground with spaces between and provided with bellows over the tops. The tops were extended upwards to take the charcoal and finally must have looked very like the Ming furnaces.

The bell-founder's window in York Minster seems to show that a different type of furnace was in use in the 14th century (*see* Fig.57). In this case two men are treading bellows behind, and it would seem that a forced draught rather than a natural draught was being used. Perhaps the charcoal fuel was mixed with the metal charge. Similar furnaces where the charcoal and fuel are mixed are used at the bell-foundry at Villedieu-Les-Poëles in Normandy but here the draught is obtained by a chimney. A wood-fuelled reverberatory furnace with a separate firebox has been found at Keynsham, near Bristol. Since the hearth is paved with tiles from the chapter house of the adjacent abbey this must date to after 1540.

In the early part of this period the alloys used were mainly tin bronzes and gunmetals, as shown in Table 44. Brasses enter the picture later in the period and it is almost certain that most European brass originated in the Ardennes, especially near Dinant and Aachen. This was calamine brass made in the manner described by Theophilus, and it seems from the analyses of the memorial brasses that the calamine invariably contained lead.

The calamine was first calcined, ground up fine and mixed with charcoal, and placed in hot crucibles until they were one-sixth full. These were then filled with pieces of copper and covered with charcoal. When the 'copper' had become molten (it would of course have been a brass by this time and have had a melting point well below the 1 084°C of copper) it was stirred with an iron rod to make sure there was no segregation. More calamine was put in, and was covered with fresh charcoal. The contents were then poured into sand moulds in the ground and the crucibles refilled.

It is clear that gunmetal was not used for guns in this period. Guns were made of straight tin bronzes containing between 8 and 14%Sn and little lead. But true gunmetals appear in castings such as fonts, effigies, and spouts where castability rather than ductility would be the more important property (*see* Table 44).

For the ordinary run-of-the-mill articles, small amounts of metal were melted in crucibles in a simple hearth with bellows or induced draught. We have quite a number of the crucibles and their capacity varies from about 5 to 400 cm^3. Some are triangular, others circular, and by late Saxon times the larger ones were bag shaped and capable of melting about 300 cm^3 or 3.7 kg of bronze. Many of the medieval crucibles had pointed bottoms and few were flat bottomed; such a crucible was not common until the 16th century. Examples of Migration period and medieval crucibles are shown in Fig.58; some have been found on the coast of East Africa in an Arab context.[88]

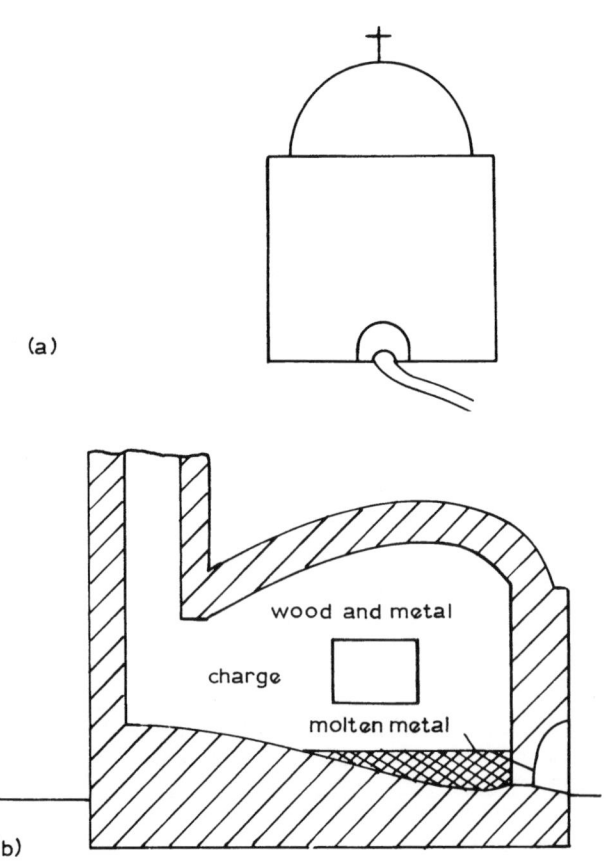

(a)

(b)

a front view as seen in York Minster; *b* reconstruction based on above

57 *Fourteenth century bell-founding reverberatory furnace based upon a stained glass window in York Minster*

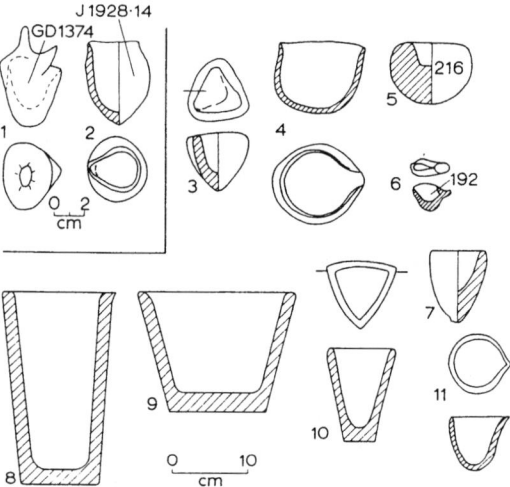

1 integral lidded, type E, from Garryduff, Eire, early Christian; 2 pointed bottom with interned rim, type G, from Wadsley, near Sheffield, medieval; 3 deep triangular crucible with pointed bottom, type A2, from Lough Faughan Crannog, Ireland, early medieval; 4 bag-shaped crucible, type F, from Winchester, England, 12th century; 5 globular stone crucible, type B1, from Garranes, Ireland, 5th–6th century; 6 horizontally pinched type of crucible, D1, from Balinderry, site no.2, Ireland, early Christian period; 7 conical round crucible, type B3, from the Tanzanian coast, Arab period; 8 flat-bottomed tapered crucible from the R. Nene, Northampton (not stratified); 9 flat-bottomed tapered crucible from Tábor, Czechoslovakia, 15th century; 10 triangular flat bottomed crucible, type A4, of the 16th century (after Agricola); 11 hemispherical crucible, type B1, from Corinth; Byzantine, AD 1150–1300; (for typology *see* Table 12)

58 *Some examples of medieval crucibles*

Table 44 Composition of copper-base alloys from the Migration and medieval periods

Object and provenance	Date	Sn	Pb	Zn	Fe	Others	Ref
Crucible contents;							
L. Rea, Ireland	5-10th century	3.1	-	15.5	-	Ni, 1.0	144
Dunshaughlin, Ireland	5-10th century	11.6	-	-	-	-	145
Lagore Crannog, Ireland	7-11th century	4.2	-	nil	1.3	Ni, 2.2	144
Bells;							
Winchester	10th century	18.75	4.35	0.1	-	-	91
Theophilus	12th century	20	-	-	-		56
Cheddar	12th century	20	nil	-	-		89
Thurgarton	12th century	22-26	3.5-6	0.5-2			92
Wharram Percy	1617	24	1.0	-	-		92
Cauldrons;							
Scandinavian	1200	14.1	12.1	tr.		Ni, 0.012	98
Scandinavian	1300	6.2	20.9	tr.		Ni, 0.012	98
Scandinavian	15th century	8.9	1.8	0.68		Ni, 0.86	98
Downpatrick	13-14th century	7.08	13.47	-	-	As, 0.54; Sb, 0.5; P, 0.07;	Brownsword Brownsword
Appin C.		12.2	10.8	0.14	0.91	Ni, 0.23; Sb, 0.29; As, 0.9; Ag, 0.05	Brownsword Brownsword
Lochmaben B		5.48	11.3	1.40	0.08	Sb, 0.3; As, 1.03 Ag, 0.08; Ni, 0.18	Brownsword Brownsword
Lockerbie A		5.20	13.0	0.24	0.16	Sb, 0.46; As, 0.96; Ni, 0.31; Ag, 0.04	Brownsword Brownsword
Foil, Kirkstall	1100-1500	6.07	1.17	4.89			146
Foil, Kirkstall	1100-1500	4.21	-	-			146
Foil, Kirkstall	1100-1500	4.82	0.51	6.44			146
Tap, Kirkstall	1100-1500	7.5	0.50	5.0	0.5		147
Font, Liège	Medieval	5.0	1.7	15.3	0.2		152
Charity spout,							
Dymchurch	13-14th century	4.4	6.6	14.4	1.9		148
Effigy, Warwick	1453	3.6	1.2	8.2	2.6		149
Brass plate, Flanders	1496-1504	3.0	3.5	29.5			142
Memorial brass	14th century	tr.	-	23.3	0.08		150
Memorial brass	1456	1.16	7.14	24.2			150
Memorial brass	1470	2.56	2.13	28.5			150
Guns;							
Tower of London*	1451	10.2	-	-	-		143,95
Turkey	1521	12.3	2.2	-	-		151
Italy	1530	13.9	1.3	-	-		151
Sweden	1535	7.6	2.6	0.4	-		151

* Turkish; considerable variation in composition

THE FOUNDING OF BELLS, GUNS, CAULDRONS, AND STATUES
Bell founding was the largest metallurgical operation of the medieval period. The mould consisted of two parts, (*a*) the core which formed the inside surface of the bell; and (*b*) the outer part, or cope, which enclosed the bell and the core. According to Theophilus,[56] the core was moulded on a tapered oak spindle and the pattern of the bell plastered on in tallow. It was thus an example of the *cire perdue* process. The tallow was then covered with the cope or mould proper which, like the core, was usually made of a fibrous clay mixed with peat, grass or dung. The oak spindle was then withdrawn and some of the clay of the core was removed to facilitate drying. In clay moulding all the uncombined water must be removed before casting, as the moulding material has not sufficient permeability to allow the steam evolved on casting to leave the mould without causing trouble. After withdrawing the spindle an iron ring was inserted to support

the clapper, and the cannons (supporting gear) and a feeding head were moulded on top. The mould was strengthened with iron bands and then moved to the bottom of a pit and warmed to run out the tallow and complete the drying (*see* Fig.59).

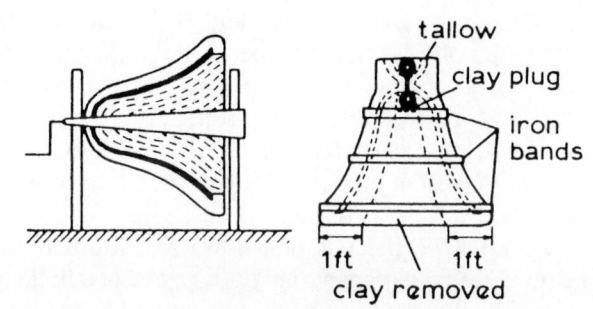

59 *Processes of bell moulding based upon Theophilus and others*

A bell-founding pit dated to the 13th century was discovered at Cheddar in Somerset,[89] and was probably intended for a bell for a chapel built in about AD 1220. This pit was 1.56 m dia. and 1.06 m deep, and the bell would have been about 60 cm in dia. and weighed about 340 kg. In the pit, pieces of the fired mould of a fibrous clay, which had been heated to about 500 °C, were found as well as scraps of bell metal containing about 20%Sn, the composition recommended by Theophilus. This method of bell production became well established and is shown by Biringuccio in use in the 16th century.[90] But the casting was not always done in a pit: sometimes the bell was cast at ground level and the metal was ladled out of the furnace, or carried up a ramp to the top of the mould. Recent excavations at Winchester unearthed the site of a bell-casting place which had probably been used for the bells of the late 10th century minster, i.e. the church before the present one. The bell had been cast at ground level amidst the remains of an earlier church.[91] All that remained was the trench in which the fire for melting out the wax or tallow and drying the inside of the mould had been placed. Part of the mould was still in position across the sides of the trench. There were also pieces of metal which had probably broken away from the parting line between core and cope. One piece of the mould seems to be inscribed *SC* or *ISC*, which is more likely to be part of a Saxon inscription than of a Latin one. This is the earliest find of its type: no Saxon bells survive in Britain but it is probable that the bell cast here bears some resemblance to those illustrated in an illuminated page from the Benedictional of St. Ethelwold (*c.* AD 980).

The remains of a similar structure, which had consisted of a stone foundation with the remains of the base of a fired clay bell-mould supported across it,[92] were found at Thurgarton in Nottinghamshire. The mould consisted of clay containing charcoal, faced internally and externally with pure clay. The mould was about 60 cm dia., and contained some pieces of bell metal (22–26%Sn). This had been surrounded by a thin clay wall acting as a screen to retain the heat while drying. The inside of this had been burned red to a depth of 8 mm. In no case had the clay been fired above 580°C. A sample, probably from the mould, was found to contain much fine organic matter— probably pond mud.

Three pieces of metal had come from within the mould (*see* Table 43 for analyses). They were of slightly different composition: one was analysed more thoroughly and found to contain 0.4%Sb, 0.3%As, and 0.02%Bi, in addition to Sn, Zn, and Pb. It is possible that more than one cast is represented, but one cannot rule out segregation effects and the use of a number of crucibles with slightly different contents. The remains of a furnace, used perhaps for melting, were found about 2 m south west of the mould but had been considerably disturbed by later activities.

By AD 1372, the use of tallow for large bells was proving rather expensive and its use declined in favour of the 'strickle board'.[75] After the core had been turned, either horizontally as described by Theophilus and as shown in the York window, or vertically as described later by Kricka, it was merely greased with tallow. The pattern of the bell was then built up with clay (the 'case') and accurately shaped with a strickle, a thin board cut to the profile of the bell and rotated against the core and finally, with a slight change in shape, against the pattern. This permitted a more accurate wall thickness and hence a better tone to be obtained. The optimum shape was not developed until the 14th century when bells started to be rung together and became rather more musical than noisy.

After the shape of the bell had been accurately built up by this method, any inscription or decoration was now added to the pattern in wax. After drying, the pattern was smeared with more tallow and the cope was built up with the same moulding material as that of the core. The cope was strengthened by the moulding in of iron wire, and finally it was reinforced externally with iron bands. The mould was then baked, and when hard the cope was lifted off with ropes fastened to the iron bands inserted during moulding. The pattern came away from the core with the cope which was then laid on its side and the pattern and decoration broken away from the inside piece by piece.

Apart from the destruction of the pattern, this technique is more like modern engineering moulding practice. Location of the cope and the core was ensured by making the cope fit over the base of the core, and no chaplets were required. Towards the end of this period the cast bronze gun was beginning to replace the inefficient wrought iron guns. There is no doubt that the moulding and casting technique was derived directly from bell founding which had been well established over the length and breadth of the civilized world by the 10th century.

We have a good description of the casting of a gun in 1451 which must be very similar to that of Muhammad 11, dated to 1464, which can still be seen in the Tower of London.[76] This gun consists of two parts screwed together, as were many large guns, for ease of carrying (*see* Fig.60). The smaller part contains the 25 cm dia. chamber and is about 18 m long, while the barrel is 63 cm bore and about 3 m long. Each part weighs about 8–9 t. Like the bells, these pieces were cast on the spot, some actually outside the walls of Constantinople. The core for the chamber and barrel seems to have been made with solid clay mixed with linen, hemp, and other fibres. The cope or mould proper was made in a pit and consisted of an inner lining of fibrous clay about 13 cm thick surrounded by a reinforcement of timber, clay, and stones, both to reinforce it and support it in a vertical position in the pit. The core was inserted with the muzzle end downwards, and presumably located in the ground. The dead-head or feeder at the top was 76 cm high and cut off with axes when still hot.

The two melting furnaces must have been very like the Chinese ones used in the Ming Dynasty, blown with bellows situated round the circumference. They were made of bricks and clay and was fuelled with wood and charcoal. The molten metal ran down 'earthen pipes' straight into the mould. The metal was virtually free of

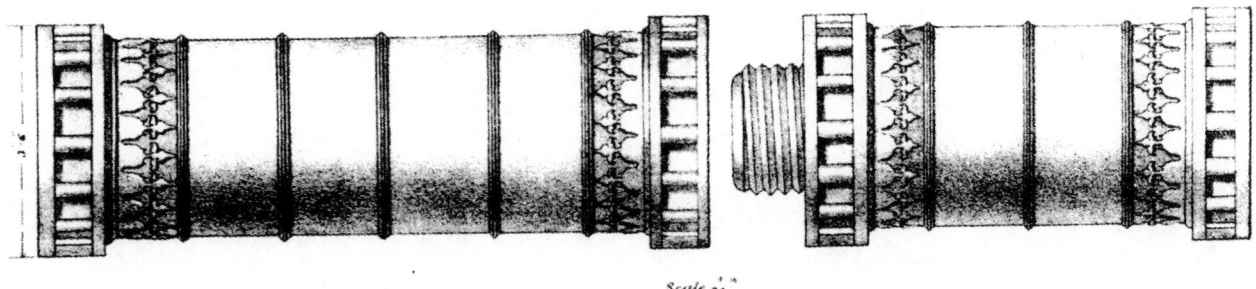

CANNON OF MUHAMMAD II, *cast A H 868/A D 1464*
PRESENTED TO
HER MAJESTY BY THE SULTAN ABDUL AZIZ KHAN, 1866
NOW IN THE MUSEUM OF ARTILLERY, WOOLWICH.

Scale ¼

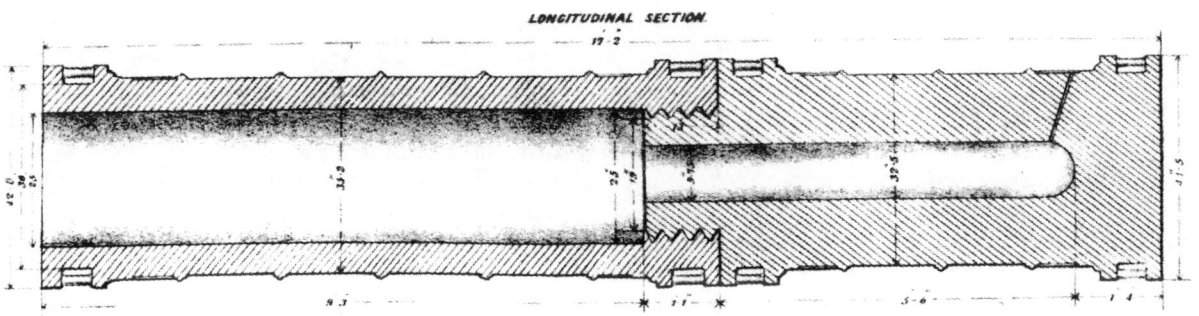

LONGITUDINAL SECTION

60 Bronze cannon with chamber and barrel cast separately and screwed together (from Lefroy [94])

Zn, Pb, Sn, As, Au, and Ag but the tin content varied from 4.8% at the muzzle end, to 10.15% elsewhere: this would be consistent with casting with the muzzle end down.

Another sophisticated product of this period was the cast bronze cauldron. Several foundries for the manufacture of bells and cauldrons have been found in Britain, two of them in Exeter dating to 1560–70. The remains of reverberatory furnaces with deep hearths capable of burning wood have been found together with casting pits. Many cauldrons and skillets have been found in Ireland and Scotland,[97] some in England, and a few in Scandinavia.[98] They were probably made by itinerant bell founders and many of them are poor examples of the Metallurgical arts. Some examples of Irish provenance are shown in Fig.61. Essentially, they were made with a one-piece core and a split mould. The handles were probably made by cutting two holes to meet inside the mould, while the legs were made by ramming separate patterns into the mould in the right positions and withdrawing them from the inside when the core had been removed (*see* Fig.62). Unlike the early bells, these were not examples of the lost-wax technique, and one can say that they provided some of the earliest and largest examples of the two part mould-with-core technique. As seen from Fig.63, the core and the mould were kept apart with chaplets. In many cases these chaplets did not bond in, so that in the course of use they were corroded out. If care had been taken to see that they were clean, of the right composition, and that the cast metal was hot enough, the corrosion might have been

prevented, but perhaps one should not be too critical as such problems are still with us.

Three cast cauldrons of South-West Scottish provenance have been examined by the author and more recently analysed by Brownsword.[99] Two of them, from Lochmaben and Lockerbie showed the 'as cast' structure of a low tin bronze with a little delta. The hardnesses were 72 and 122 HV5. The other, from Appin, had 12.2% tin but retained none of its cast structure and had been

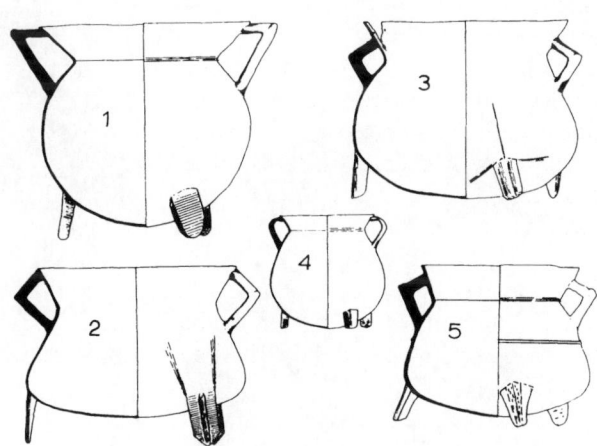

1 from near River Erne near Belturbet, Co. Cavan;
2 from near Ballymena; 3 from Co. Antrim; 4 from Ireland; 5 from River Erne near Belturbet; scale 1/10

61 Sections of bronze cauldrons from Belfast Museum (from Marshall [97])

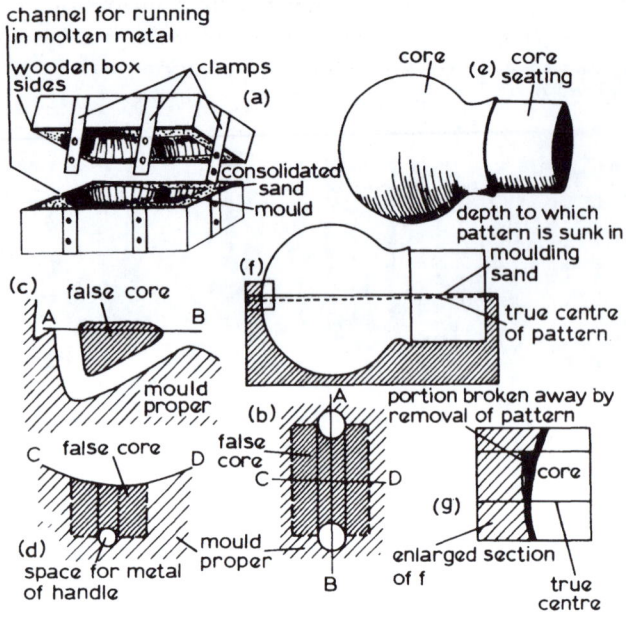

62 *Cauldron moulding (after Marshall[97])*

heated to 400–500°C after casting; this had a hardness of 90 HV5. All showed signs of final cold work, probably resulting from a planishing operation. The final wall thickness varied from 0.8 to 2.5 mm. All contained appreciable lead and variable porosity analyses are given in Table 44.

Buddhism, with its requirement for large statues, provided the incentive for making large bronze castings, examples of which can be seen in Japan and China today. The art of casting was brought to Japan from China by way of Korea in about 300 BC. In the early stages this art was applied mainly to weapons and bells but, by the 6th century AD, we begin to see the production of large Buddhas, such as that at Asuka in Japan which was cast in AD 605. This is about 3 m high, 1 cm thick, and

weighs 16 t. The great Buddha at Nara was built in AD 749 and is 16 m high, 40 m in circumference, more than 5 cm thick, and weighs about 250 t. Of course, these enormous statues could not be cast in one piece: the mould was made in pieces rather like concrete shuttering, and each section was cast separately and in such a way that the last piece interlocked with the previous piece. The statue at Kamakura was made in AD 1252, is 11.6 m high and weighs 110 t. It was made with a wooden pattern from which inner and outer mould pieces were made. These pieces were finally baked and assembled on site, rather like the piece moulds of the Chinese Bronze Age.[79–100]

INTERNATIONAL TRADE IN COPPER

We have already alluded to Germany's early place in the revival of the production of copper after the fall of the Roman Empire, and its premier position in the 16th century is well known. For this reason it is worthwhile examining the position in the medieval period, when the foundations were laid for this situation. Following the opening up of the mines of Saxony and Mansfeld, copper mining was started in Sweden at Stora Kopparberg in Falun in about 1200. This was almost certainly undertaken by German miners, and this mine seems to have supplied most of the requirements of Northern Europe through the Hanseatic League.[101] In due course this trade competed with that of the south through Nürnberg. The latter had been extended with the reopening of the Hungarian mines where miners from Saxony had become the founders of the 16 towns in the Zips—the Slovak Ore Mountains—in 1243.

In 1502, an alliance developed between Mansfeld and Nürnberg, while at about the same time the Fugger family of Augsburg, who had a monopoly of the Hungarian and Tyrolese copper, were trading it through Leipzig. The Fuggers seemed to have relied at first on the more primitive techniques of mining and smelting, which no doubt had persisted in the areas outside Germany

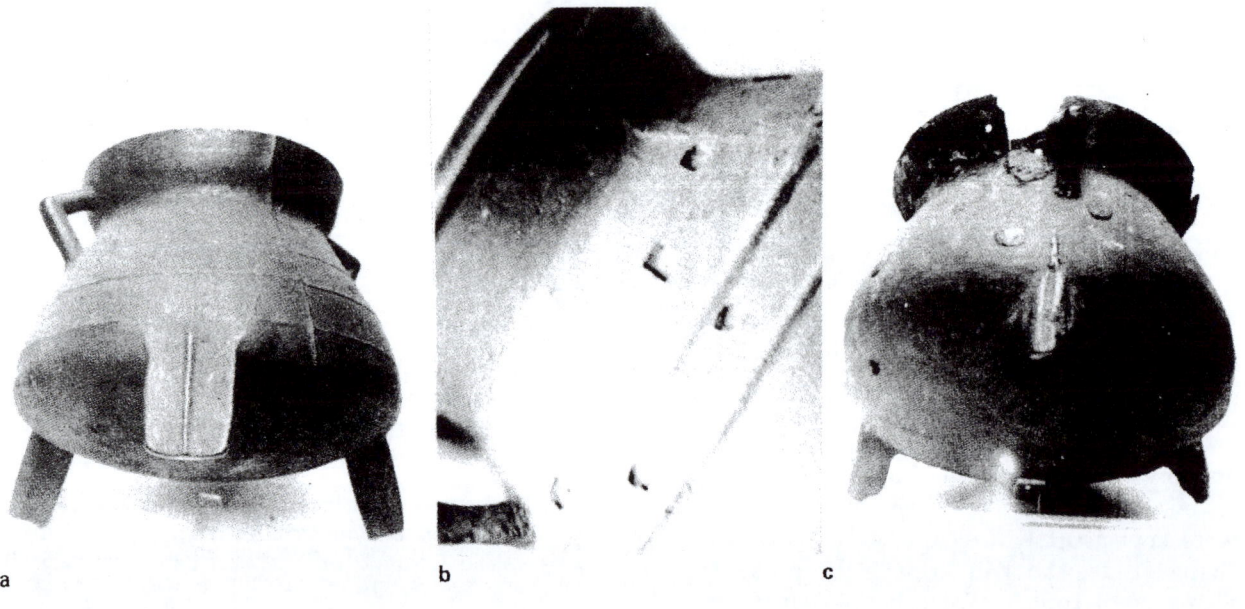

a   b   c

63 *Details of cauldrons from Belfast*

such as the British Isles. By 1500 Jakob Fugger stood on the threshold of a world monopoly in copper, but for political reasons a decline set in and by 1546 the Fugger family abandoned the Hungarian side of the business.

The existence of a European copper trade centred on Germany assisted the exchange of copper for spices and other material from the East and Africa. It is probable that this trade, carried out mainly by the Portuguese, was responsible for the increasing use of brass and leaded brass in the sculptures of Nigeria from the 16th century onwards. West Africa was receiving copper-base alloys from the north as early as the 12th century when a caravan, apparently from Morocco, abandoned a load of 80/20 brass in the Western Sahara. However, it is probable that some of the alloys, especially the bronzes, had a more local origin.[102]

But it must be remembered that Nigeria and its neighbour Niger have copper and lead-zinc deposits of their own, and there is now increasing evidence that these were used at an early date and might even be the source of the metal for the famous 'Benin Bronzes'.

Lead and silver

There is no doubt whatever that by this period most of the silver was extracted from lead and copper ores by cupellation. In many of the coinages, silver gradually replaced gold from the seventh century onwards, and the greater part of this silver would have been derived from lead rather than copper. It is therefore very surprising that we do not hear more about lead mining in the early period. Of course after the destruction of the Roman towns a good deal of scrap lead would have been available but this would have been of little use as a source of silver as it would have been low in this element.

It is possible that the Spanish mines which had been so intensively exploited during the early period of Roman occupation continued to be worked but, apart from the mercury mine at Almaden, there is little evidence of this during the Arab occupation. In most of Europe the metal trade came temporarily to a standstill, but we can surmise that there was intense exploitation of local sources to supply local mints. There is evidence of considerable international trade by the 10th century.[103] For example, the total of 40 t of silver necessary to pay the English Danegeld on four separate occasions in the 10th and 11th centuries could represent the production of 400 000 t of lead with a mean silver content of 0.1%. It is not suggested that all this silver was from current English lead production: a lot of it was from coins that had been in circulation for some time and some of it was from other sources.[104] England must have been a very wealthy country, and we know from the Sutton Hoo treasure that Anglian kings were able to amass considerable personal wealth.

What was the reason for all this apparent wealth and what was the source of the silver? We know that Germany was paying for something obtained from England at this time by silver remitted to England. Rammelsberg in the Harz was the probable source of the metal but the

Saxon mines no doubt made their contribution. The most likely import would have been wool. But there is no doubt that the English lead production was high and most of it was being cupelled for silver, although the mean silver content was low. During the two centuries after the conquest, English mineral exports increased by a factor of ten.[105] Lead ingots were exported to France for roofing purposes and a good deal of tin also went abroad. In addition, a large amount was used for monasteries in England. In the 12th century 100 cartloads went from Newcastle to Rouen, and 241 cartloads were exported from York. The latter probably came from the Yorkshire mines. Metal from the Derbyshire mines was exported via Boston and King's Lynn. This trade continued into the 16th century.

Mines yielding native silver are on record. Biringuccio refers to one at Schio, near Vicenza in Italy, which used the process of amalgamation in a grinding mill, with additions of vinegar, vitriol and verdigris in difficult cases. This is clearly a reference to what was to become the 'Patio' process in South and Central America. Agricola[107] mentions the application of this process to gold but not to silver, although he was familiar with Biringuccio's work. It seems that other processes such as the *Saiger* process (*vide inf.*) were being used on the ore bodies of the Erzgebirge and that amalgamation was in decline in Europe.

The mine near Beinsdorf, Saxony, where 'pure silver and native copper were exposed to the daylight' and which was opened in 922, was unusual and can be compared with the silver mine at Hildeston in Scotland,[108] which produced some native silver up to 1873. There were lead mines in Derbyshire at work in 835, and seven *plumbaria* were mentioned in the Domesday survey of 1086.[20] But of course, Derbyshire lead is low in silver and it is unlikely that it made much contribution to the supply of this metal, although we know that a group of manors sent 40 lb (18 kg) of silver to the mint at Derby.

The Swedish copper from Falun contained about 0.1% Ag and 0.005% Au, and this mine was certainly at work by 1200, if not before.[81] The mines at Alston in Cumberland were worked for silver and lead between 1100 and 1307, and the bishops were allowed by the Crown to retain the silver for their mint at Durham. By the 14th century, lead mining and smelting was in full swing all over Europe and great numbers of people were employed: mention is made of 10 000 in the Mendip area in Edward IV's time. It is unlikely that the numbers were as great as this although, while the miners and smelters would have been few, there would be a considerable number of women and children employed as ore dressers. The smelting processes used had undergone little change since Roman times. Much of the lead was smelted in 'boles' or induced-draught furnaces in exposed places, which gave slags with high lead content (*see* Fig.64). It is almost certain that these slags were worked up with a forced draught in slag hearths, which were like blacksmiths' hearths and bloomery furnaces.

To recover the silver from the lead the cupellation

process was used, again with no change in principle but merely with an increase in scale. However, for the recovery of silver from copper, we have definite evidence of the processes of lead soaking and liquation for the first time. Lead soaking involves placing the silver-bearing copper in a furnace or cupel with several times its weight of lead. The lead dissolves the silver from the copper, and the silver can be recovered from the lead. By the beginning of the 16th century nearly all the copper was being desilverized either by smelting it together with lead followed by liquation of the lead, or by the extraction of the silver from the matte by the addition of lead to the furnace. This involves mixing the copper with three times its weight of lead and melting in a blast furnace. The lead–copper alloy is tapped out and, during solidification, the two metals separate into a finely dispersed alloy with all the silver in the lead phase. The alloy is reheated in a separate hearth at a red heat to melt out the lead, leaving porous copper behind. The lead is then cupelled for silver recovery. The method was introduced by the Portuguese into Japan in 1591 where, by the 19th century, it had undergone a slight change in that it was manually squeezed so that the lead would be forced out.[79]

This method is usually referred to as the *Saiger* process after its German name. It appears to have been known from the earliest times.

Zinc

The Middle Ages saw the introduction of metallic zinc, certainly into China and possibly into Europe. The use of metallic zinc was unknown to Theophilus and, consid-

ering the ease with which calamine brass could be produced, there was not much incentive to use metallic zinc for this purpose. Metallic zinc is not an easy metal to produce, as zinc oxide cannot be reduced by charcoal at temperatures below 1 000°C. While such temperatures can be relatively easily obtained by means of a bellows blast, there is a complication in that zinc boils at 923°C and the metal is therefore produced as a vapour which requires special precautions before it can be successfully condensed. This is why zinc took a much longer time to appear on the historical scene than lead, copper, tin or iron.

There are one or two examples of the appearance of zinc before our period,[109] but it is possible that these are accidental deposits of zinc condensed from zinc vapour given off during the making of calamine brass. It is claimed that China first produced zinc in the period 200 BC–AD 200.[110] By the Ming dynasty, coins containing 97–99%Zn and 1–24% Cu[111] were being made and there was a large increase in the zinc content of brasses[112] (*see* Fig.39). In 1585 slabs of zinc weighing 60 kg were being produced and exported.[113] A Dutch East Indies ship foundered off Mauritius in 1609 bearing a load of zinc ingots from China.[114]

We have two 14th century references to the production of zinc as oxide in Iran.[115,116] The first, from Marco Polo, refers to the heating of a zinc mineral to produce zinc oxide, or 'tutty', near Kerman. The second is more detailed and is dated to 1340. The zinc-containing ore was ground and moistened and formed into bars about 50 cm long and 2 cm dia. These were heated in a furnace and the zinc volatilized and collected as oxide in a type of condenser in the higher levels of the furnace. Enormous heaps of the spent bars, which are now mainly iron oxides, can be seen near Deh Qualeh, north of Kerman in Eastern Iran.[117] Some of the ZnO was used for eye ointments, but it is clear that the production was very large and most of the Near Eastern ZnO must have gone into brass.

In India, at Zawar near Udaipur in Rajasthan, there was a considerable production of zinc, possibly as metal, between the 10th and 16th centuries.[119] Great heaps and walls of small retorts, consisting of pointed elliptical masses of vitrified clay 25 cm long and 15 cm dia., have recently been found. These were closed at one end but open at the other, with the remains of 2.5 cm tubes inserted in them.

It is now clear that 36 of these retorts were placed in a square furnace with their openings projected through a square perforated plate, so that the zinc vapour condensed and ran downwards into saucer-like vessels beneath. The retorts were charged with zinc mineral and charcoal.[119] This process was used from early medieval times up to the 18th century and it is estimated that the heaps at Zawar represent the extraction of 100 000 t of metallic zinc.

As far as China is concerned, there is a reference in a text book on metallurgy, dated to 1637, to the production of zinc in sealed crucibles.[110] The charge consisted of a mixture of calamine and charcoal and the metallic zinc

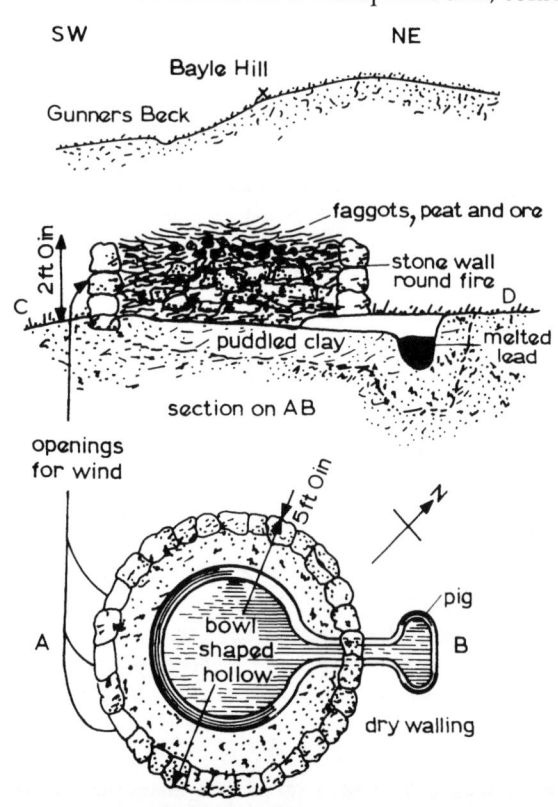

64 *Early lead smelting furnace of 'bole' from Yorkshire (from Raistrick[141])*

was deposited on the upper part of the crucibles and recovered by breaking. In 1785 a parcel of zinc ingots on its way from China to Sweden was lost in Gothenburg harbour. This was salvaged in 1872 and found to be of 98.99% purity.

We now know that a vertical distillation process has been in use in South West China up to the present time. This was done in heated crucibles 1 m high, and the zinc vapour was condensed in saucers fixed near the top.[120]

Coinage

The Migration period saw the decline of gold as the main coinage alloy and its replacement by silver. In Britain the Anglo-Saxons established mints in the seventh century at London, Canterbury, and Winchester. By Alfred's time there were additional mints at Bath, Exeter, Gloucester and Oxford. In 928 a single coinage was established and dies were distributed from London to the provincial moneyers. By this time there were about 85 mints.

Large quantities of silver coins have been excavated in Scandinavia and the Baltic Islands and most of these came from England and were, no doubt, largely the remains of the Danegeld of the 10th and 11th centuries. In four years, 991, 994, 1002 and 1007, a total of 40 t of silver coins left the country. The efficiency of the apparatus which minted these coins and collected them must have been remarkable.[121]

Some of the Baltic silver was of Moslem origin from the Emirates east and north east, of the Black Sea. This source ceased in the 11th century when these mints were disbanded, but there is little doubt that the Moslem occupation of Asia Minor resulted in a continued and intense exploitation of their silver deposits. The coinage tended to deteriorate in the 13th century owing to clipping and other causes, and a recoinage was ordered in Britain in 1299, when the mint was placed in the Tower of London.[122] In fact, British minting only left the area of the Tower in 1970 when a new mint was established in South Wales: a continuous series of trial plates against which the quality of the coinage was tested exists from 1279.[123]

The overwhelming bulk of the medieval coinage of Europe was struck free-hand, but in the early days a fixed relationship of the die axes did exist, and the abandonment coincided roughly with the transition from the gold tremisses to the silver penny. The top die (or trussel) consisted of an iron punch, and the lower (or pile) of a tapered block of iron embedded like an anvil in a block of wood. Both contain engraved steel or bronze dies projecting slightly from the iron blocks. A set of iron dies have recently been found in Anglo-Scandinavian York. Although not yet fully examined, these would appear to have been steeled on their surface like the Roman dies found at Trier.[124]

There are some exceptions to free-hand striking. We have a set of steel dies from Algiers which are said to have been used in the years 1115–16 at the mint of Nul in the Maghreb. These were pegged so that a consistent die alignment could be obtained if desired.[125] However,

it is possible that pegged dies were not introduced before the 18th century. Furthermore, it is believed that most Arab dies were still of bronze and made by cutting the design into lead, impressing this into clay, and making cast dies of bronze.[126] Certainly, steel dies would have to be engraved directly unless they were made like modern ones by engraving on to soft carbon steel, hardening it, then pressing it into a soft steel die and hardening the die. It seems that high-tin bronze was quite hard enough for the relatively soft noble metal coinage of this period. The blanks were either pieces of sheet trimmed to size with shears, or slices cut from bars. In Britain, in 1280, the silver was cast in the form of square-section bars which were sliced up and forged round.

It seems that no official coins were made by casting in this period, although this method no doubt continued to be used by counterfeiters. Machinery was not introduced until the Renaissance, and blanking punches and the screw press were used about 1530.[127,128] The composition of early Byzantine gold coins and of some other early issues is 98% Au[129] and one finds in no early gold coinage a copper content of more than 5%, probably because of the hardening effect of such an addition. The hardening of the gold coinage was done with the aid of silver which seems to have reached about 30%. A good deal (perhaps all) of the gold would have been natural, as it only has about 1% Cu and a maximum of 20% Ag. However, towards the middle of the seventh century we have examples of coins containing 59% and 69% Ag which must have looked very white. Lower denomination copper and bronze coins were also made in the Byzantine and Arab mints.[130] At the moment it is not known where the copper came from but there are some large slag heaps dating from Romano-Byzantine times on the south shore of the Black Sea,[117] in Cyprus and in Israel. It is possible that some of the metal came from the deposits north of Constantinople, and now in Bulgaria.

In Britain, the change from a gold to a silver coinage took place in 660–670. The silver *sceattas* of AD 700–710 were about 95% Ag, with 1–3% Au and about 4% Cu. These could be non-alloyed silvers or the result of cupellation with a copper addition. The gold is not lost in cupellation but it is more probable that the gold was introduced by the addition of late gold–silver alloy coins. There was no evidence of surface gilding.

The British silver trial plate of 1279 contained 6.19% Cu, 0.81% Pb, and 0.3% Au.[123] The lead content is typical of silver from cupelled lead but the copper would have been an intentional addition. This sterling standard was maintained with occasional exceptions (1542 and 1600) into modern times. The earliest gold trial plate, dated 1477, contained 99.35% Au, 0.515% Ag, and 0.135% Cu, and the gold must have been purified by chemical 'parting', i.e. by the solution of the silver in acid.

Tin alloys and tinning processes

We have virtually no information on the techniques used for the production of tin in this period. The early tin ingots ('Jews' house tin') were plano-convex like the EIA and Roman ones, and there is little doubt that they were smelted in the same manner. Rectangular ingots seem to have appeared by the end of the period: these were probably the result of 'coining', the process of quality control imposed on the Cornish tin industry in about AD 1200. But the earlier date of AD 700 has been found in the lower levels of a tin smelter at Week Ford on Dartmoor in Devon.[131]

The principal use of tin was for pewter and, for this purpose, it was often leaded to harden it and make it go further. Different countries had different standards in this respect and, as we have seen from the Roman material, the lead content could vary widely, but it was usually about 20%.[132] The tin or pewter was cast into metal, sand, or stone moulds not unlike the Roman ones, and finished by beating it into shape. Some vessels were made by 'raising', i.e. beating out an ingot or sheet, and others were spun to the finished shape. Of course, one of the great advantages of the metal was its great ductility and its low work-hardening tendency.

The joining of high grade (high-tin) pewter, such as Britannia metal, could be carried out by soldering with tin–lead solders of the 60/40 type which melt at temperatures well below the melting point of tin. The lower grades could only be joined by the localized melting of the sheet metal itself. Today, more complex bismuth-containing solders are available. These have lower melting points and can therefore be used on pewters with high lead content. At present there is no evidence that such alloys were used in early times.

In Europe, the end of the 15th century saw the invention of printing. This involved the use of tin–lead alloys as type metals. Cast bronzes had been used for Korean characters as early as 1403–43 so there was nothing new in the idea.[133] It is almost certain that Gutenberg's first type was the normal pewter of the day, i.e. Sn–20% Pb. Very soon, however, it was found that additions of antimony, and possibly of bismuth, conferred improved properties such as lower surface tension on the original pewter, which already had some of the good properties required of a satisfactory type metal, i.e. high specific gravity and low melting point. It seems that Biringuccio[90] knew of the good effect of antimony on the tin–lead alloy. The compositions of modern type metals fall in the ranges 20–25% Sn, 50–60% Pb, and 19–25% Sb.

The tinning of bronze appears to have started in the EBA as there are a number of flat axes that have a tin-rich surface. However, recent work has shown that this is probably due to inverse segregation and that the tin has not always been intentionally applied in the liquid state.[134,135] It was, however, normal practice to tinplate iron spurs in the medieval period,[136] but the process was not applied to the stirrups. Presumably, this was because the spurs were a personal adornment to be taken indoors while the stirrups, like the spearheads and other iron-work, would remain outside and be polished when necessary, or be made of brass. The swords, of course, were protected by their scabbards. The plating material was pure tin or tin–lead alloy, and it is thought that the spurs were first fluxed and then sprinkled with powdered tin.

Other variations of the tinning process were bronze plating on iron, gilding on bronze, and gilding with gold leaf on bronze. Theophilus has something to say about the processes used.[137] Copper plates were first scraped and dip tinned: the undersides of the heads of studs were tinned with an iron and iron bindings for an organ were tinned on both sides to protect them from rust. He advised cleaning iron with a file and throwing it into a pot of molten tin covered with tallow, and he warned against touching it with the hands after filing. Here we have something much more like the modern tinning process. He also described a special vibrating tool for roughening and cleaning surfaces before applying inlay, and he made it clear that inlaying was a pressure-welding process in which the applied metal was cold hammered on to the prepared surface.

Mercury gilding was initiated with a warm mixture of argol (crude alkali tartrate), salt, water, mercury and freshly-milled gold. This was stirred and applied to the silverware with a linen cloth and brush of hog's bristles. When amalgamation had started to take place, the work was heated over charcoal and gold leaf applied with a copper gilding tool. Theophilus warned against trying to gild leaded brass or leaded bronze with mercury.[138,139] Recent analyses of copper base objects of this period show that his recommendations were generally followed.

The manufacture of tinplate, i.e. the tinning of iron sheets, was a medieval development. It probably started in Wunsiedel in the Fichtelgebirge in the 13th century.[140] It is known that a merchant of Nürnberg supplied the Netherlands with 28 *tonnelets* of tinplate in 1428. Wunsiedel was the site of an early tin mine: such a large use of tin would naturally require a local source and it seems that, as the tinplate trade developed, the Nürnbergers bought up the local tin mines. The sheet iron came from the Oberpfalz, and these two sources seem to have been sufficient for much of the European requirement until the 16th century.

Tinning was carried out by first immersing the iron sheets, in groups of 300, in a bath of fermenting bran placed near the side of the furnace for three days or more. They were then washed and one by one placed into a bath of tallow and then pretinned in a hot tin bath where they remained for 15 or 30 minutes. The iron tanks containing the tin measured 50 2 40 cm across and were 50 cm high. The French used tin containing 1.5% Cu to thicken the coating. After this pretinning stage, the sheets were finished in a cooler bath just above the melting point so that a thicker layer could be applied. Finally, they were cleaned in a mixture of bran and flour.

References

1   H. R. ELLIS DAVIDSON: 'The sword in Anglo-Saxon England', 1962, Oxford, Oxford University Press
2   A. K. ANTEINS: *J. Iron Steel Inst.*, 1968, 206, 563
3   E. SALIN: 'La civilization Merovingienne', part 3, Les Techniques, 1957, Paris
4   A. M. ROSENQVIST: 'Sverd med Klinger ornert med figurer i Kopperlegeringer fra elders jernalder i Universitets Oldsamling', 1967–8, Oslo, 1971, Universitets Oldsaksamlings Arbok

5   E. M. JOPE: *Ant.J.*, 1946, **26**, 70

6   G. BEHRENS: *Mainzer Zeit.*, 1946–48, **41–43**, 138

7   E. E. EVANS: *Ulster J. Arch.*, 1948, **11**, 58

8   A bloom from Carrigmuirish in Co. Cork, Ireland weighed 3.6 kg; this was kindly provided for examination by Professor M. J. O'Kelly

9   M. S. GUISEPPI: *Arch.*, 1913, **64**, 145

10   I. SERNING: *Durrer Festschrift*, 73–90

11   R. PLEINER: *Slovenska Arch.*, 1961, **9**, 405

12   At Baysdale in N. Yorkshire for example: personal communication from A. Aberg

13   W. GUYAN: *Durrer Festschrift*, 163–194

14   D. W. CROSSLEY and D. ASHURST: *Post-Med. Arch.*, 1968, **2**, 10

15   R. PLEINER: *RHS*, 1962–3, **3**, 179

16   G. HACKENAST *et al.*: A Magyarországi Vaskohaszat Története a Korai Kozepkorban, 17, 1968, Budapest, Akademie Kiado

17   VITRUVIUS: 'Ten books on architecture', (Trans. M. H. Morgan), 1914, Cambridge, Cambridge University Press

18   Gregory of Tours: 'History of the Franks.' (ed. O. M. Dalton), 1927, Oxford, Oxford University Press

19   M. T. HODGEN: *Antiquity*, 1939, **13**, 261

20   SIR H. ELLIS: '*A general introduction to Domesday Book*', 1833, London, Commissioner on The Pipe Rolls of the Kingdom

21   G. T. LAPSLEY: *Eng. Hist. Rev.*, 1899, **14**, 509

22   Taccola's furnace of 1440 is illustrated in H. R. Schubert: 'History of the British iron and steel industry', 135

23   J. R. SPENCER: *Tech. Cult.* , 1963, **4**, 201

24   G. HACKENAST: *RHS*, 1967, **2**, 73

25   J. PERCY: 'Metallurgy; iron and steel', 278, 1864, Murray

26   W. G. COLLINGWOOD: *THSLC*, 1901, **53**, 14

27   OLE EVENSTAD: *Bull. HMG*, 1963, **2**, (2), 61

28   R. SCHAUR: *Stahl u. Eisen*, 1929, **49**, 489

29   J. PERCY: *op. cit.*, 321

30   R. F. TYLECOTE *et al.*: *J. Iron Steel Inst.*, 1971, **209**, 342

31   G. MAGNUSSON: 'Lapphyttan — an example of medieval iron production', In: 'Medieval Iron in Society' (ed. N. Bjorkenstam *et al.*), Jernkontorets Forskning, H 34, 1985, 21–60

32   I. SERNING, HANS HAGFELDT and P. KRESTEN: 'Vinarhyttan', Jernkontorets Forskning, H 21, 1982

33   R. F. TYLECOTE: 'The early history of the iron blast furnace in Europe; a case of East–West contact ?' In: 'Medieval Iron in Society', (*see* Ref. 31), 158-173

34   N. BJORKENSTAM and S. FORNANDER: 'Metallurgy and technology at Lapphyttan', In: 'Medieval Iron in Society' (*see* Ref. 31), 184-228

35   C. S. SMITH *et al.*: *Tech. Cult.*, 1964, **5**, 386

36   E. BOHNE: *Stahl u. Eisen.*, 1928, **48**, 1577

37   J. NEEDHAM: 'The development of iron and steel technology in China', 1958, London, Newcomen Society

38   R. D. SMITH and RUTH R. BROWN: 'Bombards; Mons Meg and her Sisters', Monog. No. 1, Royal Armouries, Tower of London, 1989

39   J. W. ANSTEE and L. BIEK: *Med.Arch.*, 1961, **5**, 71

40   R. F. TYLECOTE and B. J. J. GILMOUR: 'The metallurgy of early ferrous edge tools and edged weapons', *BAR. Brit. Ser.* 155, Oxford, 1986

41   'TO-KEN': A catalogue of an exhibition of Japanese swords held in the Ashmolean Museum, Oxford, 1968, London

42   H. H. COGHLAN: 'Notes on prehistoric and early iron', 166, Oxford, Pitt-Rivers Museum (quoting M. Chikashigo, Alchemy and other chemical achievements of the ancient Orient, Tokyo, 1936)

43   H. O'NEILL: *Trans. Inst. Weld.*, 1946, **9**, 3

44   C. S. SMITH: 'History of metallography', 1960, Chicago, Chicago University Press

45   H. W. VOYSEY: *J. Asiatic Soc. Bengal*, 1832, **1**, 245

46   F. BUCHANAN: 'A journey from Madras through the countries of Mysore, Canara, and Malabar', 1807, London

47   K. N. P. RAO *et al.*: *Bull. HMG*, 1970, **4**, (1), 12

48   J. NEEDHAM: 'The development of iron and steel in China', Newcomen Society, 1956, Plate 15, Fig.25, (AD 1334)

49   MARTIN RUSSEL: Personal communication, August 1981

50   J. P. FRENKEL: *Tech. Cult.*, 1963, **4**, 14

51   W. ROSENHAIN: *J. Roy. Anthrop. Inst.*, 1901, **31**, 161

52   B. BRONSON: 'Terrestrial and meteoric nickel in the Indonesian Kris', JHMS, 1987, **21**(1), 8–15. (see also letter by A. Maisey in *JHMS*, 1988, **22**(1), 58-59)

53   P. WHITAKER and T. H . WILLIAMS: *Bull. HMG*, 1969, **3**, (2), 39

54   Stewart Rowe of Brisbane, Australia, has found that he can make excellent Damascened blades from old mine haulage rope in which the outer layers of the strands of decarburized steel weld remarkably well.

55   R. E. OAKSHOTT 'The archaeology of weapons', 143, 1960, London

56   J. G. HAWTHORNE and C. S. SMITH: 'On divers arts; the treatise of Theophilus', 1963, Chicago, Chicago University Press

57   C. PANSERI: 'Ricerche Metallografiche Sopra, Una Spada da guera del XII secolo', 1, Quad. I, 1954, Milano, AIM

58   H. E. BUHLER and C. STRASSBURGER: *Archiv. Eisenh.*, 1966, **37**, 613

59   Sir F. Stenton (ed.). 'The Bayeux Tapestry', 1957, New York, Phaidon; plate I showing the death of Harold; H. H. Coghlan, *op. cit.*, 191

60   C. FELL: *TCWAAS*, 1956, **56**, 67

61   A. R. WILLIAMS: 'Four helms of the 14th century compared' *J. Arms and Armour Soc.* 1981, **10**(3), 80–102

62   H. H. COGHLAN and R. F. TYLECOTE: 'Medieval iron artefacts from the Newbury area of Berkshire', *JHMS*, 1978, **12**(1), 12-17

63   E. E. P. COLLINS and H. H. BEENY: *Man*, 1950, **50**, 114

64   E. M. BURGESS: *Ant.J.*, 1953, **33**, 48, 193

65   A. R. WILLIAMS: 'The manufacture of mail in medieval Europe: a technical note', *Gladius*, 1980, **15**, 105–134

66   A. SNODGRASS: 'Early Greek armour and weapons', 1964, Edinburgh, Edinburgh University Press

67   C. S. SMITH: *Tech. Cult.*, 1959–60, **1**, 59, 151

68   G. W. HENGER: *Bull. HMG*, 1970, **4**, (2), 45

69   J. E. REHDER: 'Ancient carburization of iron to steel', *Archeomaterials*, 1989, **3**(1), 27–37

70   J. STEAD: 'The uses of urine', Old West Riding, 1981, 1(2), 12–18

71   R. THOMSEN: *J. Iron Steel Inst.*, 1966, **204**, 905

72   Peter Crew, who has been smelting bog iron ores from Wales, has found that these may give more than 1% As in the forged bloom

73   O. JOHANNSEN: *Stahl u. Eisen*, 1910, **30**, 1373

74   J. C. ALLAN: *Bull. HMG*, 1968, **2**, (1), 47

75   W. GOWLAND: *J. Inst. Metals*, 1910, **4**, 4

76   R. HARTWELL: *J. Econ. Hist.*, 1966, **26**, 29

77   M. L. PINEL *et al.*: *Trans. AIMME.*, 1938, **5**, Tech. Pub No. 882, 20

78   K. KUBOTA: 'Japan's original steelmaking and its development under the influence of foreign technique', 6, 1970, Pont à Mousson, Int. Co-op. Hist. Tech. Committee

79   W. GOWLAND: *Arch.*, 1899, **56**, 267

80   K. KIRNBAUER: In 'Copper in nature, technique, art and economy', 40, 1966, Hamburg

81   S. LINDROTH: 'Gruvbrytning och Kopparhantering vid Stora Kopparberget', 2 vols., 1955, Uppsala, Almquist and Wiksells Boktryckeri AB

82   P. S. DE JESUS: 'A copper smelting furnace at Hissarcikkayi near Ankara, Turkey', *JHMS*, 1978, **12**(2), 104–107

83   N. CUOMO DI CAPRIO and A. STORTI: 'Oridinamenta Super Arte Fossarum Rameriae Et Argentariae Civitatis Massae', In; 'The Crafts of the Blacksmith', (ed. B.G. Scott and H. F. Cleere), Belfast, 1984, 149–152

84   J. G. CALLANDER: *PSAS*, 1931–32, **66**, 42

85   N. BARNARD: 'Bronze casting and bronze alloys in ancient China', 1961, Canberra

86   To be seen in York Minster in the north aisle near the transept

87   B. J. LOWE *et al*: 'Keynsham Abbey excavations; 1961-1985', *Proc. Som. Arch. Nat. Hist. Soc.*, 1987, **131**, 81–156

88   J. R. HARDING: *Man*, 1960, **60**, (180), 136

89   P. RAHTZ: *Med. Arch.*, 1962–3, **6–7**, 53

90   V. BIRINGUCCIO: 'Pirotechnia', (ed. C. S. Smith and M. T. Gnudi) 1943, New York

91   M. BIDDLE: *Ant.J.*, 1965, **45**, 230; see also *Foundry Trade J.*, 1964, **117**,460

92   P. W. GATHERCOLE and B. WAILES: *Trans. Thoroton Soc.*, 1959, **63**, 24

93   J. G. M. SCOTT: *Trans. Devon Assoc.*, 1968, **100**, 191

94   J. H. LEFROY: *Arch.J.*, 1868, **25**, 261

95   A. R. WILLIAMS and A. J. R. PATERSON: 'A Turkish bronze cannon in the Tower of London', *Gladius*, 1986, **17**, 185–205

96   C. G. HENDERSON: 'Archaeology in Exeter, 1983-84', Exeter Museums Arch. Field Unit, 1985

97   K. MARSHALL: *Ulster J.Arch.*, 1950, **13**, 66

98   A. OLDEBERG: 'Metallteknik under Vikingatid och Medeltid', 1966, Stockholm, Victor Pettersons, Bokindustrie AB

99   R. BROWNSWORD and E. H. H. PITT: 'Alloy composition of some cast "latten" objects of the 15th/16th century', *JHMS*, **17**(1), 44–49

100   ANON: 'How the Diabatsu of Nara was made', Japan Info. Bull. 1974, Feb. 21(2), 9–11 and J. W. MEIER: 'Non-ferrous metals casting; history and forecast', Aug. 1970, Information Circular IC 239, Mines Branch, Department of Energy, Ottowa

101   W. TREUE: 'The medieval European copper trade', in *Copper in Nature, Technics, Art & Economy*, Hamburg Centenary Volume, 1966, 95

102   T. SHAW: 'The analysis of West African bronzes; a summary of the evidence', 'Ibadan', 1970, (28), 80

103   V. E. CHIKWENDU, P. T. CRADDOCK *et al.*: 'Nigerian sources of copper, lead and tin for the Igbo-Ukwu bronzes', *Archaeom.*, 1989, **31**, 27–36

104   P. H. SAWYER: *Trans. Roy. Hist. Soc.*, 1965, **15**, 145

105   M. CARUS-WILSON: 'Medieval England', (ed. A. L. Poole), 230, vol.1, 1958, Oxford, Clarendon Press

106   D. KIERNAN: 'The Derbyshire lead industry in the 16th century', Chesterfield, 1989

107   G. AGRICOLA: 'De Re Metallica'

108   H. AITKEN: *Trans. Fed. Inst. Min. Engrs.*, 1893–94, **6**, 193

109   M. E. FARNSWORTH *et al.*: *Hesperia*, 1949, Supp. 8, 126

110   L. AITCHISON: 'A history of metals', 480, 1960, London, Macdonald and Evans

111   E. T. LEEDS: *Num. Chron.*, 1955, **14**, 177

112   N. BARNARD: *op. cit.*, 194, Fig.52.

113   E. BROWNE: *J. Roy. Soc. Arts.*, 1916, **64**, 576

114   M. L'HOUR and LUC LONG report the finding of zinc ingots probably of Chinese origin off the island of Mauritius

115   MARCO POLO: 'The Travels', (Trans. R. Latham), 1958, Harmondsworth

116   G. LE STRANGE (Trans.): 'The geographical part of the Nuzhatel-Qulub by Hamd-allah Mustafi of Qazvin in AD 1340', 1919, Leiden

117   R. F. TYLECOTE: *Metals and Materials*, June 1970, 285

118   S. W. K. MORGAN: Avonmouth Digest, 1969, **22**,(25), 4pp

119   LYNN WILLIES, P. T. CRADDOCK, L. J. GURJAR and K. T. R. HEGDE: 'Ancient lead and zinc mining in Rajasthan, India', *World Arch.*, 1984, **16**(2), 222–233

120   P. T. CRADDOCK, L. J. GURJAR and K. T. R. HEGDE: 'Zinc production in medieval India', *World Arch.*, 1983, **15**(2), 211-217

121   G. BROOKE: 'Europe in the Central Middle Ages', 228, 1964, London, Longmans

122   J. H. WATSON: *Trans. Inst. Min. Met.*, 1959, **68**, 475

123   J. S. FORBES and D. B. DALLADAY: *J. Inst. Metals*, 1958-59, **87**, 55

124   PATRICK OTTAWAY: Personal communication. The dies were on exhibition in the British Museum in 1984

125   P. GRIERSON: *Num. Chron.*, 1952, **12**, 99

126   P. BALOG: *ibid.*, 1955, **15**, 195

127   F. S. TAYLOR: *TNS*, 1954, **29**, 93

128   F. C. THOMPSON: *Edgar Allan News*, 1949, **27**, (322), 275

129   S. C. HAWKES *et al.*: *Archaeom.*, 1966, **9**, 98

130   T. PADFIELD: 'Methods of chemical and metallurgical investigation of ancient coinage', (eds. E. T. Hall and D. M. Metcalf), 219, 1972, Special Publication no.8, Royal Numismatic Society

131   B. EARL: 'A note on tin smelting at Week Ford, Dartmoor', *JHMS*, 1989, **23**(2), 119

132   H. J. L. J. MASSE: 'Pewter plate', 1904, London, George Bell and Son

133   C. S. SMITH: 'Metal transformations', 2nd Buhl. Int. Conf. on Materials 1966; (eds. W. W. Mullins and M. C. Shaw), 1968, New York

134   N. D. MEEKS: 'Tin-rich surfaces on bronze; some experimental and archaeological considerations', *Archaeom.*, 1986, **28**(2), 133–163

135   R. F. TYLECOTE: 'The apparent tinning of bronze axes and other artefacts', *JHMS*, 1985, **19**(2), 169–175

136   E. M. JOPE: *Oxoniensia*, 1956, **21**, 35

137   J. C. HAWTHORNE and C. S. SMITH: *op. cit.*, 187

138   THEOPHILUS: 187 (*see* Ref. 56)

139   W. A. ODDY, S. LA NIECE and NEIL STRATFORD: 'Romanesque Metalwork', Brit. Mus. Publ., 1987

140   A. LUCK: *RHS*, 1966, **7**, 141

141   A. RAISTRICK: *TNS*, 1927, **7**, 81

142   H. HAINES (ed.): 'A manual for the study of monumental brasses', 1848 Oxford Archaeological Society

143   P. J. BROWN: *Foundry Trade J.*, 1960, **108**, 163

144   R. J. MOSS: *PRIA(C)*, 1924–7, **37**, 175

145   H. HENCKEN: *ibid.*, 1950, **53**, 1

146   R. HAYNES: *J. Iron Steel Inst.*, 1956, **183**, 359

147   L. ALCOCK and D. E. OWEN: *P. Thoresby Soc.*, 1955, **43**, 51

148   L. R. A. GROVE: *Arch. Cant.*, 1956, **70**, 268

149   J. M. FRIEND and W. E. THORNEYCROFT: *J. Inst. Metals*, 1927, **37**, 71

150   W. GOWLAND: *ibid.*, 1912, **7**, 23

151   *Tin and its uses*, 1959, (49), 4

152   Fr. BOUSSARD: 'La Fonderie Belge', 16, 1958

Chapter 8
Post-medieval metallurgy

This period, which can be conveniently dated from about 1500, laid the foundations for the Industrial Revolution which started in Western Europe, with the use of coal for metallurgical processes, in about 1700. Its beginnings are well documented by a number of books, of which the most important are those by Biringuccio[1] and Agricola.[2] These contain a wealth of detail on metallurgical techniques following the tradition of Theophilus. This is the period of the Renaissance in Europe when an increasing number of people were taking an interest in technology and the arts, and many had the ability to write about it. Before this period, with one important exception, the artisans had not been literate and historians were neither familiar with, nor interested in, technological processes.

Biringuccio wrote in Italian and it is probably for this reason that his work on metallurgy, which preceded that of Agricola, was not so well known. Biriguccio was not as familiar with non-ferrous smelting methods as Agricola, and it is clear that Agricola borrowed certain sections of Biringuccio's work and incorporated them into his own to make up for his deficiencies on the ferrous side. The publicity given to German metallurgy by Agricola's work, together with the reputation of German bankers and traders such as the Fuggers of Augsburg, led various governments, including that of Elizabeth I of England (1533–1603), to invite German workers to develop their mineral resources. As far as Britain is concerned, it is diffcult to understand why this move was necessary since there had been a very considerable exploitation of lead and silver before this period. However, we know that Elizabeth was somewhat apprehensive[3] of the power of Spain and wanted to err on the safe side as far as self-sufficiency in metals was concerned. In addition, there was also no doubt the feeling, common in Britain, that foreigners knew more about everything than the natives.

On the other hand, the Germans were looking at Britain from the point of view of the colonialist and merely wanted the raw materials to send back to Germany; these were to be smelted with superior German knowledge to satisfy the requirements of an expanding trade. This was, of course, scotched by Elizabeth's ban on the export of strategic materials, which finally led to the over-production of copper and the return of some disappointed German workers.

Germany was not the only European power to be interested in the exploitation of metals. During the Middle Ages, Spanish ferrous metallurgy had been quietly developing particularly in the more stable north: the southern part had been disorganized by the struggles of the latter part of the Moslem occupation. Apart from Catalan ironmaking and the Toledo tradition of swordmaking very little is known about Spanish metallurgy in the Middle Ages, but there is little doubt that there was a local tradition upon which the colonial expansion of Spain and Portugal could be based: it is this tradition that was to blossom in South and Central America, recorded in the work of Alonso Barba in 1640, and which showed evidence of independent development.

Our period, therefore, tends to divide into two parts: (a) the 16th century or late Renaissance period, where the increase in demand and the dissemination of knowledge are the main features; and (b) the 17th century, a more confused period in which the new ideas were being put into practice and an increased effort was being made in the more highly populated parts of Europe to use coal and coke as metallurgical fuels, rather than charcoal.

The fuel crisis, which should have been a problem, was temporarily obviated in Britain by the opening up of the American colonies: in some parts of Europe the timber resources were to prove sufficient for a moderate level of metallurgical activity, and techniques based on charcoal were able to continue into the early years of the 19th century. In Britain, and later in France, the Netherlands and Germany, the problem was solved by the increasing use of coke for iron smelting from about 1720 onwards. To a metallurgist, this marks the beginning of the Industrial Revolution.

Ironmaking

THE FIRST PHASE; AD 1500–1600

We have described the development of the blast furnace in the last chapter. In the 16th century it spread to most areas of Western Europe with the exception of the Iberian peninsula.

The original incentive was probably a purely military one. Yet the demand for cast iron for guns was peripathetic and much of the iron went to the finery for conversion. Wrought iron could be made more cheaply by the bloomery process. For a furnace of a given height, say 2.5 m, cast iron required a greater fuel/ore ratio than the bloomery but it produced slags with lower iron content. The increase in capital and fuel requirements outweighed the reduction in the cost of ore, especially when wrought iron was the end product, since this required further expenditure on fuel.

A cast iron gun was very much cheaper and, from some points of view, better than a bronze gun, so blast furnaces were built for this purpose and amortized on the basis of gun casting. Surplus cast iron could go to the finery to be converted to wrought iron. Later, as the height increased, the fuel efficiency increased and the blast furnace became the normal process for the production of all iron in the more advanced countries. However, the advantages as far as wrought iron was concerned were slight, as seen from the prolonged use of the Stücköfen in Austria, and the Catalan hearth in Spain.

The second description of any blast furnace is by Nicola Bourbon[4] and dates from 1517. The furnace was probably situated in the Ardennes region. It had a square stone structure with a sandstone lining and two large leather bellows at the back of the furnace which were worked by a waterwheel. The ore was washed and roasted and charged without flux, and with charcoal made near the furnace.

The slag was not free-running as it was removed with an iron hook: the iron was cast into moulds and the ingots went into a finery and finally to a chafery where they were forged into long rods. The length of a furnace smelting campaign was about two months.

The main impurities in the iron ore—the gangue—were silica and alumina, as we see from the analyses given in Table 45. These would combine at high temperatures to give a slag which could be tapped out of the furnace at temperatures about 1 300°C. This was presumably the case with the slag from Low Mill, Yorkshire, shown in the table.

The addition of limestone to the charge took place fairly early on. If this could not be done by the digging out of limy ironstones as in the Weald (Cyrenae limestone) then limestone was added. Analyses of slags from the area around Bray in Northern France—the area from which the Weald got its blast furnace technique—shows limy slags at the outset.

The early blast furnace slags often held quite a lot of iron but not as much as the bloomery slags, and it is certain that limestone was not always added (*see* Table 45). Some of the ores were self-fluxing and produced

free-running slags with some lime, but it is clear that the temperatures were often not high enough to run the slag. The iron content of the slag could, of course, be adjusted by altering the fuel/ore ratio, but sometimes it was judged better to have a free-running slag with an appreciable iron content rather than a semi-solid slag with no iron. Manganese had a marked effect in this respect and, if available, could replace iron, as we see in the slag from Duddon.[10]

A painting of an early blast furnace, by the Flemish painter Blés, refers to the period 1511–50 and the same place as Bourbon. The furnace was powdered by an overshot waterwheel and seems to be about 4.6 m high to the charging floor. The tuyere is at the side rather than the back, as in the Bourbon furnace. The product went to the finery and chafery, which is also shown in the picture. Between about 1496 and 1520, blast furnaces were being started in the Sussex Weald in England, at Newbridge, and Steel Forge, and the foundations of the latter have recently been identified.

At this point, it is necessary to discuss the evolution of the blast furnace.

We now know that the earliest true blast furnaces in Western Europe date from about 1345.[12] This was in the County of Namur where the word 'fonderie' is taken by Awty as meaning a producer of cast iron and not a forge. The same word was used in the Pays de Bray in Northern France between 1486 and 1563. A map of 1508 shows a furnace with a charging ramp. Another works in the area, at Hodeng, supplied a neighbouring forge with sows (gueuses) for fining in the last decade of the 15th century.

The input of French terminology, and the workers themselves, clearly link the Weald with the Pays de Bray. It is equally clear that the indirect process of blast furnace and finery were introduced to Britain together.

Nearly all the early furnaces were square externally and built of stone, and in some cases this type lasted into the 19th century. The stone furnace depicted in a Sussex fireback in 1636 shows the use of timber lacing, and very soon iron rods were introduced as reinforcement. The biggest changes, however, were in the internal shape. The early Blas- and Stücköfen were square-sectioned and parallel-sided like the small German blast furnaces

Table 45 17th–18th century charcoal blast-furnace slag composition, %

Element	Sharpley Pool (Worcs.)[7] 1652	Coed Ithel (Gwent)[8] 1651–?	Melbourne (Derbys.)[9] 1725–c.1780	Duddon (N. Lancs.)[10] 1736–1866	Low Mill (Yorks.)[11] 1761–?
FeO	2·7	4·75	–	2·6	16·2
Fe$_2$O$_3$	nil	–	2·6	–	–
SiO$_2$	49·3	62·8	41·6	56·4	57·8
Al$_2$O$_3$	11·4	7·3	22·7	12·4	18·6
CaO	22·8	15·9	14·1	14·6	0·7
MgO	12·0	8·4	14·2	3·6	–
P$_2$O$_5$	tr.	0·13	0·023	–	–
S	tr.	0·01	c.0·1	–	–
MnO	0·84	0·40	3·01	9·8	–
K$_2$O	2·0	–	–	–	2·7
TiO$_2$	–	0·3	–	–	1·2

– Not determined

used for non-ferrous smelting. The Siegerland Blauofen of the 16th century was about 4.5 m square externally, and 1.7 m square internally,[5] while the English furnace of 1542 at Panningridge was 5.2 m square at the base, and 1.2–1.5 m square internally.[6] Of course, a lining could have been built inside the latter to give a crucible not more than about 0.3 m square. We do not in fact know precisely when the shallow low-angle bosh came into being, but it seems that the Siegerland Blasofen (Hoherofen) had this form by about 1550. In England this form is typified by the late 17th century furnace at Allensford.[14]

Most 16th century blast furnaces have two openings in the sides; one for blowing and the other for tapping. The blast was supplied by a waterwheel operating two bellows which were alternately compressed by cams. The waterwheel was large and narrow and usually overshot (*see* Fig.65). The power required would not be high, probably no more than 1 hp, and this could be obtained from a wheel 3 m dia. by 0.30 m wide, running at a speed of 6 rev/min and consuming 1.7–2.5 m³/h of water. The important thing was continuity, partly because a large stone structure such as a blast furnace could not stand the frequent heating and cooling involved in bloomery operations, and partly because intermittent operation of such a large unit would, thermally, have been very wasteful. The campaign of the Bourbon furnace of 1517 was two months during which, unlike the bloomery, it was in continuous use, running day and night. This usually meant the building of a large pond with adequate storage to cover periods of low rainfall.

Some of the blast furnaces had only one opening, the front opening, e.g. that at Pinsot near Allevard in Savoie.[15] In this they were like the Stücköfen which was blown, and the bloom extracted, from the front (*see* Fig.70). Some furnaces were blown with a device known as a trompe which used air drawn into a falling jet of water. When the water with its entrapped air had reached the bottom of its fall it yielded up its air to the furnace through a pipe which led to the tuyere in front. This principle was also used in the Catalan hearth (*see* Fig.72).

The output of a mid-16th century furnace seems to have been only about 4–5 t of cast iron per six days (a founday). A furnace could not store such large quantities of cast iron and this limited capacity led to the early development of the double furnace such as those at Worth in Sussex which, in 1549, were capable of producing gun castings of up to 2 200 kg weight.[16]

The lining of the hearth built into the shaft needed renewal at the end of every campaign, and it was for this reason that almost all blast furnaces have been built with a self-supporting shaft or stack and a separate hearth, which was built up from the bottom to meet the shaft at the bosh (*see* Fig.65). All the earliest blast furnaces were essentially of this type but probably of square section, and we know that some 17th century furnaces carried on this tradition of two truncated pyramids meeting at the bosh, but by that time the hearth had become conical to avoid the problem of material accumulating in the corners. In fact, at Coed Ithel (*c*. 1650) we can witness the difficult problem that the builders had when a conical hearth met a pyramidal shaft.[17]

It would seem that some 16th century furnaces consisted of three portions; a square-sectioned crucible, a short inverted pyramidal bosh section, and a pyramidal shaft (*see* Fig.66). In the course of operation the crucible and the pyramidal bosh section gradually became one: presumably, the restricted 0.30 m wide crucible led to easier working with a single tuyere. The tuyere seems to have been a simple iron tube (tue-iron) about 5 cm dia. grouted into a tapered brick or masonry opening in the side. This opening was corbelled out or arched like the tapping opening to take the ends of the bellows, which

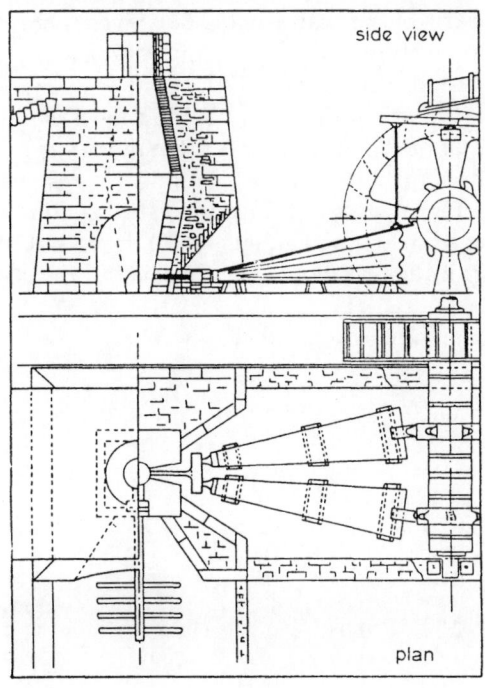

65 *A blast furnace of the steep-bosh type (after Morton* [20] *with slight modification*

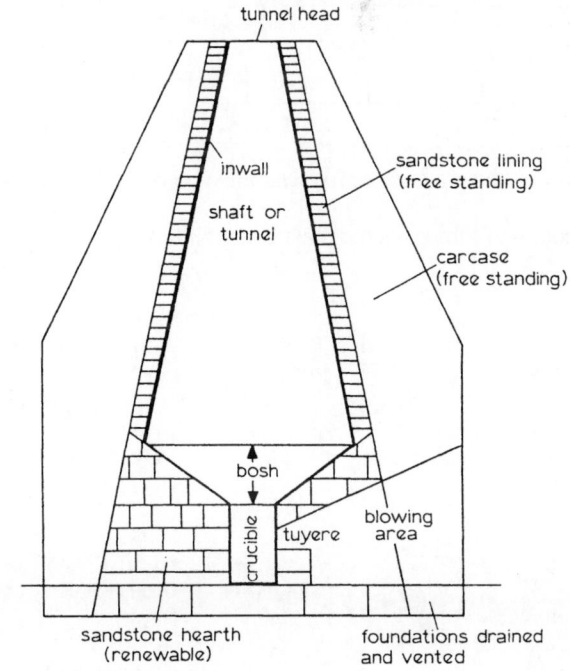

66 *A blast furnace of the shallow-bosh type (after Swedenborg* [19])

were from 3.6 to 6 m long. The tuyere would be about 0.30 m above the bottom of the hearth and its level set the limit to the level of the metal and slag.

It would appear that the earliest blast furnaces were tapped directly from the square crucible by making a hole in the side and stopping it with clay when the furnace was empty of liquid. But, by the mid-16th century, the forehearth had appeared in non-ferrous smelting furnaces and the principle was soon to be applied to blast furnaces to give the tymp and dam (*see* Fig.67). This meant an effective enlargement of the capacity for storing molten metal and slag, and removed the tapping holes ('notches') a convenient distance away from the hottest part. The surface of the forehearth would usually be covered with a layer of solidified slag. Tapping was effected by cutting a notch in the corner of the dam stone and afterwards temporarily stopping it with clay. The tymp was probably originally arched but would quickly burn away: it was soon to be modified with a bar and plate of wrought iron.

We know something of the consumption and production of at least two 16th century furnaces, at Robertsbridge and Panningridge in the Weald of Kent.[18] Unfortunately, the units are sows of iron and loads of charcoal and there is some doubt as to the meaning of these. It would seem that a sow weighed 10 cwt, or 500 kg. The figures shown in Table 46 are consistent and indicate a fuel/ore ratio of about 1:1, and a fuel/iron ratio or 'coke' rate of about 5:1 if it is assumed that a load

Table 46 Charcoal and ore requirements per unit of iron in 16th century blast furnaces (after Crossley[18] and Schubert[45])

Material	Robertsbridge 1559–62	Panningridge 1546	1546	1551–6	Newbridge
Cast iron	1	1	1	1	1
Charcoal	4·6	5·0	4·9	4·7	5·5
Ore	5·7	6·0	5·9	5·7	7·0
Fuel/ore	0·8	0·83	0·83	0·82	0·79

Assumptions: 1 load of ore or charcoal = ½ t; 2 sows of iron = 1 t

of charcoal and ore weighs 500 kg and a sow the same. There is no mention of limestone or roasting of the ore and we can assume that the burden was self-fluxing, which is in keeping with the analysis of ore and slag found at Panningridge[6] (for ore type, *see* Table 28, p.50).

The iron, in spite of its low silicon content (*see* Table 47), which is typical of cold-blast charcoal iron, was usually grey owing to its slow cooling rate. Wealden iron had comparatively high phosphorus, as did most of the later coke iron of Britain.

THE SECOND PHASE: AD 1600–1720

After 1600 there is a good deal more information to call on. In Britain, three furnaces are still standing and the foundations of two more have recently been excavated. It is clear that the furnace size has been increased; the base of that at Sharpley Pool[7] was 7.2 m square instead of the 5.1 m at Panningridge,[6] and the heights by inference have increased from 6 to 9 m. According to Swedenborg,[19] the furnace at Lamberhurst in Sussex was 8.4 m high. This was clearly one of the narrow crucible shallow-bosh type and seems to have had a tapered exterior to reduce the amount of masonry in the upper parts of the shaft (*see* Fig.66). Swedenborg's furnace is really typical of a 16th century type which, as we know, lasted well into the 17th century. But we know from the furnaces at Sharpley Pool,[7] Cannock,[20] and Coed Ithel[17] that the bosh angle was often much steeper. The latter three furnaces give 80°, 78° and 77°, respectively, and from what we see of Coed Ithel are essentially more like that from Schmalkaldens[13] (84°). It did not require German influence, however, to develop this type in

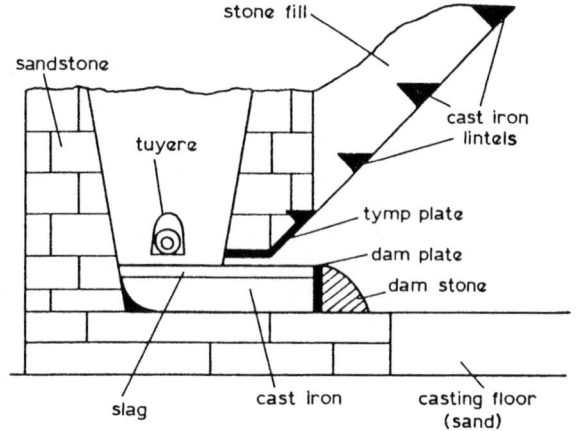

67 Details of a blast furnace hearth

Table 47 Composition of cast iron artefacts

Provenance	Object	Date	Total C	Si	Mn	S	P	Type of iron	Reference
Liége, Belgium	Fireback	1548	3·59	1·14	1·58	0·03	0·62		45
Hastings Museum, UK	Fireback	1586	3·65	0·52	0·42	0·086	0·56		"
	Fireback	1642	3·65	1·00	0·92	0·06	0·55		"
	Fireback	1683	3·58	0·56	0·86	0·074	0·61		"
	Fireback	1707	3·99	0·65	0·82	0·048	0·47		"
Saugus, USA	Kettle rim	c.1640	3·67	0·77	0·37	0·094	1·21		23
	Crane hook	1600±60	3·70	0·74	1·15	0·05	0·72	Grey	132
Quincy, Mass.	Fragment	1644–7	3·59	1·73	0·51	0·04	0·85	Grey	132
Sharpley Pool, UK	Pig	1652	3·9	0·49	0·05	0·068	0·31		7
Duddon, UK	Pig	1736–c.1866	4·3	0·65	0·10	0·023	0·124	Mottled	10
Nibthwaite, UK	Pig	18th century	3·73	0·85	0·05	0·029	0·11		44

Britain as it was the natural conclusion to be drawn from the wear of the crucible-type hearth.

The furnaces at Gunns Mill in the Forest of Dean (1683), and Rockley in Yorkshire (1652), are two of the remaining furnaces of the pyramidal shaft type. In both cases the hearth is missing but, from the position of the bosh above ground level, we can safely say they are also examples of the narrow crucible shallow bosh type. The furnace at Rockley[21] has a third opening which may have been designed for a second tuyere but it is more likely to be an inspection hole. By the middle of the 17th century the declining fuel reserves in Britain had led to the fostering of the iron industry in the British Colonies and in Ireland, and the works at Saugus, Mass., were started in the 1640s by Winthrop.[22] Previously, in 1584, Walter Ralegh had found ore deposits off the coast of North Carolina, while in 1609 ore was supplied from Jamestown, Virginia, for a trial smelt in Britain. The works at Saugus[23] were based on bog ore which, because of its porosity, is comparatively easy to smelt. The burden was fluxed by the addition of rock ore from Nahant. The blast furnace was 8 m square at the base and about 6.4 m high. It was powered by an overshot wheel 4.9–5.2 m dia., containing 50 buckets 30 cm apart, and capable of giving 2 hp at 8 rev/min. The crucible was 46 cm square and is believed to have had a height of 1.07 m and a capacity of 1 t which was tapped once every 24 hours. Sows have been found 71 cm long, 11 cm thick, and weighing up to 230 kg.

Although coke smelting was beginning to be introduced into England about 1710–20, some charcoal-fuelled furnaces were still being built. Again we see the use of the two distinct types. The furnace at Melbourne in Derbyshire was built in 1725 and blown out in c. 1780: it seems to have been of the vertical-crucible shallow bosh type.[24] On the other hand, the furnace at Low Mill in Yorkshire,[25] which was blown-in in 1761, has a bosh angle of 80°.

In the development of European furnaces we see the two basic types in competition. We have examples of both the shallow bosh furnace typified by Jars[26] at Johangeorgenstadt on the Saxon side of the Bohemian border, and the high steep-sided bosh furnace that was common in Britain. The shallow bosh type has been criticized as being 'demodé and the construction a little inappropriate'. It must have been very difficult to build and maintain, yet it still had its adherents in Bohemia in the 19th century.[27]

The charcoal furnaces of the 16th and 17th centuries continued in use in some parts of Europe well into the 19th century, until the availability and price of charcoal led to the inevitable closing down or change to coke.[20] However, by this time most of the furnaces were too small for economic coke working and were finally blown out. There were occasional exceptions, e.g. a charcoal furnace at Backbarrow[29] in North Lancashire carried on with charcoal until 1920, and then had a short period on coke until it was finally blown out in 1966. But the general change to coke smelting started in about 1720, and will be discussed in the next chapter.

Far eastern blast furnaces

The priority of the Chinese in the use of cast iron and the development of the blast furnace cannot, apparently, be questioned. But by the Ming period (1368-1644) all development seems to have stopped, just when there was a steady increase in size in Europe. The Chinese blast furnace was wide and stubby with a wide and open top, if we accept the four well-known drawings by artists of the period 1313–1637.[30] These are very like the furnaces used to melt bronze in the LBA (*see* Fig.56). In the drawing of 1334 we see a short furnace about 1.8–2.4 m high tapering near the top but with a maximum diameter about $^{3}/_{4}$–$^{7}/_{8}$ of the height. The tuyere is nearer the bottom than in the early bronze melting furnaces. In the drawing of 1637, the diameter of the furnace is as large at the top as it is high, and the taphole is nearly halfway up which makes one suspect that it is being operated as a cupola, i.e. melting iron and not smelting it.

The blast furnaces used in Chiangsi in about 1900 were still little more than man-height (1.8–2 m); they could be tipped through 30° to facilitate tapping. Using blackband-type ironstone and coke they could produce 600 kg/day of cast iron with high sulphur and silicon. The traditional blast furnace of Yunnan (1879) was 3.3–3.7 m high, of masonry construction, and could make 800 kg of cast iron per day. This furnace had a steep-sided crucible with an angle of 80°, but no bosh in the sense that there was no reduction in diameter (throat) at the top of the furnace. This was blown by an inclined tuyere at the back opposite the taphole (*see* Fig.68). The effect of widening at the throat would reduce the rate of gas flow and avoid the blowing out of the fine charge of magnetite sand which was often used in these furnaces.

The long standing tradition of the blast furnace in the East was not confined to China. We have a detailed record of the making of cast iron in the Philippines in 1903.[31] The external appearance of the furnaces is just like the Chinese type depicted in the Ming paintings. The internal diameter increased from 0.7 m to 0.82 m at the top (*see* Fig.69). The internal height was 1.8 m and the tuyere entered from the back with a slight downward inclination like the Yunnan furnaces. The height from the bottom of the hearth to the 38 mm bore tuyere was only 10–15 cm. The tuyere was of best fireclay, 15 cm o.d. and 80 cm long. The furnace was made of sun-baked fireclay, sometimes from bricks and sometimes monolithic, but better grade material was used near the hearth. The similarity with the Roman bloomery furnaces is striking.

The ore was broken to 4 cm size by means of a hammer and was smelted unroasted, with charcoal, and without flux. The volume ratio of fuel/ore was from 8:1 at the start, to 4:1 when going well. This would suggest that the fuel/ore ratio was about 1.5:1 by weight. The slag was free running and contained some iron; it was tapped every 2–3 min. The iron was tapped into a forehearth every 2–3 h and then ladled into moulds for cast implements. The output seems to have been around 2–3 t/month of castings. This sort of tradition was the basis of the recently reintroduced 'backyard' furnace

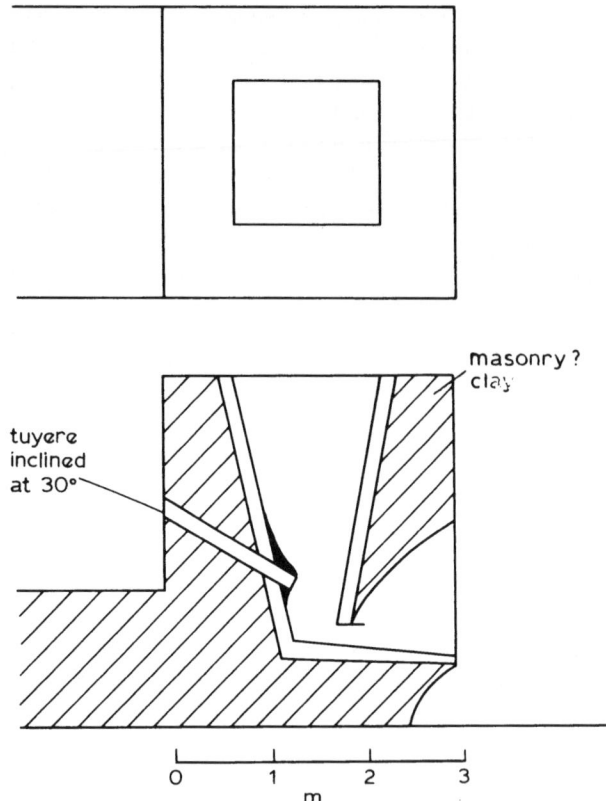

68 Yunnanese blast furnace of 1879 (after Needham[30])

which tided the Chinese over a difficult period in the 1950s.[54] Coal was used as fuel in these: Chinese history shows a considerable knowledge of the metallurgical effects of the various fuels, coal being used exclusively or blended with charcoal when the product required it.

The fact that development tended to stop in the Ming period has given us many useful relics of early Chinese

practice. The need for wrought iron and steel has given us details of fineries that survived into the 1950s.[32]

The development of the bloomery

The introduction of the blast furnace did not at once render the bloomery obsolete. This was partly because the majority of metal consumed was still wrought iron, which was required because of its malleability, and this meant that another step—the finery—was needed to convert cast iron into wrought iron. This required a further consumption of charcoal and the loss of $^1/_4-^1/_3$ of the cast iron in the finery process. In theory, the slag responsible for this loss could have been put back into the blast furnace, but it was often in such unwieldy pieces and the finery so far from the furnace that this was not practicable. In most countries it was still worthwhile to produce a large proportion of the malleable iron by the direct process.

There is documentary evidence for the high bloomery furnace at St. Gallen as early as ad 1074.[33] In Austria, and to a certain extent in some other countries, the developments which gave rise to the blast furnace were used to improve the bloomery, and by the 15th century the Stückofen had been evolved. This was as high as the blast furnace (3 m) but, by reducing the fuel/ore ratio, wrought iron was produced instead of cast iron. It was thus more economical to produce a tonne of malleable metal in this way as there was no additional cost of fining. However, the method of working was more complicated. Instead of the 90 kg blooms of the low bloomery, by the 15th century the Stückofen was producing 'Stücke' or blooms weighing up to 400 kg each day. In order to remove these the bellows were first removed, the front of the furnace opened up, and the Stück dragged out with the aid of a winch.

The furnace was first square in section with parallel sides like the blast furnace, but later it became 'D' shaped and the shaft tended to narrow towards the top (see Fig. 70). The disadvantages of the two stage or indirect process became less with the gradual improve-

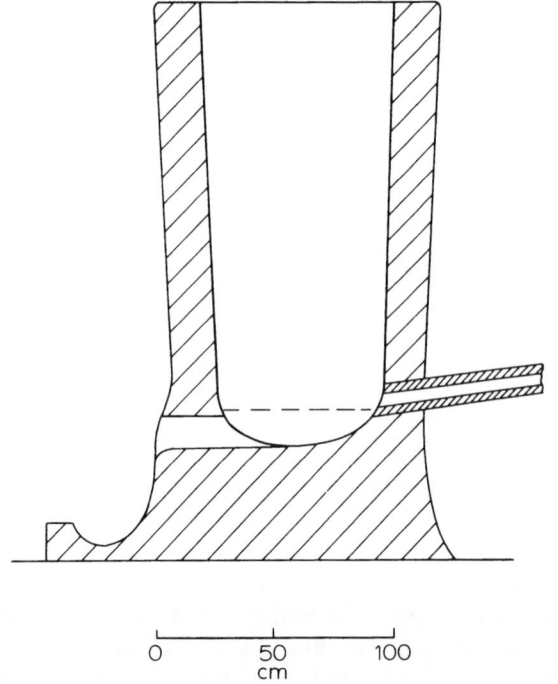

69 Philippine blast furnace of 1903 (after McCaskey[31])

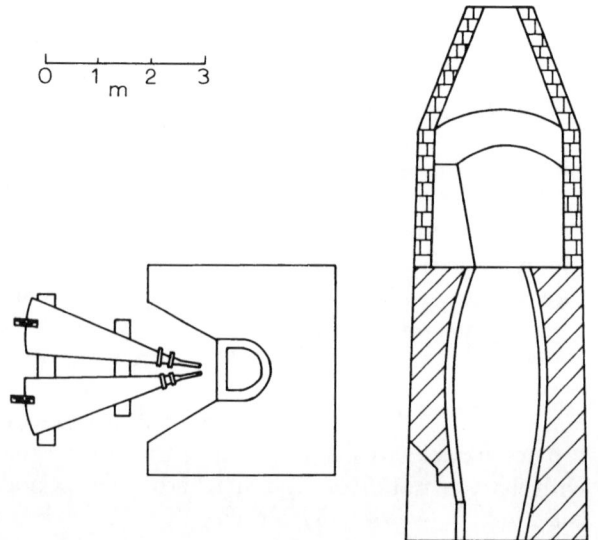

70 Styrian Stückofen (after Jars[26])

ment of the blast furnace, and between 1750 and 1767 all the remaining Styrian Stückofen were converted into blast furnaces.

In Britain and Spain, and in some parts of the Continent, the low bloomery continued in use after the introduction of the blast furnace, perhaps because of the capital required to build the more efficient Stückofen. This type of bloomery therefore had considerable advantages over the high furnaces and the blast furnace in particular. It developed in Britain into quite a deep, round furnace[26] up to about 0.7 m in internal height (*see* Fig.71), and such bloomeries were still being commissioned as late as 1636 in the more remote parts of the British Isles.[35] Of course, the height was limited by the need to withdraw the bloom through the top, rather than the side in the case of the Roman-type bloomery and Stückofen. In Catalonia and the northern flanks of the Pyrenees it was powered by the trompe, a form of blower in which air was injected into a falling stream of water (*see* Fig.72). The entrapped air and the water were separated in a container at the bottom and the air fed under pressure to the tuyere. The Catalan hearth was

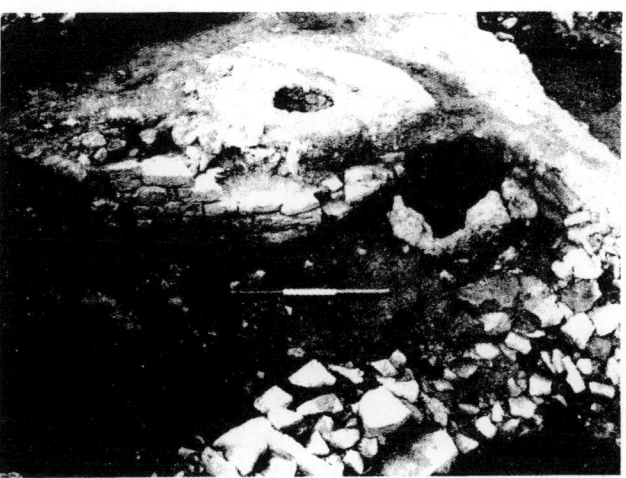

71 *Blookmery at Rockley, near Sheffield, after excavation (courtesy of D. Crossley)*

low and square in section, i.e. about $0.6 \times 0.4 \times 0.52$ m high.[36] This type of plant only went out of use in the early years of the 19th century in Europe, and perhaps rather later in the northern parts of New York State.[37]

72 *Catalan hearth and trompe (Crown Copyright, Science Museum, London)*

In Norway and Sweden the bloomery process was in use into the 19th century and we have very detailed descriptions.[38] Swedenborg[19] shows a relatively high bloomery furnace from which the bloom was extracted via a hole at the bottom. Evenstad,[39] later in the 18th century, shows a somewhat lower bloomery furnace from which the bloom and the slag were removed through the top (see Fig.73).

The finery and the chafery

The finery carries out the first stage of conversion from cast iron to wrought iron.[40] The aim here is to reduce the carbon content of the cast iron (usually in the range 3–4%) and the silicon when present, to very small amounts, usually <0.05%. This is done by melting the cast iron and oxidizing it in front of a tuyere and for this purpose sows, pigs, or 'plats' were cast at the blast furnace. In about 1750 a sow would have weighed about 500 kg and would have to be broken into smaller pieces. The plats were smaller pieces intentionally cast for this purpose.

Two versions of the finery process are known. The forges in the greater part of Germany, Austria, and Italy carried out the whole process in one hearth. In Belgium, Luxembourg, France, and Britain two hearths were used; the first, the finery proper, did the conversion work and the second was reserved for most of the reheating for forging and named the 'chafery' from the French verb chauffer, meaning to heat. This two-hearth process is generally known as the 'Walloon' process.[41] It would appear that it was developed so that charcoal could be saved by using coal in the chafery hearth. This is possible owing to the reduced area of contact between the fuel and the iron in the final stages of the process, which reduces the risk of sulphur pick-up.

There is little doubt that the single hearth process was the earliest and was developed from the bloomery. After oxidation of the impurities in the cast iron the consolidation of the 'fined' metal was continued in the same hearth, which was wasteful since some of the time the 'loup' or semifinished bloom was being forged under the hammer. Even in the Walloon process, use was made of residual heat in the slag at the bottom of the finery hearth to reheat the loup while a new pig was being oxidized in the top part of the hearth near the tuyere. However, all later reheating was done in the chafery.

Not much is known about the detailed mechanism of the finery process. In Britain, a finery hearth has been

excavated[42] and it is therefore possible to build up a fairly reliable picture from the actual evidence (see Fig.74), and such descriptions as that of Plot in his Natural History of Staffordshire in 1686.[43] To this may be added more recent descriptions of l9th century fineries.

The Walloon finery consisted of a charcoal hearth blown with bellows and a tuyere. The bellows seem to have been water powered right from the beginning, and the air required was probably as much as that required by the blast furnace. Charcoal or other low-sulphur fuel was needed: coal could not be used for the finery although it was often used for the chafery from about the 16th century onwards. The finery and chafery were usually placed together, probably so as not to waste the residual heat in the fined bar (loup).

The burning of the charcoal in front of the tuyere produced two zones, one oxidizing and the other reducing. The maximum temperature was obtained immediately in front of, and above, the oxidizing zone. Here the pig iron was melted and, on entering the oxidizing zone, the silicon in the pig would be rapidly oxidized (see Fig.75). The major portion of the carbon was in the form of graphite and the removal of silicon allowed some of this carbon to replace it as iron carbide. The desiliconizing step was later to be carried out in a separate ' refining' hearth or 'running-out fire'.

The partially oxidized, still molten, iron fell into a slag bath below, composed of the silica and iron oxide (i.e. fayalite–$2FeO.SiO_2$). The pasty mass of iron was raised on the end of a bar remaining from the previous heat, into the oxidizing zone when further oxidation or 'fining' would take place.[44] In the course of an hour's working the oxidized metal would become still more

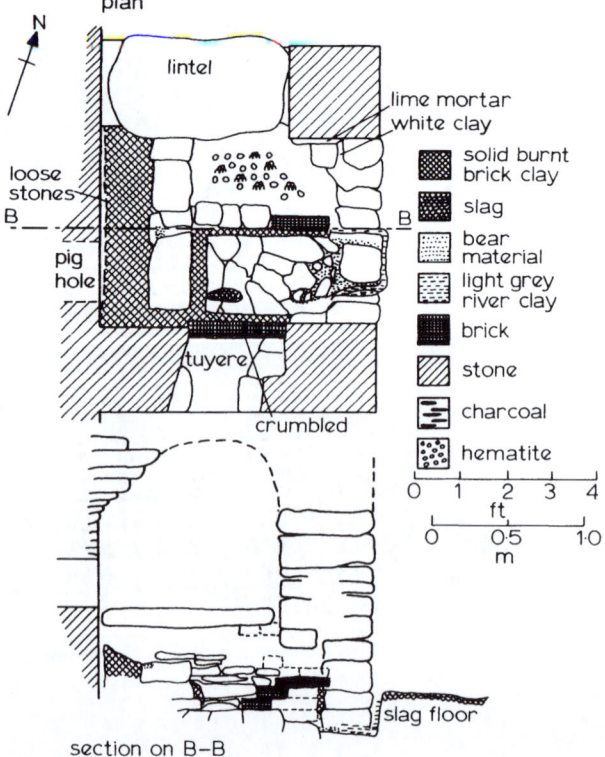

plan

N

lintel

loose stones

pig hole

tuyere

crumbled

B — B

lime mortar white clay

- solid burnt brick clay
- slag
- bear material
- light grey river clay
- brick
- stone
- charcoal
- hematite

0 1 2 3 4
ft
0   0·5   1·0
m

section on B–B

74 Details of finery hearth excavated at Stony Hazel, North Lancashire (from Davies-Shiel [42])

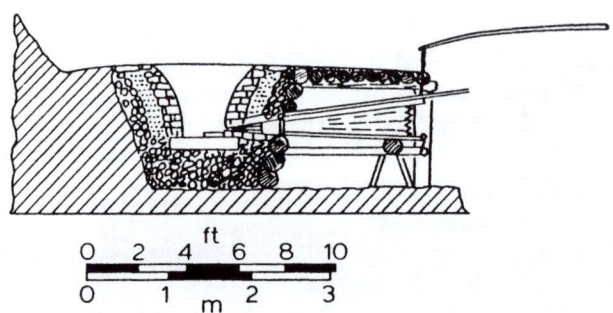

0   2   4   6   8   10
ft
0   1   2   3
m

73 Scandinavian bloomery furnace (after Evanstad [39])

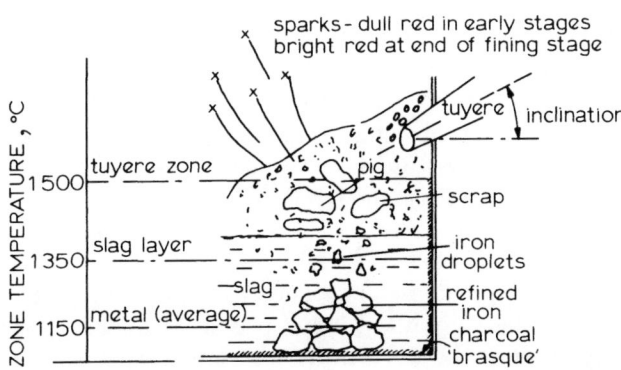

75 Hearth conditions during the refining stage of the fining of grey cast iron (after Morton [44]; courtesy of The Metallurgist)

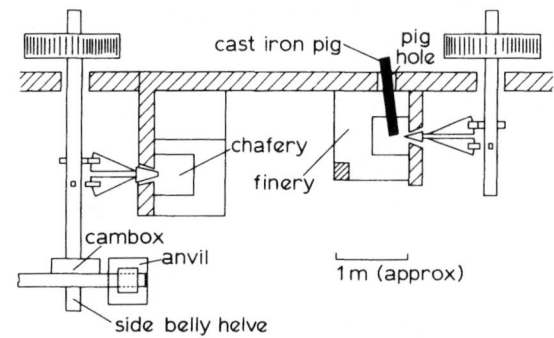

76 Reconstruction of a Swedish Walloon forge from Hillestrom's painting of 1793

solid as its melting point increased with decreasing carbon content, and by this means the finer would know when the process had been completed.

The amount of slag formed depended on the silicon content of the pig and, when this was low, or when starting up, additional slag or slag-forming material would be necessary. This would come from hammer-scale, sand, or even added iron ore. These fayalite slags (*see* Table 48) are very active and it was soon found that the best way of protecting the hearth was to use cast iron plates often coated with some charcoal-clay refractory. From time to time surplus slag would be run off, producing plano-convex slag cakes.

By the 17th century the finery was usually combined with the chafery and the two together, along with a water-driven hammer, made the 'forge' (*see* Fig.76). Before fining was finished the loup was forged into a thick square 'halfbloom', and then returned to the finery for a further hour for reheating. This was then forged a second time, resulting in a dumbell-shaped bar with one end, the 'mocket-head', being much greater than the other which was called the 'ancony' end.[45] The bloom was then transferred to a chafery where the ancony end was first heated and forged into a bar. The mocket head was then forged down in two stages with intermediate reheating.

The chafery fire was of a quite different form from the finery. Of course, its function was entirely that of heating

and no chemical changes were intended, and for this reason coal could be used. The heating zone was made into a hollow beehive shape above the hearth which was much larger than that of the finery, and the fire was made of small fuel, slag, and clay: the whole could be likened to a muffle furnace. The heating zone was much larger than the finery and heat was produced by oxidation of the fuel in front of the tuyere and by combustion of the hollow walls of the muffle which would radiate downwards on to the work. It is estimated that temperatures in the region of 1 400–1 450°C were obtained in these fires.

In construction, the finery and chafery hearths were very similar and both required cast iron plates. The volume of slag produced in the former was very much greater than in the latter, and this was mainly due to the oxidizing nature of the process. The overall loss of metal during conversion was as high as one-third. In about 1630, at a South Hereford forge, 1 300 kg of cast iron was needed for every 1 000 kg of bar iron produced.[46] The slag produced in the chafery is of similar composition to the finery slag, but very often one comes across evidence of the use of coal because of the high sulphur content (*see* Table 49). This slag was run out into hollows in the floor to form plano-convex lumps; those with runners attached have often been described as 'ham bones' and weigh as much as 25 kg.[47] Sets of plates for the hearths were cast at the blast furnaces. In 1591, Rievaulx furnace in Yorkshire supplied two finery bottoms weighing a total of 200 kg.[48]

The mechanical hammer

Agricola[2] showed the use of a wooden sledgehammer for the removal of adhering slag and charcoal from the bloom, but a mechanical hammer was shown for the working-up of pieces of the bloom into wrought iron. It is unlikely that a hammer large enough to break up or work a 100 kg bloom existed until the 18th century, and it is almost certain that the job of breaking up blooms was done hot with axes, as shown by Swedenborg in connection with the so-called Osmond process in Sweden.[19] It is often thought that the only type of mechanical hammer in use before the 16th century was the 'oliver' or treadle-operated tilt hammer.

Two olivers were in use in Yorkshire in 1368, and a large hammer is referred to in North Wales in 1335

Table 48 Composition of finery slags (after Morton[44])

| | Amount, % | |
Element	Stony Hazel N. Lancs.	Lowick, N. Lancs.
FeO	56·20	65·10
Fe$_2$O$_3$	3·20	9·10
SiO$_2$	24·20	11·20
CaO	2·50	3·10
MgO	0·90	0·29
MnO	0·13	3·83
Al$_2$O$_3$	8·90	3·50
P$_2$O$_5$	0·71	2·59
S	0·22	0·10
TiO$_2$	–	0·23
Fe (metal)	1·90	0·75

Table 49 Composition of chafery slags (after Morton[44] and Morton and Gould[46])

Element	Amount, %		
	Little Aston, Staffs.	Ipsley, Worcs.	Nibthwaite, N. Lancs.
FeO	45·20	58·80	65·53
Fe$_2$O$_3$		9·90	9·43
SiO$_2$	22·62	17·80	16·16
CaO	Nil	2·60	2·00
MgO	0·15	0·22	1·20
MnO	0·06	2·50	–
Al$_2$O$_3$	13·38	4·90	4·30
P$_2$O$_5$	–	2·01	0·23
S	0·08	1·50	0·47

which may have been an oliver. The oliver is merely a device for transferring the power from the hand to the foot, thus freeing the hands for other purposes, and also power is more easily applied by the foot. The smith can then hold the work with tongs in one hand and a drift or set in the other the tool being hit by the foot-operated hammer.[49] It is thought that the trip, tail or tilt hammer was the first water-driven hammer. The tip of the shaft was hit by cams in the same way that the boards of the bellows were, thus raising the head and allowing it to fall by gravity. However, it is known that the belly helve or lift hammer was in use in the Siegerland as early as 1467,[50] and it is this type that Agricola shows in connection with his illustration of the bloomery process. This type of hammer has its waterwheel shaft extended parallel to the hammer shaft and terminates in cams which raise the hammer shaft midway between its pivot and its head. It seems to have had a lower rate of working

than the tail helve but was capable of lifting a larger head (300–320 kg) at a rate of 70 strokes/min (*see* Table 50 for examples).

With water driven hammers the speed could be regulated by controlling the water. The normal working rate was inversely proportional to the weight of the hammer head. On the whole, tail helves were used for rapid blows—up to 200/min—with light heads of 50–250 kg, and strokes of 0.25–0.6 m. These were often used for plating, i.e. making plate or sheet from blooms or bars. The simple tail helve was used in the Catalan smithy into the l9th century and could then raise a head of 600–670 kg at a rate of 100–125 strokes/min. The belly helves or lift hammers were intermediate and had head weights of 200–400 kg, and strokes of 0.55–0.80 m. Usually, they had only 4 cams and worked at a rate of 100–120 strokes/min.

A variation of the belly helve is the side belly helve in which the wheel shaft is at right angles to the hammer shaft, and the cam box raises the hammer shaft by means of a projection on its side. The effect of the falling head was enhanced by the drome beam against which the head of the hammer struck on its return stroke. This acted like a spring (*see* Fig. 76). Another was of increasing the blow used in the tail helve was to insert an iron recoil block on the floor under the tail.

Finally, the nose helve was developed in which the cams raised the nose of the shaft, i.e. the end remote from the pivot, and the hammer head was placed a short distance back along the shaft. The nose helves were the heaviest and had head weights of 50–400 kg.[51] All hammers had to have extensive foundations for the anvil as well as for the frame of the hammer. The anvil was usually supported on cross baulks of timber to

Table 50 Water-powered hammers

Date	Site	Type*	Head weight, kg	Cams	Max. rev/min	Blows/min	Reference
1540	Germany	B		4			Agricola[2]
1791	Cramond	T	150			120–160	Schubert[45]
1591	Rievaulx		250				Schubert[45]
	Staffs.	B	300				Schubert[45]
1713(?)	Wortley	B		6	36	200	–
c.1765	Swalwell	B	280	4	–	–	Jars[26]
c.1767	Larvik (battery)	T	320–360	4	20	66–80	Jars[26]
1773	Périgord	N		4	32	128	RHS[49]
c.1765	Swalwell	B	280	4	–	–	Jars[26]
c.1784	Pontypool	T	150	4			Schubert[45]
18th C.	Luxembourg	B		4			RHS[49]
1831	English	B	360–410	4		150	Lardner[134]
1840	Pyrenees	T	600–670	4	30	100–125	Percy[36]
1850	Wortley	B		4	30	100	–
19th C.	Kirkstall	N		4			RHS[49]
19th C.	Périgord					90–120	"
1858	Dahn	T					"
1858	Dahn	T		12			"
	Defrance (painting)	B		4			"
19th C.	Cotatay	T		15			"
19th C.	Bracco (Tende)	T	45	6			"

*Types B; Belly helve: T; Tail or tilt hammer: N; Nose helve

spread the load over a large area of ground. No doubt additional timbers were added from time to time to make up for any sinking of the foundations. These hammers were worked by beam engines in the 19th century. The final form of the belly helve was the Krupps beam hammer of 1852, in which the cam wheel was replaced by a fast acting steam piston placed between the hammer shaft and the floor.[52]

Slitting and rolling mills

While the hammer was the main implement for working iron up to end of the 18th century, it was not the only one. Although the rolling mill was not very common, rotary action in the form of the slitting mill was often applied for the cutting of plates or flat bars into rods or nails. This was essentially a continuous multiple rotary shear. It may have developed from the small handmill used for rolling lead cames for glass windows, the earliest description of which is dated to 1568. It would appear that rotary shearing was introduced by the beginning of the 17th century.

The first description of a slitting mill is that given by Plot[43] in his Natural History of Staffordshire of 1686. However, the Saugus Iron Works of 1647 has recently yielded a piece of partly slit bar which is the first product of a slitting mill to have been metallurgically examined.[53] The bar had a section of 6.5 × 0.74 cm and had been partly slit into 10 pieces varying in width from 0.5–0.8 cm. The material was poor quality wrought iron with slight surface carburization on one side and segregation of phosphorus leading to banding. The bar would not seem to have been heated above 910°C in some places, thus providing good reason for the accident which gave rise to its rejection. This piece of iron gave some indication of the poor condition of the slitting discs and the mill in general.

According to Plot[43] the bars were first cut into short lengths by pivoted cold shears and heated in a furnace, which was probably coal-fired. They were flattened and reduced to some extent 'between rollers', and then put between the cutters of the slitting mill. The first drawing of a slitting mill was made by Anton Schwab in 1723 but was not published until 1966.[53] This was used for strips for copper coin blanks. The first publication of a drawing of a slitting mill for iron was that shown by Swedenborg in 1734. Sven Rinman made a drawing, from the period 1750–60, which was not published and was followed by the drawing in Diderot's Encyclopédie of 1765.

It is clear that the two sets of discs or rollers were originally worked independently from separate waterwheels. There was no screwdown or method of adjustment of either mill; initial adjustment of the flat rolls was made by wedges. According to Schwab's drawing, three discs were placed on a square-sectioned shaft with two spacer discs, all discs being clamped together by four bolts. The lower assembly consisted of four slitting discs with spacers assembled in the same way. The three top discs meshed in the spaces between the four lower discs, and guides were inserted to make sure that the slit rods did not get entangled with the discs

and wrap themselves round the spacers between the discs.

The limit of size was largely governed by the size of the wrought iron members for the housing. From the scale of Schwab's drawing it would appear that the overall height was about 1 m. The discs were about 30 cm dia. and must have been made of steel, although it has been suggested that they were merely steel-ringed. The Saugus bar shows that there was considerable wear on the slitting discs and also that they were loose on the square shaft. As for the rolling mills, plain rolls for a pair of rolls and forming part of a slitting mill near Wetzlar, were cast at the local furnace in 1606–7.[54] The rolls of the same mills working in 1758 were 20 cm dia., while a mill in Birmingham in 1755 was capable of hot rolling 7.5 cm wide bars with a reduction of 75%, i.e. with an increase of length from 0.3–1.2 m. Such a reduction would need considerable power and a strongly built mill.[54] Smeaton's design for Kilnhurst in 1765 included waterwheel(s) of 5.5 m dia. 2 1.3 m wide, similar in size to those used for hammers.[45] The power available from two such wheels would be about 20–30 hp (50 hp is now considered sufficient to power a two-high mill with rolls 25 cm dia. and 40 cm long). Obviously, developments of this sort depended on the substitution of metal for wood whenever possible. There is little doubt that the housings of the 18th century rolling and slitting mills were metal, and therefore illustrate another use for that newly developed metal-cast iron.

The production and use of steel

During the medieval period, steel was produced either by cementation or direct from the bloomery process. Biringuccio[55] tells us of a different process, that of so-called 'cofusion'. First, some pieces of broken cast iron are melted in a bloomery or smith's hearth with an inclined tuyere, under a slag cover. Marble ($CaCO_3$) is one of the constituents of this slag. Three or four forged blooms of 14–18 kg weight, each of wrought iron, are added to the crucible and kept covered for 4–6 hours, during which time they absorb carbon from the cast iron. The temperature must be about 1 150–1 200°C, and after a certain time the pieces are taken out, quenched, broken up, and then returned to the bath. In this way a more even carburization is ensured instead of the surface carburization of large lumps. The completely carburized pieces are then removed and quenched into water to demonstrate their hardness. Agricola repeats this description almost word for word. Rehder[56] found that the time of immersion must be severely limited and the temperature kept as low as possible, otherwise the carburized surface is redissolved by the cast iron. Clearly, the smaller pieces, the more effective the process would be, and 14–18 kg seems much too large and 4-6 hours much too long.

Steel was still used sparingly in most countries and, because of the need for a low phosphorus and sulphur content, had to be made from pure charcoal iron which was often imported from the Scandinavian countries. Naturally, we see its use mostly in edge tools such as swords, knives, and scythes. The finery process could produce steel by stopping it before the complete decarburization of the cast iron.[57] For this purpose, the cast iron was cast into plate or flat bars and fined. To produce 57 firkins of steel (1 firkin = 330 kg) 27.5 t of cast iron plates were required, representing a yield of 70%.

The main steelmaking process was cementation, and during this period we witness the move from localized cementation of the individual tools in the smith's hearth to the production of large quantities of steel by cementation in specialized works. Swedish iron was obviously ideal for this and in some cases Swedish steel itself was imported. In Britain, this was supplemented by imports of charcoal iron and steel from Russia.

Cementation was now carried out in large kilns, like pottery bottle kilns in which the imported iron bars, interspersed with charcoal, were stacked in chests. These chests, or 'coffins' , had to be made of good clay, originally from the Stourbridge area, and coal was used as fuel. This was the basis of Elliott and Maysey's patent of 1614, but in this case it would seem that a reverberatory furnace was used. At Bromley in Staffordshire, John Heydon used Spanish and Swedish iron in his 'tile house' or tile kiln. According to Plot[43] this resembled a baker' s oven with a grate at the bottom: on each side of the grate were laid coffins of Amblecote clay containing bars 1–1.5 m long. The fire was kept going from 2 to 7 days and the iron broken into short pieces 2–5 cm long called 'gadds'. By the 16th century the makers of edge tools were buying their steel in firkins and using it in conjunction with local wrought iron. In the probate inventory of 1608 of Richard Smith of Drayton, a barrel of steel worth £23 is recorded.[58]

Cemented steel was usually worked many times by piling to even out the carbon gradient across the section, and became successively shear steel and 'double shear' steel.[59] Even so, it was often too brittle by itself and, being expensive, it was sometimes used as a sandwich with wrought iron plates on either side, such as in scythe making.[60,61] It is clear that the craft of pattern welding was still very much alive and there is essentially no difference in principle between a scythe and a Japanese sword. In Toledo in Spain, the traditional method of swordmaking which lasted until 1733 was revived in the 19th century. Here, two Swedish steel bars were welded to an inner (Biscayan) iron core with a flux of flint clay powder.[60] An example of an 18th century sword of the Toledo type showed an average of 0.39%C with 0.45% Cu and low phosphorus. This had been heat treated to give a hardness of 455 HV30. Clearly it had been made from an iron mineral associated with copper ores.[62]

Besides being used in the Black Country, cementation furnaces were being used in Sheffield and near Newcastle upon Tyne. This latter area was visited by Kahlmeter in 1719[63] and by Jars in 1764. Nineteenth century brick furnaces of this type still stand in Sheffield, and a stone furnace at Derwentcote, near Newcastle, has recently been examined[64] and could well date from the early 17th century when German steelworkers and swordmakers started work at Shotley Bridge nearby.[65,66]

Steel products

Sword blades and edge tools were the chief uses to which steel was put. The steel from the cementation furnaces was heterogeneous and, owing to its surface appearance, took the name of 'blister steel'. A sword required twenty heats and first a trip—and then hand hammering to finish it. Fullers (grooves) were then forged in, using a suitably shaped hammer and anvil. The temperature was such that there must have been considerable diffusion of the carbon from the steel into the iron core. The blade was coated with soap and hardened by quenching from a dull red heat into a pan of water held between 14 and 24°C. The blade was held at an angle of 15–20° and quenched from point to tang. Once in, it was gently moved up and down with the flat side horizontal until cold, and then tempered until it was slightly purple.

In the 17th century, mass-production techniques were introduced, and it was said that Cromwell was buying swords from Solingen at 7s 6d. each, scabbard and belt included.[67] These blades were probably rolled wrought iron strips, case hardened, and oil quenched. Many of the blades required at this time were for civilian use and might more correctly be termed swords of fashion. Solingen bladesmiths devised a light triangular blade with concave faces which they called a 'hollow' sword blade. German workers were brought to England in 1685 for the manufacture of such blades at Shotley Bridge. Whether they made civil blades or not, wars in Ireland kept them busy on military weapons and a flourishing business was established for a time.[68] At the time of Jars' visit to the area in 1764,[26] Crowley's Winlaton Works in the same valley was making files, and the method of making these was not all that far removed from case-carburized sword blades. The files were first annealed and then had the teeth cut in with a chisel. They were then soaked in a trough full of beer dregs, covered with a thin layer of a carburizing compound consisting of salt and cow hooves and horns, and were slowly heated in a bellows-blown furnace. When they had been at temperature for long enough (neither time nor temperature are indicated by Jars) they were quenched vertically in water.

Knife blades were usually steeled. A recent example of a 16th century blade from Goltho in Lincolnshire, showed a very neat technique which in some ways resembled scythe making.[69] A piece of steel (0.8% C) had been covered on its two sides and the back by piled wrought iron so that the steel was exposed only at the cutting edge.[61] The welding was well done with a good joint between steel and iron and yet little diffusion: the steel was very symmetrically positioned. After welding, the whole blade must have been heated to 800°C and quenched to harden the edge. It was finally tempered to give a hardness of 557 HV at the edge and 151 HV in the wrought iron back. Such a hardness would be more than sufficient for domestic use and it might have been used as a butcher' s knife.

Iron casting

Most of the cast iron produced in the blast furnace in this period was destined for conversion into wrought iron in the forges but, at Rievaulx in 1591–92, where no guns were being made, 3.5% of the iron went for miscellaneous castings.[70] These comprised plates for fineries and forges, anvils, hammers, and hursts (parts of hammer shafts). These were cast 'at the furnace' and double furnaces were erected for large castings, such as guns, which were the largest castings of the period. The other military item was shot and it would seem that a good deal of this was cast at the furnace by ladling metal from the forehearth, between the tymp and the dam, into moulds. Only a limited amount of this metal could be used for this purpose as the furnace had to be taken off blast while the ladling was in progress.

Other items were firebacks and grave slabs cast in open moulds, firedogs, and railings. A large amount of ironworking machinery was cast iron, as can be seen from the Rievaulx accounts. In addition, there were lintel beams, iron rolls, slitting mill discs, and presumably the stands (frames) of rolling and slitting mills.

Iron cauldrons began replacing those of bronze right from the beginning of our period, but they were not very thin walled until the period of Abraham Darby at Coalbrookdale (after 1707). Up to about this time all iron was tapped or ladled from the blast furnace: the reverberatory furnace or air furnace had not yet been applied to cast iron.

The technique of gun casting was borrowed from bell founding. The guns were at first cast round a core which was reinforced with a 'core iron'. Biringuccio[55] described the method of making the core between two trestles, like the core of a bell. Clay mixed with cloth chippings, ashes, and dung was then applied to the core which had first been wrapped intermittently with a rope of hemp tow. The core was built up in layers and allowed to dry between each layer, finally finished with a strickle board, then dried and baked (*see* Fig.77).

Meanwhile, a mould for the outside of the gun was made on a solid wooden pattern to which pieces representing the feeder head, the decorations, and the trunnions were applied. The pattern was coated with ashes or tallow and then a coat of fine loam (clay) was applied with a brush: the mould was then built up layer by layer, the outer layers being reinforced with longitudinal iron wires. More clay was applied and, after drying, a jacket (armature) made of iron rods and bands was put on to reinforce the whole mould. This was dried and warmed to release the pattern, which was withdrawn from the muzzle end, leaving behind the decorative parts which had only been loosely stuck to the inner pattern. These were then withdrawn from the inside of the mould. The trunnions could also be withdrawn through the inside or the outside, in which case the holes had to be closed on the outside. An iron ring with arms like spokes was then placed in the breech or chamber end to support the core and maintain concentricity. The mould for the breech was usually made separately and formed the third component of the mould. A wooden pattern was again used, and the mould was built up in the same way as the barrel, but with the pattern supported at only one end. The breech mould was offered up to the barrel mould and its outer framework fastened to the rein-

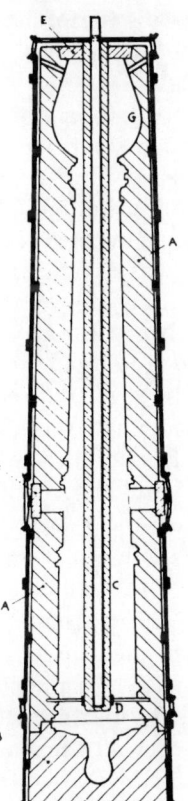

This drawing (which serves to illustrate Biringuccio's description of gun founding, Book VI, chapters 3 to 11) is based on one by Saint Rémy (*Memoires d'artillerie*, 1697) but has been modified slightly to agree with Biringuccio's account The mould consists of the main body, *AA*, closed at the bottom by the fitted breech mould, *B*. The core, *C*, built up of clay upon an iron core-rod, is held centrally in the mould by the iron chaplet, *D*, at the bottom and by the clay disc, *E*, at the top. The trunnion cavities are closed by bricks, *F*. The whole mould is reinforced with an armature of heavy iron bands and iron wire and is buried in a pit for casting with the muzzle and feeding head, *G*, uppermost.

The pattern on which the clay mould is built up is made of either wood or clay, but in either case the cornices, trunnions, and other projecting parts are applied over a layer of wax so that the pattern can be driven lengthwise out of the mould, leaving these parts to be later removed by hand. The ends of the pattern are extended to act as journals which rest and rotate on suitable trestles. Clay, applied as the pattern is rotated, is shaped by a strickle board securely held on the trestles.

There is some doubt about the size and shape of the feeding head. Biringuccio's drawing of the pattern shows a very small hemispherical head *above* the core-disc print. To have been effective it must have been below this, as shown in the present drawing. The importance of the feeding head in giving a sound casting is emphasized by Biringuccio, though he gives no dimensions for it. Saint Rémy said that the feeding head on a cannon should weigh no less than 4,000 pounds. It was, of course, sawn from the gun after removal from the mould.

(Courtesy of C. S. Smith (ed.) and M. T. Gnudi (trans.))

77 Gun moulding (after St. Remy [1])

forcement jacket of the barrel. Finally, a clay disc was needed for locating the core iron of the core at the muzzle or top end, and the runners and risers were cut into the feeder head mould.

Biringuccio follows his description with all sorts of admonitions and points of detail showing without any shadow of doubt that he was well-versed in the art from personal experience. There is no reason to believe that the making of moulds for iron guns was any different. The higher temperatures would mean more attention to detail, such as the quality and drying of the moulding material. After casting and removal from the pit the guns were 'bored' : actually, at this time it was merely a broaching operation to clean out the bore. It was not until after 1700 that guns were cast solid and bored out. This seems to have produced a sounder gun, no doubt owing to the porosity that arose from residual moisture in the low permeability clay core of the hollow-cast gun.[71]

Drilling was done horizontally, according to Biringuccio in 1540, with the bit rotated by water power and the gun 'fed' towards the bit on a sliding table. But Kricka (1550) shows a gun being bored vertically; the gun is suspended by a block and tackle and the bit rotated by horse power below.[72]

A boring bar—more correctly a reamer—with part of its carriage, has been found in the Weald, not unlike that shown by Biringuccio.[55] The cutting end of the wrought iron bar has had four inserts of steel welded onto iron.[73] Turnings from a smaller set-up have been found on another Wealden site.[74]

Cast iron is not malleable in the normally accepted sense and its brittleness has often been a hindrance to its use. However, the brittleness depends to a great extent upon the purity of the iron and the geometry of the artefact, as can be seen from the high strength of some of the early cast iron cannon balls. Prince Rupert, grandson of James I of England, is usually credited with the invention of malleablizing cast iron (1671), i.e. of making white cast iron malleable by controlled graphitization. However, we have already seen in Chapter 5 (p. 56) that the Chinese, not surprisingly in view of their prowess in the cast iron field, had actually done this in Han times.

In fact, malleablizing was reinvented in Western Europe by the French scientist Réaumur, who wrote about it in 1722.[75,76] For this purpose he built massive brick kiln-type furnaces about 2 m high, not unlike the furnaces he recommended for cementation, in which he would heat the cast objects to temperatures in the range 950–1 000°C. He heated his iron in bone ash to provide a neutral environment and give 'black heart' iron, and also in iron oxide (hammer scale) to give an oxidizing atmosphere and therefore 'white heart' iron. His main object was to permit the engraving of architectural castings. Of course, the degree of malleability is only relative and such iron cannot be deformed to any considerable extent. It was, therefore, not really in competition with wrought iron, although today the process is in active competition with mild steel castings.

Non-ferrous metals

Unlike the larger ironmaking blast furnaces, the furnaces for producing non-ferrous metals were merely adaptations of the small shaft furnaces used for ironmaking in the Roman period. Agricola[2] shows a series of furnaces used for the smelting of various non-ferrous metals. These were square externally, about 0.8 m wide, and 2 m high. The inside walls appear to be almost parallel and the section nearly square. The main difference between the furnaces for various metals was in the hearths, where there are different arrangements for separating and tapping slag and metal. Some furnaces worked with an open taphole, others with the taphole closed.

As far as copper was concerned, much of the ore of this period was pyritic (sulphide), and required preliminary roasting. Agricola recommended the heap roasting of broken ore with the finer fractions on top, the heaps being contained by low walls. The smelting furnaces were built in banks of up to six against a stone wall, behind which the bellows were driven from a single waterwheel-operated shaft. The lower parts of the square shaft furnaces were brick built, the upper was of stone and widened somewhat at the top. The tuyere was placed in the back wall, 22 cm above the hearth, and there was an arched opening in the front wall for tapping. Underneath the furnace and forehearth there were channels for venting ground moisture.

COPPER SMELTING

According to Agricola, copper smelting was done with the taphole always open. The furnace linings were made of a clay-charcoal mixture rammed into place at the bottom and smeared on to the inner walls: the forehearth pits were similarly treated and the tuyeres were not allowed to project beyond the lining.[77]

The three types of hearths are shown diagrammatically in Fig.78. Hearth No.2 has the open type of sump or continuously running taphole.[78] Hearth No.3 is a syphon arrangement, which allows the metal or slag to leave the furnace as it accumulates but effectively keeps the bottom closed. The second type was recommended for copper. Agricola gave all the metallurgically obvious advice about the furnace being slowly preheated before charging, and then the furnace was 'tempered' by charging slag. Oxidized ores could be smelted direct to copper in this furnace, but it is clear that Agricola is dealing with ores which were mainly sulphides. In this case, the result of the first smelting would be a matte (a mixture of copper and iron sulphides) and a slag which separated in the forehearth, and possibly some copper which would have contained any precious metals present. It took 12 hours to produce 300 kg of matte and 25 kg of metal. The process was repeated three times with the object of enriching the matte in copper and fluxing away the iron as slag. After the fourth smelt in the blast furnace the matte cakes (now all Cu_2S) were roasted four times, resmelted, and finally roasted three more times. The final smelting to impure (black) copper then took place; any precious metals present were sepa-

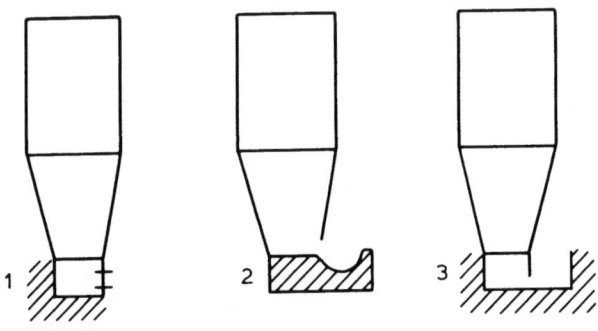

78 *Tapping arrangements in Agricola's blast furnaces (after Krulis[78])*

rated out by adding lead to the earlier furnace charges, which finally produced a Cu-Pb-Ag 'bullion' below the matte. This was later treated in liquation furnaces in which the silver-rich lead was melted out of the solid copper.

The prowess of the German smelters was such that this process was adopted in many parts of Europe and in Britain. German workers were responsible for the organization of mining in the Lake District (Cumberland), and the setting-up of a smelter at Keswick.[3] The technique seems to have been very much as Agricola described it. In 1564, 500 kg of dressed ore yielded about 150 kg of matte, which finally gave 50 kg of copper. The smelter had only six furnaces, of which not more than four were in use at any one time. The matte was then roasted with peat, which took eight days, and was reduction smelted once a month to give an impure copper (black copper). The roasting operation could be done inside in shaft furnaces or outside in 'stalls' , (*see* Fig.117). Up to this stage, peat and coal were the fuels and limestone was needed as a flux. For the final stages, carried out eight times, charcoal was needed since the elimination of sulphur was one of the aims. Again, many shaft furnaces were used for this purpose.

The silver was extracted from the matte by smelting with locally-mined lead ore. The silver was retained in the lead liquated from the Cu–Pb–Ag alloy, which was finally subjected to the cupellation process; in this way the silver was recovered from both the lead and the copper. The litharge made in the process could be reduced back to silverfree lead and copper.

The total output from the Keswick smelter over the period 1567–1584, was about 500 t, and the maximum was 60 t/a. After 1584, the ore was smelted near Neath in South Wales. Silver output totalled 5 000 troy ounces, which came from both the copper and lead. The average silver content was therefore about 10 oz/t of copper (0.03%), and its value represented some 4% of the total income. Since the relative values of Ag: Cu: Pb were 1000: 10: 1, it was always profitable to extract the silver.

The wood and charcoal was obtained from the immediate area with considerable local opposition. Coal came from Bolton nearby, and 1 000 t were used per year, giving a fuel/metal ratio exceeding 17:1. Two problems hindered output; the first was the recurrent shortage of working capital for mining development, and the sec-

ond was the absence of markets for the raw material. The producers were mostly forbidden to send the metal abroad, as England was then at war with Spain, but there was little demand for raw copper in Britain. The Crown could not afford to purchase it for the war stockpile. Therefore, in 1572 the producers were forced to enter the battery business to produce beaten copperware for which there was more demand. This needed still more capital and, as the extraction process itself was slow (stock-in-hand in 1572 represented three years output of copper), a considerable amount of capital was tied up in the process.

The great problem in copper smelting is the enormous amount of fuel needed for the various smelting and roasting processes. During the 16th century, about 20 t of fuel, either charcoal or coal, were needed for every tonne of copper. Even in the 19th century, in spite of the improved techniques, this figure did not change because of the decrease in the copper content of the ore.

Although it is clear that coal was used in Cumberland whenever it was technically possible, it must have been relatively expensive. Also, as the area was not convenient for cheap transport and communications were bad in general, it is not surprising that we see a gradual move to the south west of Britain where advantage could be taken of cheap coal and seaborne transport. The move had already started in 1584 when Cumbrian ores were smelted at Aberdulais near Neath in South Wales.[79,80] These works had been established a few years previously by the Society of Mines Royal, also with the aid of German workers. Here the scale of operations was large by the standards of the day, and blast furnaces and a reverberatory furnace were used, the latter being able to smelt 1 200 kg of ore per day. Unfortunately, owing to irregular deliveries of ore, this smelter was forced to close, and by about 1660 there was virtually no smelting of copper in Britain.

Meanwhile, let us have a look at Sweden where iron and copper were responsible for 50% and 30%, respectively, of Swedish exports in the first part of the 17th century.[81] The proportions changed to 75% and 10% by 1720, largely owing to the disastrous cave-in at the important Falun mine in 1687. In the 1690s, Sweden still supplied about 50% of European copper consumption and almost the whole of the British. But Japanese copper was predominent in Asia and, in the 17th century, the ships of the East India Company were bringing increasing quantities to Europe.[81] It is doubtful whether the worlds copper consumption exceeded 500 t/a at the beginning of the 17th century, but by 1712 it had reached about 2 000 t/a.

The copper ore deposit to the west of Falun supplied a large proportion of Northern and Central European requirements during this period.[82] It is a typical primary sulphide deposit but with no oxidation nor enrichment zones. By 1490, it was producing 300 t of copper a year from ore containing about 5% copper. The broken ore was roasted with wood in large pits containing as much as 180 t of ore. The roasted ore was put into low shaft furnaces which were in private hands, and which were scattered along the water courses all around the Kopparberg.

The Swedish furnaces stood individually between side walls, and by 1520 they were about 3 m high, i.e. larger than the Agricolan type. The back and the sides were permanent but the front wall was easily replaceable. The hearth was raised from the ground by means of an arch to ensure ventilation, and was lined with a mixture of charcoal and clay. The slag tapping hole was in the front wall but the metal was tapped at a lower level through one of the side walls; this seems to have been the main departure from the Agricolan furnace. The early 16th century furnace shown by Månsson had a distance of about 1.5 m between the tuyere and the bottom of the hearth. By the end of the century, however, the hearth bottom had been raised, and by 1760 it had the position shown in Fig.79. This change resulted in a larger output, probably owing to the larger cross-section of the higher hearth.

Unlike the German technique, in which a large number of separate furnaces were used for different stages of the smelting programme, the Swedish process relied upon one furnace for all stages. The larger furnace could of course deal with larger quantities at a time, but several charges of matte were worked before the roasted matte was smelted to copper in the same furnace. The final process was carried out separately as a third operation. The furnace was filled continuously with roasted ore and charcoal, and matte containing about 10% Cu was tapped every second day. One campaign took 12-18 days, after which the furnace was left to cool. Meanwhile, the matte was roasted in rows of stalls and after each roasting it was broken up and reroasted 6-9 times with gradually rising temperatures. This part of the process took 5-6 weeks.

The oxidized matte was then resmelted in the shaft furnace, again with charcoal to give black copper, but the furnace hearth was reduced in size with the aid of stones. One charge took four days to work and the copper was tapped twice during this period, giving about 380 kg of 90% grade black copper on each occasion. The copper was refined with charcoal by oxidation before an air blast in a smithing hearth. The tuyeres were set at an angle of 25–30°: there was a loss of 12.5–20% and

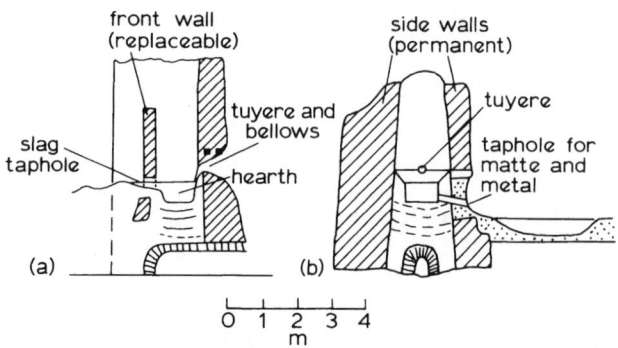

a vertical section through tuyereline; *b* vertical section through axis

79 Swedish shaft furnace (after Lindroth [82])

Table 51 European copper production

Country	Date	Annual output, t
Germany (Mansfeld)	1530	2 000
	1770	400
	1790	750
	1800	750
Sweden (Falun)	1650	3 000
	1655–60	2 100
	1665	<2 000
	1710–20	1 000
	1770	< 700
Sweden (Östergötland)	1781	100–150
Britain (approx.)	1560	100
	1712	1 000
	1800	7 000
	1850	22 000
	1880–90	80 000 (zenith)*

*Much of this was refined imported blister copper

the finished copper had a purity of 97–99.7%. Green wood was pushed in to reduce the oxide content (poling back) and the metal was removed with tongs, slice by slice, after water had been thrown upon it.[83]

At the end of the 17th century, when the output of the German and Swedish mines and smelters was decreasing, a revival was taking place in Britain based on the use of Welsh and Bristol coal and Cornish copper ores, which were being found below the tin deposits in the Cornish mines (*see* Table 51). These mines were using coal from South Wales for steam pumping engines, and it was, therefore, cheap to bring back the lean and bulky copper ores as a return cargo.

At Upper and Lower Redbrook in the Wye Valley near Bristol, the smelting of Cornish ores with local Forest of Dean coal was started in 1690–1692.[84] In 1690, 50 t of ore were sent from St. Ives, and this increased to 986 t in 1697. By 1693, 30 t of copper ore were sent to Bristol and Gloucester. There is evidence of Swedish management by 1708. At the Lower Redbrook Works there is a reference to the lease of a 'cupilo', or reverberatory furnace, in 1716. This is not surprising since we have evidence of one at Aberdulais in 1584, but we know that, by the beginning of the 18th century, reverberatory furnaces were gradually replacing blast furnaces for tin in Cornwall, and elsewhere for lead smelting. Our information about this period mainly comes from the descriptions of visits made by the Swedish observers, Cletscher in 1698, and Kahlmeter in 1724–5. Swedenborg used their reports for his book De Cupro, published in 1734, but he seems to have confused the two sources.[85]

Cletscher[86] describes the works at Bristol (Hotwells) in 1698, where rich ores containing 10–30% Cu were being smelted entirely in reverberatory furnaces. First, the ore was roasted and then smelted to give matte, which was broken up by stamping, and resmelted (oxidized) in reverberatories to give black copper which was finally refined, again in a reverberatory furnace. These operations, apart from stamping, required little water power.

Another Bristol works, at Shrewshole, was described by Kahlmeter[86] in 1724–25. Here the ore, a mixed oxide and sulphide which rarely needed preroasting, was smelted in reverberatory furnaces to matte which was roasted and resmelted up to 20 times in the same type of furnace to concentrate it; it was finally oxidation-smelted to black copper and then refined.

At Upper Redbrook in 1724–25 the matte was separated from the gangue in ten reverberatories and then crushed.[86] The fines were then roasted in more reverberatories, and the roasted matte was taken to so-called 'clotting' furnaces which agglomerated it before it was reduced to metal in blast furnaces. These latter were worked with bellows and were 1.2 m high, 56 cm wide, and 61 cm deep: they were clay lined and built of stone with two tapholes, one for copper and another a little higher up for slag. The charge of roasted matte and coke went through a hole in the side, presumably at the 1.2 m level. The furnace had a 4.3 m high stack which was 46 cm wide.

These are slightly shorter than the Agricolan furnaces, but otherwise there is no change apart from the tapholes. There is little doubt that the more controlled conditions of roasting in the reverberatories was a great improvement and would speed up the process. But it was still, essentially, the German process and not the later and more sophisticated English or Welsh process. The consumption of coal was still high—18 t of coal to 1 t of copper.

We are not given any details of the reverberatories, but they were capable of calcining 180 kg of matte at a time (12 hours), and the temperature was kept below the melting point (about 1 100°C) by the dilution of coal with ashes. The 'clotters' were also reverberatory furnaces, and sintered the powdered ore into suitable pieces for the blast furnaces, sometimes with the addition of sand to stiffen the lump. The lumps were broken up to give an 'egg size' piece which would be very suitable for the small blast furnaces.

At the Lower Redbrook Works, also seen by Kahlmeter in 1724–5, blast furnaces are not mentioned and it seems that here they had been entirely superseded by reverberatories. The process was similar to that at the Bristol smelting works, and Swedenborg gives a drawing of one of the reverberatories which he had taken from Cletscher. The inside measurements were 1.7–1.8 m long by 1 m wide, widening to 1.2 m near the centre. The bottom of the hearth was about 0.6 m from the ground. The hearth was made of sand (perhaps with some clay to form a ganister) and the ore was fluxed with a little lime. English copper was rarely refined if its purity was greater than 92%, and this purity was not suitable for brass, yet a good deal of the Bristol copper was granulated for brassmaking and this must have been carefully refined. At this time there are references to the use of suphur in refining, and it would seem that this was added to ease the removal of iron, which was the main impurity, as a superficial matte which could be oxidized more easily into a slag.

Copper refining

The techniques of copper refining practiced in Britain are rather glossed-over in most of the accounts.[87] However, Japan had been an exporter of copper to the West since about 1650 and the traditional processes in use in the late 19th century, which are almost certainly those used earlier, have been recorded.[88] The furnace was simply a hemispherical hole in the floor of the melting shop lined with a mixture of clay and charcoal, into which a crucible with an outside diameter of about 60 cm was placed, resting on powdered charcoal. Lumps of ignited charcoal were placed in it and piled above in a low conical heap. An inclined tuyere, connected to hand-powered box bellows, was placed over the top of the crucible and pieces of copper were placed on top of the hearth and melted. As they fell past the blast of the tuyere, impurities were oxidized out and the refined metal collected in the crucible, displacing the charcoal. This must have operated very like a finery fire in an iron forge.

When all the copper had been melted to give 40 kg of molten copper in the crucible, it was 'poled back' with a piece of charcoal inserted into the molten metal. Then the crucible was lifted out with tongs and the contents poured into canvas moulds placed in hot water 14 cm below the surface. The time taken to process 40 kg was about $1^{1}/_{2}$ hours and 8 charges were worked per 12 hour day. An addition of 0.1–0.2% Pb was made to the copper after refining to give the traditional brilliant colour to the cast copper. We do not know whether this refining technique was ever used in Europe, but it would seem to be a very convenient one for the 16–17th centuries or earlier.

Brassmaking

Since, at this time, a good deal of copper was used for brassmaking, it is important to discuss the technique used. Brass was still being made by the calamine (ZnO or $ZnCO_3$) process and hitherto had been centred in the Liège-Dinant-Aachen area of what is now Belgium and the Rhineland. As has already been explained, this process involves heating pieces of copper and zinc ore with some charcoal in a crucible. Originally, the fuel was wood and charcoal but there is little doubt that in Belgium, and in Bristol where most of the English brass was made, coal came to be used very early.[89]

A good deal of the output of English copper smelters went into brassmaking, and the Elizabethan Society of Mines Royal were making brass at Isleworth in Middlesex and at Tintern in Monmouth.[90] In 1568 it is recorded at the latter site that the uptake of zinc reached 31%, the normal limit for a ductile alpha brass. The calamine came from the Mendip hills in Somerset, Gloucester, Nottinghamshire and Flintshire near Holywell. All this had rather a high lead content and good calamine had to be imported.

One description of the technique of brassmaking in use in this period is to be found in Neri's *The Art of Glass*, dated to 1662.[91] The furnace, which is probably German, had its floor about 2 m below ground and at that level was 1 m wide, narrowing towards the top (*see* Fig.80). It was blown with bellows and contained eight to ten crucibles which were put in with the fuel through the hole at the top.

This type was suspended by the 18th century when natural draught was used. There would almost certainly be a canopy and chimney, as shown in Diderot's 18th century French drawing of brassmaking[92] (*see* Fig. 122, p.151).

At Baptist Mills near Bristol, 36 furnaces were in use which were 20 years old in about 1700.[93] Each of these held eight crucibles made of Stourbridge clay, each containing 18 kg of granulated copper and 20–30 kg of calamine which were cast twice every 24 hours. It is recorded that the machinery was worked by water; most of this would be for the grinding mills and tilt hammers, but it is possible the furnaces were bellows-blown, probably like those shown in Fig.80. A reverberatory furnace was used for preheating or annealing in the battery process.

In 1697, brass plates were made by pouring into moulds consisting of two sloping sandstone slabs 1.5 m long, 23 cm broad, and 7 cm thick, held together by iron bands. The plates weighed 30 kg and were probably the starting material for the battery process which involved reduction by, and shaping with, water-operated tilt hammers.[94]

THE MAKING OF COPPER-BASE ALLOY ARTEFACTS

By the 16th century there was almost a spate of works dealing with copper-base metallurgy. Less well known than the works of Biringuccio (1540) and Agricola (1556) is that of Kricka, a Czech who died in 1570. His treatise (really a practical guide for his successor) deals wlth bells, cannon, vessels, and even pumps.[71,92] His bells were larger than those described by Theophilus and were moulded vertically with a strickle board in the casting pit, much as they are today.

Biringuccio prefers the horizontal method of moulding bells, with horizontal strickle boards for shaping (*see* Fig.59). The pattern, or 'case', was not made in wax but in clay, which was prevented from sticking to the core by a parting medium of ashes. The case was cut from the core after moulding the cope, and weighed. The weight was multiplied by a factor of 7 or 8 to give the weight of the metal to be cast; 10% was added for melting losses and further additions were made for the cannons (the supporting gear). Since the specific gravity of damp clay is about 2, it would seem that either there was much fibre in the clay or that Biringuccio was making sure that he had plenty of metal to spare. A warning about the danger of having too little was made many times.

Gun founding practice, as far as we can tell, was the same for bronze as for iron, but here the metal was melted in reverberatory furnaces fuelled with wood. Kricka shows such a furnace with a grated hearth. The compositions were not strictly gunmetals, in the sense that we use the term today to describe the hybrid alloy of copper, zinc, and tin, but normal tin bronzes in which the tin varies from about 5 to 14%.[95] Lead was probably present as an impurity but it would help to make the

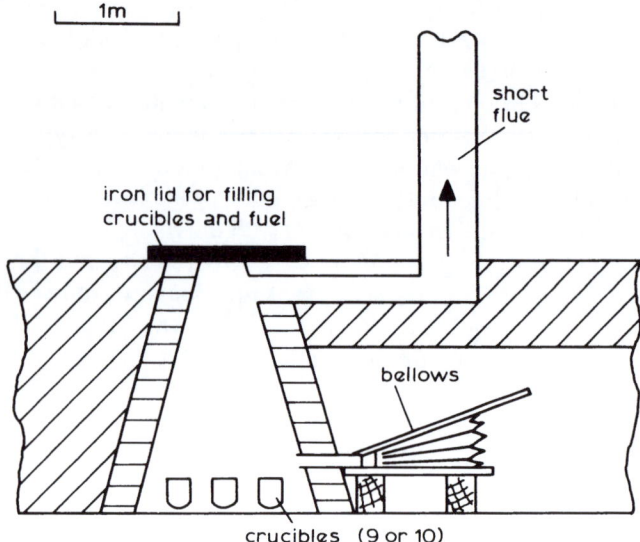

80 *Bellows-blown crucible furnace for brassmaking (after Neri in Golsmith[90])*

casting more pressure-tight. By the 19th century the zinc content was probably intentional: the composition of the US gun is that of so-called Admiralty gunmetal (*see* Table 52).

The French guns of 1638, found in the wreck of the 'Association', have been found to contain blow holes in some areas and shrinkage 'sinks' near the muzzle.[96] A sliver of metal from the trunnion showed no coring so that it is virtually in the annealed condition, probably owing to the slow cooling allowing homogenization of the cast and segregated metal to occur. The small amount of delta-phase had been preferentially corroded and there was some evidence of non-metallic inclusions, possibly sulphides.

Kricka confirms that clay was the main moulding material of the 16th century. The permeability was increased somewhat with fibrous additions, but great care was taken to see that the clay was well-baked before casting. Since Biringuccio mentions the use of a cereal bonded sand, and since many prehistoric cores have been made with clay-sand mixtures, it seems strange that the 'green-sand' technique was not widely used

Table 52 Composition of bronze cannon

		Composition, %			
		(rest Cu)			
Date	Provenance	Zn	Pb	Sn	Reference
1521	Turkey	–	2·2	12·3	133
1530	Italy	tr.	1·3	13·9	"
1535	Sweden	0·4	2·6	7·6	"
1628	Sweden	1·5	0·9	4·5	"
1638	France	1·4	0·9	4·1	96
1700	France	0·7	–	7·8	133
1700	Japan	–	3·3	12·7	"
1819	Russia	–	–	10·7	"
1863–4	USA	1·8	–	9·5	"

until the 18th century. In fact, green-sand moulds, using a good quality local sand, were first used in Bohemia at Zbiroh in 1660.[92] Biringuccio gives a graphic description of the mass production methods used for clay-moulding buckles and small brassware. Clay mixed with cloth clippings was rolled out on a board like pastry, 1.3 cm thick, sprinkled with fine charcoal and moulded using tin or brass patterns attached to gates and risers. The moulds were baked on sheet iron trays in a furnace, assembled like a multidecked sandwich with patterns between each two layers. If each two layers contained 60 patterns, a complete mould of 20 sections would produce 1 200 small castings.

Lead and silver

Although not all the silver produced at this time was extracted from lead, it is clear that a large proportion certainly was. For this reason, we will first consider silver as a by-product of lead manufacture, and later as a metal in its own right. Very little is known about the early history of the extraction of lead on the Continent. But it is well known that half of the exports of Britain in the 17th century were tin and lead, so it is very probable that Britain supplied much of the Continental requirements, just as Sweden supplied most of the copper and iron needed by both the Continent and the British Isles. During this time the exploitation of the metals of the South American continent was taking place under Spanish auspices, and there is a good account of the techniques used in the work of Alonso Barba which was published in Spain in 1640.[97]

Agricola[2] smelted lead in one of his general purpose blast furnaces, such as that shown in Fig.78, with the taphole always open. He stated that the 'Carni' partially roasted the lead ore and crushed it before charging it into an open hearth, between low walls, which was fuelled with green and dry wood on to which the broken ore was thrown. The lead smelts easily and drips down on to a sloping hearth made of charcoal dust and clay, which extends to form a crucible outside the hearth (*see* Fig.64).

Many of the early lead-smelting furnaces were even more primitive than this. In Westphalia 'they heap up two waggon loads of charcoal on some hill-side which adjoins a level place'. A layer of straw is placed on top and on this 'is laid as much pure lead ore as the heap can bear; then the charcoal is kindled and when the wind blows, it fans the fire so that the ore is smelted. In this way the lead, trickling down from the heap, falls on to the level and forms broad thin slabs'.[98]

These are the so-called 'bole' furnaces or 'bole hills' of Britain. According to a British source[99] a 'bole' had three low walls, two walls 2.1 m long, and the back wall 6 m long. Such were used in 16th century Derbyshire and are well discussed by Kiernan.[100] Similar furnaces were used in 19th century America.[101]

Only lead responds to such a primitive technique, owing to the relative ease of the double decomposition reaction, in which partially oxidized lead (PbO) reacts with the unoxidized galena (PbS) lower down to give

metallic lead and sulphur dioxide. The yield is poor, a good deal of the lead going into the slag. But, as a way of getting metallic lead easily and cheaply, there is nothing to beat it. Almost any sort of dry fuel can be used such as peat, coal, and wood, and the rich slags can be reworked in a bellows blown 'slag hearth' . These furnaces were usually situated on hill tops, not because of the better draught—like any bonfire they will create their own—but because the fumes spoilt the vegetation for miles around and the lead deposits poisoned the soil.

Alonso Barba[97] recommends a small furnace only 30 cm high for lead oxide ores to minimize fume loss. The bottom and forehearth are made of clay mixed with iron oxide. For galena (PbS), he recommends a pit furnace which consists of a hollow in the ground 2.0 m dia. and 1.5 m deep with, apparently, no means of blowing. The hollow is mostly filled with wood and the galena is placed on top. It is difficult to see how this can work and it certainly must have been very slow-acting; its only advantage would have been the elimination of fume.

More sophisticated furnaces are discussed in detail in the State Papers of Elizabeth I of England (*c.* 1581), when William Humphrey claimed that his patent was being infringed by the Derbyshire lead smelters.[90] At this time the 'old order' was represented by foot operated bellows-blown hearths, very often erected on a platform so that they could be rotated to take the smoke away from the blowers. These appear to have been only 50 cm deep with a tuyere placed near the top of the hearth (*see* Fig.81). The 'new order' was represented by Burchard's furnace, which was in fact a pair of blast furnaces very like those of Agricola, each blown by a waterwheel and two bellows (*see* Fig.82). These shaft furnaces were about 1.5 m high and 50 cm square, blown by bellows 3 m long and 1 m wide. Their originator, Burchardt Kranich, alias Dr. Burcot, obviously had German connections and this no doubt explains the similarity to Agricola's furnaces. It was claimed that the foot-blast furnace, which succeeded the natural draught boles, had twice the fuel efficiency of the latter and made 3 t/week. Whatever the merits of the shaft furnace, the shorter furnace continued but with water-power bellows and finally became the ore hearth of the 19th century, and the Newnam hearth of the 20th century. For a time, the low hearth worked the rich slags of the bole hills; the higher temperatures available from such a hearth, which was in essence like the bloomery hearth, allowed slagging reactions involving the replacement of lead in lead silicate slags by iron to give iron silicate slags.

Barba mentions the existence in South America of primitive native furnaces, which were shaft furnaces working on induced draught like the African iron furnaces. Unlike most induced draught furnaces they had holes through the top of the shaft with 'ears' by which he claims the ingoing air was heated. This is very doubtful, but the holes could assist the oxidation of a galena charge and aid the double decomposition reaction. For his own work Barba used shaft furnaces on the Agricolan model. The furnace was preheated with charcoal some hours before charging, helped with blast until the furnace was white hot, and then charged with slag from the reverberatory furnace. The slag was used to 'glaze' the furnace lining. Gold and silver ores were then added with lead or lead ores as a collector to give a lead bullion and slag in the forehearth. He suggested that ores containing sulphur and antimony should have iron ores added as flux. This might be advantageous in cases where the gangue was highly siliceous and devoid of iron compounds.

The reference to the use of a reverberatory furnace as a new type in 1637 is interesting and Barba seems to have used them occasionally for roasting, but mainly for smelting with wood, grass and dung as fuels. Such furnaces were not in use for lead smelting in Elizabethan England and the earliest references to them here for lead is in a patent of 1678 using coal as a fuel. There is little doubt that the reverberatory or 'cupola' was first used for lead smelting in Flintshire in about 1699.[102] In 1708 one plant, Gadlys, had four ore furnaces which were almost certainly reverberatories, two slag hearths, and four refining furnaces. This plant had been built by the London (Quaker) Lead Company, who had also taken over the works at Baguilt nearby, which had been established before 1694 and must have been the earliest lead plant known to have used reverberatory furnaces successfully.

Coal was being used in large quantities in the proc-

ess, and the siting of such works was no doubt decided by the proximity of this fuel. In 1721 there was a coal mine within a mile of Gadlys with a seam 1.5 m thick, at a depth of 70 m. We are given no details regarding the reverberatory furnaces, but we know that the slag hearth was worked manually with a simple pair of bellows, which in 1704 replaced a pair that had been found to be too large. As would be expected, the refining (cupellation) furnaces used a large amount of bones which had to be brought from outside, mostly by sea from London.

The average lead content of the smelted ore ran as high as 65%. The silver content of the lead produced was 450 g/t, and the cost of extracting this quantity of silver was £3.125/t of lead which gave an estimated profit of £1/t, after allowing for the reduction of the litharge back to lead. But this shows that there would have been no profit if the silver content had fallen to less than 340 g/t. The refining bellows were also manually worked. It would appear that no water power was available at Gadlys, nor was this a disadvantage. The use of coal would have produced such a saving that the small cost of blowers' wages was of little importance. Also, the reverberatories had good chimneys and needed no assistance from bellows.

The other Welsh smelter of which we have a record at this time was close by the Cwmsymlog mine and had started working ores with 7 300 g/t silver in 1604, which

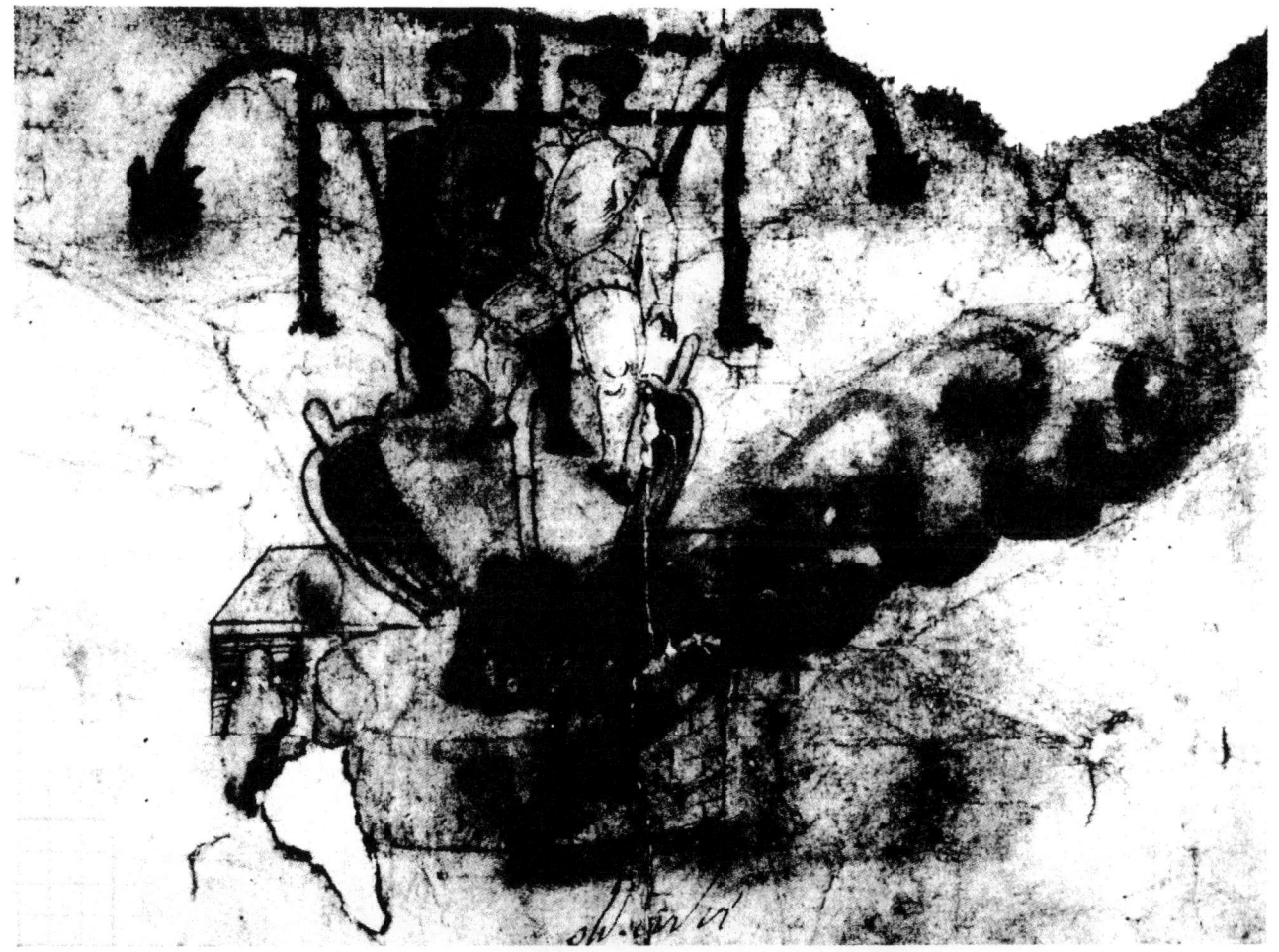

81 'Old order' lead-smelting furnace using foot-powered bellows

82 Burchard's double blast furnace for lead smelting

had fallen to an average of 220 g/t by 1710.[103] This mine had been granted by Elizabeth I to the Mines Royal Society in 1568 but had been worked long before this time. In 1588 the ore was smelted at the mine with wood and some change in the method of working resulted in an improved fuel efficiency.

Recent excavations confirmed the site of the smelter as being that of the Dyfi iron blast furnace.[104] Lead and lead slags were found below the casting house and show the use of bone ash cupels. Clearly this was the site described in detail in Pettus's Fodinae Regales of 1670.[105] By 1779 coal was being used and imported via Aberystwyth.

While the reverberatory furnace and coal was introduced into Wales and the Midlands at the end of the 17th century, the North of England seems to have been quite happy with the ore hearth. This furnace, like the slag hearth, had developed from the bloomery and the low hearth of the 16th century (the 'old order' foot blasts). Here they were also concerned with replacing wood by coal, and it is recorded that at the Brickburn (Northumberland) lead mill in 1701, coal was marginally cheaper than wood.[106] At Fallowfield in 1702, the smelters were investigating the advantages of the iron-plated hearth (like the iron finery hearth) over the old stone-lined hearth; they found that the latter gave a higher yield, perhaps owing to different temperatures or geometry which allowed more lead to escape as fume in the iron-plated hearth.[106]

As far as lead is concerned, it is clear that, by the end of the 17th century, the two competing methods were the coal-fired reverberatory and the ore hearth. The blast furnace seems to have become obsolete by the end of the 16th century; but it was to be revived in the 19th century. We have no drawings nor detailed descriptions of the 17th century ore-hearth furnaces, but we have many for the 18th century and these will be discussed in the next chapter.

Tin

The main centres of tin production in Europe were Devon and Cornwall on the one hand, and Bohemia and Saxony on the other. Brittany and Central France had produced tin in the past but their ores were poor and production was declining. The two former centres were thus the sources of tin for nearly the whole of Europe; the rest of the European requirement was made up of imports via the East India and other trading companies from the Malay Straits and China. The output in the Saxo-Bohemian area was at its peak in the 16th century but soon fell, owing to disorganization caused by the Thirty Years War. The Bohemian output is given in Table 53 but it is probable that the figures also reflect Saxon production. Their decline left Devon and Cornwall as the only producing centres in the West, and Devon production was soon to decline (as shown).

The tin ore of Devon and Cornwall was often intimately mixed with siliceous gangue, and grinding and washing were necessary.[107] Bodmin Moor in Cornwall has produced the remains of an ore stamping mill at

Table 53 Tin production in England* and Bohemia*

| Date, AD | Production, t/a | |
	English (total)	Bohemian
1200	200	–
1300	200	200
1400	300	600
1450	500	900
1500	600	1 000
1550	800	400
1600	1 200	300
1650	1 800	50
1700		–
1750		–
1800	2 522	–
1824	5 000	–
1853	5 763	–
1872	9 560	–

Devon production, t/a†

1400–50	54
1450–1500	104
1500–1550	199
1550–1600	93
1600–1650	25

*Up to 1650 according to Majer;[135] after 1800 due to Barton[117]
†Worth;[107]

Colliford Bridge.[108] Retallack in south west Cornwall has produced similar remains of more than one mill together with slags showing that it was also a blowing house[109] (see Table 54). Most of the remains of early working in Cornwall have been obliterated by more recent activity. But this was not the case in Devon nor Bohemia, where in both cases a sharp decline set in in the beginning of the 17th century. The remains of washing, grinding and smelting installations still exist on Dartmoor and many have been recorded.[107,110]

After preliminary crushing in stamp mills, the ore

Table 54 Analyses of tin smelting slags

| | Amount, % Cornish[115] | | | | Portuguese | |
Element	1	2	3	4	EIA	Modern
FeO					58·40	18
SiO$_2$					21·46	35
CaO					–	28
Al$_2$O$_3$					9·34	17
MgO					–	
Mn	0·4	0·7	0·5	0·5	0·30	
Cu	0·015	0·22	0·04	0·7	tr.	
Zn	0·17	0·10	0·5	0·5	–	
Pb	1·3	–	–	–	–	
Sn	0·8	0·5	22·0	2·8	1·60	2–3
TiO$_2$					0·96	
P$_2$O$_5$					0·30	
S					0·005	

Note; The tin in nos. 2 and 3, the Portuguese, and the modern was in the form of tin globules; that in 1 and 4 was in the form of oxide

was ground in 'crazing' mills, which were normal vertically-shafted 60 cm dia. millstones, and piles of sand can be seen outside some of the blowing houses. Some of the leats had a fall of about 4 m in a short distance; others were as much as 3/4 km long. Their capacity was as much as 1.7 m³/min, the average diameter of the overshot wheels was about 3 m, and the width 0.5 m; some of the wheels drove the crazing mills as well as the bellows. The bearings of the bellows wheels were of the open type and the axles must have been placed below the bellows.

The first furnaces were shaft furnaces and it is very doubtful whether the reverberatory was introduced before 1700, by which time Dartmoor production had virtually ceased. Yet the Eyelesborough smelter had a 21 m long horizontal flue terminating in a vertical chimney. It is quite possible that this was an improved dust catcher for the blast furnace, since there is a wheel pit in a convenient position for driving bellows for a blast furnace.[111]

Some idea of the blast furnaces used in the 16th century can be obtained from Agricola. These were 2.5–3 m high, and 60 2 30 cm internally. They were made of large sandstone blocks and were lined with clay. The tuyere was placed at the back opposite the taphole and directed downwards. The hearth was a sloping sandstone slab and the taphole was always open. The tuyere was large so that the blast should not be too fierce. The metal and slag separated out in the large (0.3 m long) forehearth and the slag was able to overflow from this: a charcoal cover was maintained over the tin to avoid oxidation.

A smelting furnace has been found at Merrivale in Devon. The remains were 1.5 m high and the internal dimensions were 50 × 60 cm. In front was a single piece of stone with a 33 cm wide depression in it which was probably the forehearth. The wheel must have been 3 m dia. × 33 cm wide. This furnace had a granite mould nearby, capable of producing rectangular 'pigs' about 30 long × 20 cm wide with a rounded bottom. A number of these mould stones have been found on Dartmoor, all of granite. The metal, smelted with water power and local fuel (probably peat), was resmelted (or refined) in the larger towns such as Exeter where it was recast into standard ingots weighing about 90 kg. The loss during this process was about 12.5%.[112]

We know a lot less about the technical side of the Cornish tin industry. The blowing houses were huts built of stone and turf covered with thatch that periodically caught fire, or were intentionally burnt to recover the tin dust. The furnace was made of large stones held together by iron clamps and was blown with a waterwheel. The metal was ladled out of the forehearth and refined by drossing in an iron 'pot', heated like a lead refining pot of the 19th century. After this, it was ladled into granite moulds; the fuel seems to have been mainly charcoal.

By the end of the 17th century there were 26 smelting houses in Cornwall and two in Devon.[113] Those near the coast probably used imported coal, although there is no

evidence that coal was used for ore smelting before the introduction of the reverberatory. But tin slags were smelted in bloomery-type furnaces like lead slags and coal was probably used for this purpose. The alluvial tin ore must have been very pure and pebbles found at St. Mawgan in Pydar assayed 78.7% Sn, which is the tin content of pure cassiterite.[114] But most tin ores containing more than 1% Sn before dressing are considered very high-grade and the bulk of the ore obtained from alluvial surface workings probably had less than this.

Smelting in a primitive blast furnace with charcoal produced a very pure tin, especially after a limited amount of drossing to reduce the iron content. The crust of a 17 kg plano-convex ingot from Tremethack Moor, Madron, contained only 1.45% of non-stanniferous material, and most of this must have been picked up from the surrounding soil. The Trereife ingot contained 99.9% Sn and is probably typical of the early production from charcoal or wood-fuelled blowing houses. Another ingot of plano-convex shape from Mabe contained 96–88% Sn.[115]

Analyses of some medieval and post-medieval slags from Devon and Cornwall are given in Table 53. These are not very different from the Bronze Age slag from Caerloggas, and carry large amounts of tin as silicate. These would have been produced by a blast furnace in a blowing house, and due to the purity of the ore charged (> 65% SnO_2), only a small amount of slag would be produced. The metal itself would be of high purity as found in the course of experimental work.[116] In order to reduce the tin content, modern smelters add iron and lime (see Fig. 53, col.6).

Precious metals

In Europe most of the silver at this time was a byproduct of lead smelting. If fuel was cheap enough or coal available it was economic to extract the silver, as long as the lead contained at least 330 g of silver per tonne (0.033%). This was done by the cupellation process which was virtually unchanged since its inception. In some cases the hearth could be moved away from the roof for charging, in others the roof was swung away from the hearth.[117] The London Lead Smelting Company smelted the cupellation furnace bottoms in a slag hearth to recover the litharge and any silver contained in them. This would be put back into the refining hearth.[104] There was to be no change in principle until the invention of the Pattinson process in the 19th century. Occasional deposits of native silver turn up, such as that at Hilderston in Scotland, but most of the silver was extracted from lead and copper ores by the time-honoured methods described by Agricola.

Gold continued to be won by alluvial and deep-mining methods all over Europe. Alluvial material merely required mineral dressing and melting and virtually no smelting. The gold-bearing sands were washed on inclined elm boards riffled by means of rough sawn slots, and the gold residue in the riffles was gathered up and amalgamated with mercury in a wooden vessel (batea). The gold was then concentrated by squeezing the excess mercury away through a leather bag and evaporating it in an alembic. This left the gold grains like sand at the bottom of the vessel. They were mixed with a flux of borax and saltpetre (potassium nitrate) and melted in a crucible.

Although Biringuccio gives a brief mention of the possible application of this technique to silver ores, Agricola does not, and it would appear that in Europe most silver ores were smelted with lead and the silver recovered by cupellation. The parting of the gold and silver resulting from melting or cupellation requires the use of nitric acid which dissolves the silver but not the gold. Nitric acid appears to have been known to the Arabian chemists of the eighth century. It was prepared in the 13th century by heating vitriol and alum with saltpetre in an alembic and condensing the distillate. The cucurbit (the lower part of the alembic) was of glass; in it was put 1 part of saltpetre, 3 of rock alum, and $1/8$th part of dried sand or the residue of previous distillation. Cyprus vitriol (copper sulphate) could be used instead of alum; the important component of the alum is the sulphate radicle. Distillation gives off water vapour, and then nitrogen oxides, which are absorbed by the condensed water vapour to form acid. The equation is:

$$KNO_3 + sulphate \rightarrow HNO_3 + KHSO_4$$

Alum (potassium alum, $K_2SO_4 . Al_2(SO_4)_3 . 24H_2O$) was often used as the source of sulphate, but iron and copper sulphates could also be used. Amalgamation processes require mercury which, is so easy to produce from its readily decomposable red mineral, cinnabar (HgS), that it must have been known from the earliest times (Vitruvius describes its preparation). Mercury gilding was practiced in the Roman period and has been descrlbed by Theophilus; the Viking site at Haithabu actually yielded metallic mercury.[119] Biringuccio[120] describes its production by grinding up the ore with stamps or with an olive-oil mill (edge runner type). The ore is then washed and distilled in a muffle furnace with condensing chambers. In fact, the ore is placed in a series of crucibles heated underneath so that the products of combustion cannot mix with the mercury vapour, which condenses on the branches of trees in the condensation chambers. Agricola[121] has used a translation of this passage from Biringuccio, but gives a better illustration. Of course, this process can be carried out in the simple alembic. In the excavations of 1955–60 at Weoley Castle near Birmingham, such a vessel of red ware with a grey core and green glaze was found;[122] it contained a metal deposit, mainly mercury, and was dated to 1450–1500. Later, parts of a similar vessel were found in the remains of a monastery at Stamford in Lincolnshire,[123] in levels also dated to the 15th century. Crucibles were found, and in the soil associated with these were droplets of mercury and quantities of red ferric oxide which could have been the products of the decomposition of an alum or iron sulphate, used for making nitric acid.

In the New World, after the Spanish conquest, an enormous amount of precious metal was removed by the conquerors. This metal, for the most part, was the accumulated product of early civilizations and was almost certainly all made from native metal. As we have shown, by the 12th century the Inca people of Peru had been accustomed to melting bronze and other nonferrous metals and would have no difficulty in converting alluvial gold to more massive forms.

However, the conquest not only removed already extracted metals from the area but traced the metals to their sources. In one case these were the mountains of the country now known as Bolivia and, in the village of San Bernardo, Potosi, arrived one Alonso Barba,[124] curate and metallurgist who lived there from about 1600–1630. He was born in Andalusia in Spain and submitted a report to the Spanish Crown in 1637, which was published. His account of the amalgamation process for the recovery of silver from silver ores is muddled in the extreme, but essentially he was describing what is now known as the 'patio process'. He was not the originator of the process but must have brought it from Seville, as it is mentioned by Biringuccio (1540). The Hoovers suggest that it may have been known in Mexico as early as 1571.[125] It was certainly another 200 years before the process was to be used in Europe.

In the patio process the ore is crushed and mixed with 5% salt solution. It is then placed on a stone floor (patio) and trampled by mules, after which a little burnt pyrites is added together with more than sufficient mercury to liberate the silver. The mixture is trampled again for another six weeks, at the end of which the silver is amalgamated with the mercury. This is washed, concentrated by squeezing through bags, and then distilled. The crude silver left is purified by cupellation, whereafter the silver and gold can be parted with nitric acid. The condensed mercury can be used again.

Barba tried to improve upon the basic patio process and, in the 1590s, developed his variation. In this, the amalgamation was accelerated by boiling the ingredients in wooden copper-bottomed tubs. But this had disadvantages; fuel was not easily available at Potosi and the process was not efficient. But in Europe, from where it is possible that the process originally came, it was revived around the 1780s in a mechanized form. Attempts to reintroduce the method into Peru failed, perhaps because of different types of ores.

There was some difficulty in producing all the mercury required locally and this was cut off from time to time by wars (e.g. 1780–1800).

Some ores responded well to the simple, dry, process; others needed the addition of suphates and/or chlorides such as verdigris to make it work.

The reactions involved are probably:

$$(FeCu)SO_4 + NaCl = CuCl + FeCl + NaSO_4$$

$$CuCl \quad Ag_2S = CuS + 2AgCl$$

$$2AgCl + 2Hg = 2Ag + Hg_2Cl_2$$

The losses which occur are mainly due to the production of Hg_2Cl_2 in the third reaction. Barba found that the loss of mercury was equal to the silver recovered. The mercury rich amalgam, suitable for filtering and distilling, contained 66% amalgam and 33% excess mercury. Considering that the amalgam itself must contain at least 60% Hg, this means a total mercury content of over 73%, most of which should have been recoverable. The process was introduced at Potosi in 1574 and, up to 1630, over 10 000 t of silver had been made. This means that a similar tonnage of mercury had been lost.

Another metal began to make its appearance at this time — platinum. We have already referred to its existence amongst the native metals of South and Central America and it was not long before its occurrence was being notified to the Spanish authorities in Seville, as its presence had been noticed in the products of cupellation. It is soluble in copper and gold and hardens and embrittles the latter; its name came from plata (silver), platina being the derogatory diminutive. The native metal is impure, consisting of an alloy of metals of the platinum group (Pt–Pd–Rh). For a time, its export to Europe was prohibited because, owing to its high density, it could be mistaken for gold when plated with the latter. Its intrinsic value had yet to be recognized.[127]

Other metals

As far as we know, zinc was not being made in its own right intentionally in Europe during this time, but it was being imported by the East India Company from about 1605 onwards. This was undoubtedly coming from China, where zinc was being used for coins in the Ming Dynasty (1368–1644). Slabs weighing 60 kg were being produced by 1585 and these contained over 98% Zn, and traces of iron and lead. Zinc ores are said to have been abundant in Hunan, Hupeh, Kweichow, and to have been extensively smelted in Yunnan, but there are no detailed accounts of the methods.[128]

By about 1736, nickel-silver, known as paktong, paitung or petong was being imported from the same source.[129,130] The analyses of this material is given in Table 55. These are nickel-brasses, sometimes called nickel-silvers or German silvers. But this is not the first appearance of this type of alloy. Bactrian coins (c. 170 BC) have been found to contain up to 42% Zn and 20% Ni, the rest being mostly copper.[128,131]

According to an official Chinese encyclopedia, compiled in 1637,[128] complex copper ores were first washed and then smelted in simple furnaces with two tapholes, copper alloy from one, and lead from the other. In 1870, paitung or white copper was produced directly from complex ores containing copper, nickel and zinc metals in varying proportions. One of the places where this was done was Hweili, Sikang, where a pyrrhotite occurs containing 1–2% Ni and 0.5–0.6% Cu, and there are similar ores in Burma and India. These are in fact very like the famous deposit in Sudbury, Ontario. To the smelted copper-nickel alloy local copper smiths added the required amounts of other metals. The manner of production of the added zinc has been described in the previous chapter.

Table 55 Analyses of Paktong[129,130]

Element	Amount, %		
	1758	1765	1822
Cu	40·9	57·9	40·4
Zn	45·0	32·2	25·4
Ni	11·1	7·7	31·6
Fe	2·5	2·5	2·6
Pb	0·23	tr.	—
Co	0·16	—	—

References

1   V. BIRINGUCCIO: 'Pirotechnia', (Trans. C. S. Smith and M. T. Gnudi), 1942, New York

2   G. AGRICOLA: 'De Re Metallica,' (Trans. from 1556 edition by H. C. and L. H. Hoover). 1912, London, reprinted 1950 by Dover Publications, New York

3   M. B. DONALD: 'Elizabethan copper', 1955, London, Pergamon

4   N. BOURBON: 'Ferraria, Nugae', 1533, Paris, (in Latin); E. Straker has published a translation from the French in his 'Wealden iron', 41, 1931, London

5   DANIELLE ARRIBET: 'La siderurgie indirecte dans le pays de Bray Normand (vallée de la Béthune) de 1485–1565', Memoire de Matrice, Univ. de Paris I, Sorbonne, année sectaire, 1985–86

6   D. W. CROSSLEY: 'A 16th century Wealden blast furnace: a report on the excavation at Panningridge, Sussex, 1964–1970', Post-Med. Arch. 1972, 6, 42–68

7   M . M . HALLETT and G. R. MORTON: J. Iron Steel Inst., 1968, 206, 689 and P. J. BROWN: 'The early industrial complex at Astley, Worcester', Post-med. Arch., 1982, 16, 1–19

8   R. F. TYLECOTE: J. Iron Steel Inst., 1966, 204, 314

9   Bull. HMG 1964, 1, (3), 3

10   G. R. MORTON: J. Iron and Steel Inst., 1962, 200

11   H. G. BAKER: TNS, 1943-5, 24, 113

12   B. G. AWTY: 'The continental origins of Wealden ironworkers, 1451–1544', Econ. Hist. Rev., 1981, 34, (4), 524–539, and 'The Origin of the blast furnace evidence from Francophone areas', JHMS, 1987, 21, (2), 96

13   J. W. GILLES: Archiv. Eisenh., 1952, 23, 407

14   S. M. LINSLEY and R. HETHERINGTON: 'A 17th century blast furnace at Allensford, Northumberland', JHMS, 1978, 12, (1), 1–11

15   The excavation of the 17th century blast furnace of La Pelousse at Pinsot was reported by M. Benoit at the conference of CPSA at Vallecamonica, Italy in October 1988

16   H. R. SCHUBERT: J. Iron Steel Inst., 1942, 146, 131

17   R. F. TYLECOTE: ibid., 1966, 204, 314

18   D. W. CROSSLEY: Econ. Hist. Rev., 1966, 19, (2), 273

19   E. SWEDENBORG: 'Regnum Subterraneum sive Minerale De Ferro (De Ferro)', 1734, Dresden and Leipzig

20   G. R. MORTON: Iron and Steel, 1966, 39, 563

21   Now preserved by the Sheffield Trades Historical Society

22   J. B. AUSTIN: J. Iron Steel Inst., 1962, 200, 176

23   First Ironworks Gazette: 1951-5, Saugus, Massachusetts

24   Bull. HMG 1964, 1, (3), 3

25   A. RAISTRICK: TNS, 1938–9, 19, 51

26   G. JARS: Voyages Métallurgiques, 1774–81, 3 vols., Lyon

27   L KRULISRANDA: RHS, 1967, 8, (4), 245

28   J. KORAN: 'Vyvoj Zelezárstvi v Krusnych Horách',

62pp, 1969, occasional paper no. 8, Prague, National Technical Museum

29   W. H. SANSOM: *Metallurgia*, 1962, 65,165

30   J. NEEDHAM: 'The development of iron and steel technology in China', 1958, London, Newcomen Society

31   H. D. McCASKEY: *Eng and Mining J.*, 1903, 76, 780

32   D. B. WAGNER: 'Dabieshan', Scand. Inst. Asiatic Stud., Curzon Press, London and Malmø, 1985

33   R. SCHAUR: *Stahl u. Eisen*, 1929, 49, 489

34   D. W. CROSSLEY and D. ASHURST: *Post-Med. Arch.*, 1968, 2, 10

35   R. F. TYLECOTE and J. CHERRY: *TCWAAS*, 1970, 70, 69

36   J. PERCY: 'Metallurgy; iron and steel,' 1864, London, 279 Murray

37   P. TEMIN: 'Iron and steel in 19th century America', London, 1967

38   HELEN CLARKE (ed. and trans.), 'Iron and Man in Prehistoric Sweden', Stockholm, 1979

39   O. EVENSTAD: *Bull. HMG*, 1968, 2, (2), 61

40   H. R. SCHUBERT: *TNS*, 1951–3, 28, 59

41   ALEX DEN OUDEN: 'The production of wrought iron in finery hearths, Part I, The finery process and its development', *JHMS*, 1981, 15, (2), 63–87, 'Part 2, Survey of remains', *JHMS*, 1982, 16, (1), 29–32

42   M. DAVES-SHIEL: *Bull. HMG*, 1970, 4, (1), 28

43   R. PLOT: The natural history of Staffordshire, 1686, Oxford

44   G. R. MORTON: *The Metallurgist*, 1963, 2, (11), 259

45   H. R. SCHUBERT: 'History of the British iron and steel industry', 1957, London

46   G. R. MORTON and J. GOULD: *J. Iron Steel Inst.*, 1967, 205, 237

47   H. C. B. MYNORS: *Trans. Woolhope Naturalists' Field Club*, 1952, 34, 3

48   H. R. SCHUBERT: *op. cit.*, Appendix X, 402

49   M. BOULIN *et al.*: *RHS*, 1960, 1, 7

50   F. OEHLER: *Stahl u. Eisen*, 1967, 87, 207

51   *Bull. HMG*, 1970, 4, (2), 83

52   R. C. BENSON: 'Forging in the Past', 1957–58, English Steel Corporation News

53   C. S. SMITH: *RHS*, 1966, 7, 7

54   R. JENKINS: *The Engineer*, 1918, 125, 445, 486; H. R. SCHUBERT: *op. cit.*, 311

55   V. BIRINGUCCIO: *op. cit.*, 68

56   J. E. REHDER: 'Ancient carburization of iron and steel', *Archeomaterials*, 1989, 3 (1), 27–37

57   D. BROWNLIE and BARON de LAVELEYE: *J. Iron Steel Inst.*, 1930, 121, 455

58   J. A. ROPER: *West Midlands Studies*, 1969, 3, 73

59   T. A. WERTIME: 'The coming of the age of steel', 1961, Chicago, Chicago University Press; *EA. News*, 1964, 43, 159

60   J. A. R. RUSSELL: *EA. News*, 1964, 44, (506),188

61   R. F. TYLECOTE and B. J. J. GILMOUR: 'The metallography of early ferrous edge tools and edged weapons', *BAR Bri. Ser.*, 155, Oxford, 1986

62   J. M. P. VALLE: 'Comentarios Metallurgicos a la technoloigia de procecesos de Elaboracion del acero de las espados de Toledo decritas en el documento de Palomares de 1772', *Gladius*, 1986, 17, 129–155

63   Cementation furnaces in Sheffield were seen by H. Kahlmeter during his visit in 1724-25

64   J. K. HARRISON: *North East Ind. Arch. Soc. Bull.*, 1968, (9), 12, 18

65   M. W. FLINN: *TNS*, 1953-5, 29, 255

66   M. W. FLINN: 'Men of Iron', 1962, Edinburgh

67   J. D. AYLWARD: *EA. News*, 1962, 41, (476), 40; (477), 66

68   R. JENKINS: *TNS*, 1934–5, 15, 185

69   G. BERESFORD: personal communication on excavation at Goltho, Lincolnshire

70   H. R. SCHUBERT: *op. cit.*, Appendix X

71   M. H. JACKSON and C. de BEER: 'Eighteenth century gunfounding', 1973, Newton Abbot, David and Charles

72   F. PISEK: *Actes Xl Congrès Int. Hist. Sci*, 1965, 6, 84

73   D. S. BUTLER and C. F. TEBBUTT: 'A Wealden cannon boring bar', *Post-Med Arch.*, 1975, 9, 38–44

74   D. W. CROSSLEY: 'Cannon manufacture at Pippingford, Sussex: The excavation of two iron smelting furnaces of c. 1717', *Post-Med Arch.*, 1975, 8, 1-37

75   R. A. F. de REAUMUR: 'L' art de convertir le fer forgé en acier et l' art d' aducir le fer fondu', 1722, Paris

76   A. G. SISCO (trans.): 'Réaumur' s memoirs on steel and iron', 1956, Chicago, Chicago University Press

77   C. S. SMITH and BARBARA WALRAFF: 'Notabilia in Essays of Ores and Metals: A 17th century manuscript', *JHMS*, 1974, 8, (2), 75–87

78   I. KRULIS: Z. dejn. vied a techniky na Slovenska, 1966, 4, 77

79   D. W. HOPKINS: *Bull. HMG*, 1971, 5, (1), 6

80   G. GRANT-FRANCIS: 'The smelting of copper in the Swansea district of South Wales from the time of Elizabeth to the present day', 1881, ed. 2

81   H. KELLENBENZ (ed.): 'Schwerpunkte der Kupferproduction und des Kupfer Handels in Europa, 1500–1650', Kolner Kolloquium zue internationalen Sozia 1- und Wirtschaftsgeschichte Vol. 3, Koln, 1977

82   S. LINDROTH: 'Gruvbrytning och Kopparhantering vid Stora Kopparberget', 2 vols., 1955, Uppsala

83   It seems that, for a period at least, the copper was exported as the 90% grade and refined in the towns of the Hanseatic League, probably Hamburg (see *Plate Money*, Stockholm, 1987)

84   R. JENKINS: TNS, 1943–5, 24, 73

85   E. SWEDENBORG: Regnum subterraneum sive minerale de cupro et orichalco (De Cupro), 1734, Dresden and Leipsig

86   Translated extracts from the travel diaries of Thomas Cletscher and Henric Kahlmeter are in the Rhys Jenkins papers in the Library of Liverpool University; I am indebted to Mrs Joan Day for this information

87   D. DIDEROT and J. D'ALEMBERT: Encyclopédie, ou Dictionaire Raisonné des Sciences, des Arts et des Metiers, 1771–80, Paris

88   W. GOWLAND: *J. Inst. Metals*, 1910, 4, 4

89   J. DAY: 'Bristol brass; the history of the industry', 39, 1973, Newton Abbot, David and Charles

90   J. N. GOLDSMITH and E. W. HULME: *TNS*, 1942, 23, 1

91   L. JENICEK: 'Metal founding through the ages on Czechoslovak territory', 1963, Prague, National Technical Museum

92   M. B. DONALD: 'Elizabethan Monopolies (1565–1604)', Oliver and Boyd, London, 1961

93   H. HAMILTON: 'History of the English brass and copper industries', 1926, London, (ed. 2) 1967, Frank R. Cass and Company

94   M. COOK: *Trans. Birmingham Arch. Soc.*, 1937, 61, 11

95   P. J. BROWNE: *FTJ*, 1960, 108, (2253), 163

96   B. UPTON: *Copper*, 1970, 1, (1), 23

97   ALONSO BARBA: 'El Arte de los Metales', (Trans. by R. E. Douglass and E. P. Mathewson), London, 1923

98   G. AGRICOLA: *op. cit.*, 390

99   V.C.H. Derbyshire vol 11, 344; M. B. DONALD: 'Elizabethan Monopolies', p.144

100   D. KIERNAN: 'The Derbyshire lead industry in the 16th century', Chesterfield, 1989

101   H. OVERMAN: 'A Treatise on Metallurgy', New York and London, 1852,

102   M. BEVAN-EVANS: *FHSP*, 1960, 18, 75; 1961, 19, 32

103   W. J. LEWIS: *Ceredigion*, 1932, 2, (1), 27

104   J. DINN: 'Dyfi furnace excavations, 1982-87', *Post-Med. Arch.* 1988, 22, 111–142

105   J. PETTUS: 'Fodinae Regales or the History of Laws and Places', 1660, (Facsimile reprint by the Institution of Mining and Metallurgy, London)

106   Blackett, Matfen MSS, Ledger of Rents, 1685–91, University Library, Newcastle

107   R. H. WORTH: 'Dartmoor', 1954, Plymouth

108   G. A. M. GERRARD: 'The excavation of a medieval tin works at West Colliford, St. Neot Parish, Cornwall', University of Wales, MA Thesis, 1984

109   G. A. M. GERRARD: 'Retallack; a late medieval tin milling complex in the Parish of Constantine and its Cornish context', *Corn. Arch.*, 1985, (24), 175–182

110   R. F. TYLECOTE, E. PHOTOS and B. EARL: 'The composition of tin slags from the south west of England', *World Arch.* 1989, 20, (3), 434–450

111   R. M. L. COOK, T. A. P. GREEVES and C. C. KILVINGTON: 'Eylesbarrow (1814–1852)', T. Devon Assn. 1974, 106, 161–214

112   G. R. LEWIS: 'The Stannaries', 1908 (reprinted 1966)

113   D. B. BARTON: 'A history of tin mining and smelting', 1967, Truro, D. Bradford Barton Ltd

114   L. M. THREIPLAND: *Arch. J.*, 1956, 113, 33

115   R. F. TYLECOTE: *Bull. HMG*, 1965, 1, (5), 7

116   B. EARL: 'Melting tin in the West of England: Part 1', *JHMS*, 1985, 19, (2), 153–161: Part 2, 1986, 20, (1), 17–32

117   This can be seen very clearly in the triptych at Annaberg, Germany

118   ALAN PROBERT: 'Bartelome de Medina: the patio process and the 16th century silver crisis', *J. of the West*, 1969, 8, (1), 90–124

119   Recently found in excavations at Haithabu; personal communication from Dr. K. Schietzel

120   V. BIRINGUCCIO: *op. cit.*, 83

121   G. AGRICOLA: *op. cit.*, 4

122   A. OSWALD: *Trans. Birmingham Arch. Soc.*, 1962, 78, 81, 82, Fig. 13

123   *The Times*, Oct. 17, 1970

124   A. BARBA: *op. cit.*, 126

125   G. AGRICOLA: *op. cit.*, 297

126   A. LANGENSCHEIDT: 'Historia Minima de la Minera en Sierra Gorda', Windsor, Mexico, 1988 (There is an extensive review of this in English by R. D. Crozier in *JHMS*, 1989, 23, (2), 130–132

127   D. McDONALD: 'A history of platinum', 1960, London, 9

128   C. F. CHENG and C. M. SCHWITTER: *AJA*, 1957, 61, 360

129   A. BONNIN: 'Tutanag and Paktong', 1924, London

130   L. AITCHISON: 'History', vol. 2, 480

131   M. C. COWELL: 'Analyses of the Cu–Ni alloy used for Greek Bactrian coins', In: Archaeometry, Proc. 25th Int. Symp. Athens, (ed. Y. Maniatis), 1989, 335–345

132   G. W. HENGER: *Bull. HMG*, 1970, 4, (2), 45

133   *Tin and its Uses*, 1959, (49), 4

134   E. STRAKER: 'Wealden iron', 1931, London, (reprinted 1969), Newton Abbot, David and Charles

135   J. MAJER: 'Tezba cinu ve Slavkovském lese v. 16, Stoleti', 216, 1969, Prague National Technical Museum

Chapter 9

The Industrial Revolution; AD 1720-1850

An agreed definition of the Industrial Revolution is not easy to obtain; one authority would define it as the transfer of at least 50% of the means of production of any item from the house to the factory, and for many items this would date its start in Britain to about 1750. In the case of iron there is no such difficulty; the Industrial Revolution began in Britain with the transition from charcoal to coke as the principal fuel. This transition brought about the release of ironmaking from the inhibition caused by the effective shortage of charcoal as a fuel.

It has long been argued that shortage of charcoal was not the real cause of the failure of the iron industry to develop in Britain in the latter half of the 17th century, and its increasing reliance on the import of bar iron from Sweden, Spain, and Russia.[1] However, it is clear that there was intense competition for wood and charcoal and, although iron masters had shown considerable foresight and made advanced arrangements for a regular supply of charcoal, they were not popular with the local inhabitants who felt that they were being deprived of their right to local supplies for domestic purposes. Furthermore, growth of the industry was also limited by water power requirements, and coal and the steam engine were soon to be used for supplying the power for blowing.

The Transition to the Use of Coke and Coal in Ironmaking

The manufacture of iron by means of coal must have been a challenge from the Roman period at least. We know that at that time coal was used for various industrial processes,[2] and there is little doubt that it was tried for iron. In Britain, the earliest patents were granted to Sturtevant (1611), Rovenson (1613), and finally Dudley.[3] There is no evidence that the first two ever succeeded, and the granting of a patent to Dudley in 1622 suggests that, if so, nobody had been convinced. Of course, the patent situation in the 16th and 17th century was not as well organized as it is today and the claims were so all-embracing that it was possible to succeed in part of a claim but not the rest. We must remember that the use of coal in lead and copper smelting was general by the end of the 17th century, and the use of coal in other industries such as glass was very extensive.

The claims of Dudley[4] have been the subject of considerable argument[5,6] and must be considered seriously since it is obvious from his book *Metallum Martis*, pub-

lished in 1665, that he had considerable technical knowledge. Dudley claimed that he made changes to his first furnace at Cradley in Worcestershire, in 1619, to make it suitable for smelting with coal but we do not know what these changes were. When the furnace was demolished in the 1830s, it was found to be a square stone furnace 2.5 m across the boshes. He does not mention the use of limestone but the low sulphur coals were not so far away and limestone was available on his father's estate at Sedgeley Beacon. After a flood that overwhelmed his Cradley furnace he built a second furnace at Himley in Staffordshire, and later at Askew Bridge in the same county. The latter was 8.2 m square overall, larger than the first two and capable of producing 7 t/week.[7] It evidently had larger bellows than were usual for such a furnace. His second (1638) patent refers to 'bellows.... and additaments'; the latter could be taken as a reference to lime.

Recent excavation on the slag heaps at Himley showed that the slag contained a lot of coked coal but in other respects was typical of the early 17th century, with only 11.5% $(CaO + MgO)$ and 9.3% Fe_2O_3, together with 24% Al_2O_3. The coke could have been charged to the furnace as coal. The sulphur content of the slag was 0.12%, and it is thus possible that Dudley produced iron with coal but could not find much sale for it, as most of the iron produced at that time was destined for the forge and this would be unsuitable for conversion. Iron produced from Shropshire coal, with a slag of such low basicity,[8] would be far too high in sulphur to make satisfactory iron fted by vandals slashing his bellows, and by riots and lawsuits. His petition for extension of his patent, after the restoration of the monarchy in 1660, was apparently unsuccessful. Thus, while many had been attempting to smelt iron with coal, and some indeed had been granted patents, it is clear that Abraham Darby I and his partners were the first to achieve this in a practical way. Amongst other things, Darby had been concerned with the making of malt mills in Britain since 1699 and knew something about low-sulphur coals suitable for the malt industry. The Shropshire coalfield was noted for these and, after making experiments with moulding and casting techniques for thin-walled pots, he moved to Coalbrookdale in Shropshire where he leased a disused blast furnace which had been built in 1638 by Basil Brooke, and owned by him until 1695.

The records show that the furnace was in blast again by 4th January 1709, but this time on coke.[9] The most important reason for this revolutionary step was the fact that the easily worked coal seams (i.e. those near the outcrop) in the Shropshire coalfield had a low sulphur content (0.50–0.55%) and good coking properties. This was unusual in Britain; some coals have high sulphur, some make good but unreactive coke, but some of the Shropshire seams produced good coke with high reactivity and low sulphur. Modern furnaces use cokes with much higher sulphur contents than these (1–2.5%), but this is taken care of by a very basic (high lime) slag which can only be obtained with hot blast and high blowing rates.

Coking does not usually reduce the sulphur content of coal very much since the loss of sulphur as sulphur dioxide is balanced by the loss of hydrocarbons. The loss in weight on coking varies from 35 to 70%, according to the method of coking.[10] The early coke was made in heaps in the open, as in charcoal making (meiler heaps). In Britain, the start of coking in ovens can be dated to about 1765 in the Newcastle area.[11]

The outer part of Darby's original furnace still stands,[12] but its present internal state is that adapted for some other purpose. Therefore, we do not know the 'lines' of the furnace Darby used for his original coke smelting. The original case iron gave a hot-short wrought iron, due to its sulphur content, and was not suitable for the best grades of forge iron, although this quality came to be accepted later. The original sulphur content of the cast iron would be about 0.1% but there was sufficient manganese present in the iron to form the more innocuous MnS. The analyses are as given in Table 56. The silicon shows a sharp increase over that normal for charcoal iron (0.7–1.0%) and tends to give a greyer iron, softer, and more easily machined in thin sections. However, this was not as good as white iron for the finery-forges, as shall be seen later.

Abraham Darby I died in 1717 when Darby II was only 6 years old, so the Company was for a time under the control of the other partners. But, by 1732, Darby II was beginning to take an active part. By this time motive power had become a problem. Horse gins had been used since 1732 for pumping water back from a low-level pond to a high-level pond so as to conserve it for its use on the wheel for blowing bellows. The first change made by Darby II was the use of the atmospheric (Newcomen) engine for this purpose in 1742. This started the gradual development of more powerful blowing engines to supersede the large and cumbrous leather bellows which required 120 kg of leather annually.

These two important changes released the blast furnace from its need to be situated in areas where water and wood were plentiful and allowed its migration to areas where coal was plentiful. It happened that most coalfields in Britain also had carbonate ores in the shale layers between the coal seams and this ore proved eminently suitable for iron production (analyses in Table 57). The growth of the industry from 1710 onwards, therefore, really depended on the availability of capital and the solution of the engineering problems involved in making use of the steam engine for efficient

Table 56 Analyses of coke-smelting Shropshire iron

Element	Horsehay 1756	Ironbridge*, 1779 Strut	Ironbridge*, 1779 Arch	Coalbrookdale 1779
Total C	3·28	3·25	2·65	3·63
Si	1·57	1·48	1·22	1·40
Mn	0·56	1·05	0·46	1·09
S	0·09	0·037	0·102	0·07
P	0·57	0·54	0·54	0·52

*After Morton and Moseley [13]

blowing. These problems had to be solved by the iron producers themselves.

The early cylinders delivered by Darby to the firm of Boulton and Watt were not satifactory, in spite of Darby I's alleged prowess in the casting of thin-wall iron vessels. All we know is that they contained holes which had been filled up with some sort of filler. This problem was eventually solved by John Wilkinson who, with his father, had moved to the West Midlands from the North Lancashire charcoal furnace at Backbarrow, adjacent to which his father (Isaac) had been casting smoothing irons for linen.[14] The Backbarrow furnace was blown-in in 1711, but Isaac Wilkinson had the vision to realise that the future of iron lay in the coalfields and not in the forests of North Lancashire, where there was already serious limitations on fuel. So, about 1753 he moved with his son to the West Midlands and Isaac bought Bersham Furnace, near Wrexham, from the Coalbrookdale Company; this became Wilkinson's principal furnace from 1753 onwards. It was sold mainly because of the need for captial to modernize its blowing equipment, which the Darby Company could not provide.

While the Wilkinsons concentrated on the cast iron side, Darby turned his attention to the supply of pig iron for the forges. A satisfactory iron was being made by about 1750, and one supposes that it was of the quality that was supplied for the Ironbridge in 1779 (i.e. 0.04–

Table 57 Analyses of Shropshire ores as mined (after Percy[171])

Percy's No.	53 Donnington	54 Wood	55 Madeley Court	56 Madeley Wood
FeO	45·12	38·9	44·19	51·45
MnO	1·78	1·31	1·0	0·54
Al₂O₃	0·35	tr.	0·41	0·13
CaO	2·80	2·54	1·63	2·13
MgO	4·08	4·65	3·40	0·42
SiO₂	8·90	19·60	13·87	9·60
CO₂	34·0	29·81	32·0	33·3
H₂SO₄	0·49	–	0·06	–
P₂O₅	0·46	0·25	0·29	0·23

0.1% S, Table 56). Its percentage of phosphorus rendered it cold-short and therefore very unsuitable for nailmaking.

When Abraham Darby II built his furnaces at Ketley and Horsehay in 1755 and 1756 he erected steam engines with cylinder bellows for blowing.[15] Knight installed similar bellows at Charlcot[16] in 1763. When Darby III took over in 1768 this type was installed at the Dale. Even so, the steam engine was not connected directly with the bellows but merely pumped the water from a lower to a higher pond, from which it turned a waterwheel operating the bellows with the aid of cams (in 1742 the difference of level was 37m). Many other engines were in use at the time for the drainage of the Dale Company's coalpits and the company were involved in making many parts of Newcomen engines for use all over the country. Smeaton seems to have been one of the first to use iron blowing cylinders in a design for the furnaces at Carron in 1768, but even these were powered by a waterwheel (*see* Figs. 83 and 84).

By 1776, a method of direct blowing had been worked out at New Willey in Shropshire using a Boulton and

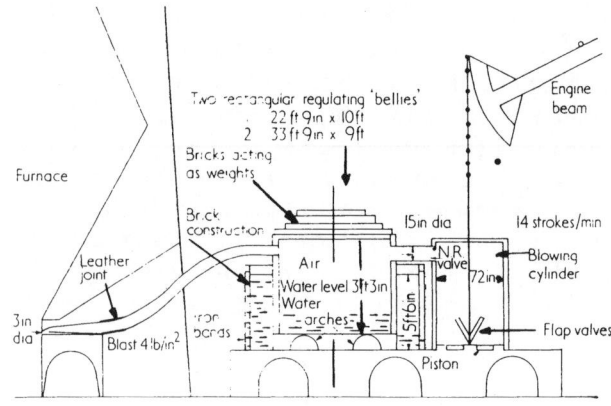

85 *Steam blowing arrangement at New Willey furnace, Shropshire*

Watt beam engine (*see* Fig. 85). A Boulton and Watt condensing engine and a sun and planet gear to obtain rotary motion had also been evolved by 1782, and these engines were applied to blowing and to all sorts of rotary action in forges such as hammers, and rolling and slitting mills. Smeaton's engine had four blowing cylinders designed to smooth out the irregularities of single cylinder blowing. But the single cylinder beam engines, such as were used at Horsehay (*c.* 1790) and Hollins Wood (1793) had double acting blowing cylinders and water regulators for this purpose.[17,18]

By the end of the century, coal and the steam engine had revolutionized the making of iron in Britain; their effect can best be shown by the production figures given in Table 58. Here we see that, by the use of coke, the annual iron production had increased from 20 500 t/a in 1720 (almost all of which was made using charcoal), to over 250 000 t in 1806, which was virtually all made with coke. We have a fairly complete set of figures for the Horsehay (Shropshire) furnace from 1755–1806, from which we can get an idea of the materials used and the increase in efficiency[19] (*see* Table 59). The ironstone was calcined at the furnace, and ore and coal would be mined at the same pits. Over this period there is little change that cannot be accounted for by change in the iron content of the ore. The smelting fuel/ore ratio varies from 0.3–0.42 while the smelting fuel/iron ratio (coke rate) varies from 1.66–2.0. It is estimated that the manufacture of 1 t of coke required about 3.3 t of coal.

Over the period, production per furnace increased from 13 t to 36 t/week. While we have no precise slag analysis for this furnace, the ratio of calcined ore to limestone at Coalbrookdale was roughly 4:1. If we assume that the CaO content of the limestone is about 54%, and the composition of the calcined ore is; Fe_2O_3, 70%; SiO_2, 20%; MnO, 1.5%; Al_2O_3, 0.6%; CaO, 2.5%; MgO, 5.0%, then we would expect the CaO + MgO content of the slag to be 48%, which would give a basicity (CaO + MgO)/SiO_2 of 1.03. This is in good agreement with the slag found on the site of Wilkinson's furnace at Bradley[20] which would have been dated to the period 1770–1800 and have been produced direct from coal or coke with a cold blast (*see* Table 60).

83 *Water-driven iron-blowing cylinders from Duddon furnace (from Morton[45])*

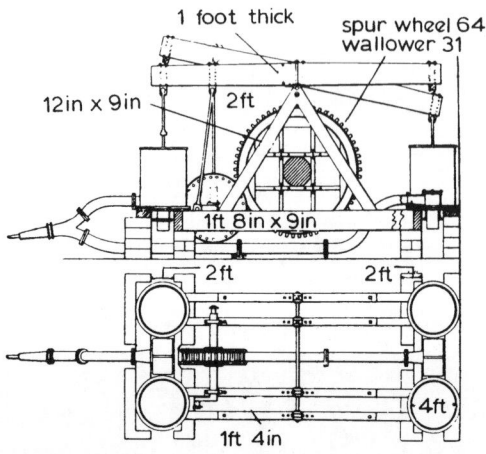

84 *Smeaton's blowing cylinders of 1768 (after Mott[15])*

Table 58 Iron production in UK 1720-1806

| | Production, t | | | | | |
| | Cast iron | | Bar iron | | Imports (Bar iron) (Charcoal) | Exports (Total) |
Date	Charcoal	Coke	Charcoal	Coke		
1720	20 500	400	14 800	nil	17 100[1]	—
1750	24 500	2 500	18 800	100	31 200[2]	c. 5 000[3]
1788	14 000	54 000	11 000	22 000	48 200[4]	—
1796	8 500	112 500	6 500	125 000	38 000[5]	24 600
1806	7 800	250 500	6 000	—	—	31 500

Notes; most of the cast iron would be converted into bar iron and therefore the bar iron total is a more realistic figure of total iron production; (1) mean 1714–18; (2) mean 1751–55; (3) estimate; (4) mean 1788–96; (5) mean 1796–1806
Source; B. R. Mitchell and P. Deane: 'Abstract of British historical statistics', 140, 1962

Table 59 Quantities used per tonne of iron made; Horsehay Works, Coalbrookdale Company (after Mott[19])

Date	Ironstone (raw)	Limestone	Coke	Coal General	Coal Calcining	Total fuel
1755	4·08	0·51	5·5	2·65	0·37	8·52
1756	4·40	0·59	4·55		0·60	7·11
1757	4·97	0·62	5·36		0·37	7·85
1758	4·95	0·92	6·25		0·75	9·82
1759	4·00	0·87	5·53		0·64	9·25
1760	4·55	1·63	5·30	2·80	0·53	8·63
1767	4·31	0·48	7·00	1·74	0·93	9·67
1768	2·86	0·69	5·87			7·88
1769	3·81	0·89	5·72			7·97
1770	3·60	0·56	5·65			7·68
1771	3·80	1·13	5·76			8·63
1772	4·44	0·83	6·7			10·33
1773	3·51	0·43	5·74	2·34	0·48	8·56
1796				1·47*		
1797				1·96		
1798				2·02		
1799			6·19	1·40		
1800			5·75	1·22		
1801		0·95	7·95	0·70		
1802	4·05	0·78	5·78	0·78	0·15	
1803	3·00	1·00	5·65	0·91	0·16	
1804	3·55	0·85	7·38	1·13	0·14	
1805	4·18	0·73	7·40	1·18	0·13	
1806		0·74	6·95	1·25		

Quantities based on the following;
  ironstone; 1 load = 18 bushels = 1 t
  bulk density of clay ironstone = 100 lbs/ft^3
  coal, 1718–27; stack = 9 loads = 36 cwt; 1728–32; stack =
  10 loads = 35 cwt lump coal; 1 dozen = 12 loads = 42 cwt
  limestone; 8 loads = 1 t
*After 1796, for fire engine only

Wilkinson's slag was found in contact with cast iron —the two having been run from the furnace together. The sulphur content of the slag was 1.37, while that of the pig iron was 0.027%. This shows that Darby and Wilkinson between them had solved the problem of sulphur. Although both had used a low-sulphur coal this was not the main reason. The main reason was the use of a high lime slag with a high basicity index, which needed a high working temperature and therefore a high blowing rate. It would appear that it was really the use of the steam engine which made this possible (*see* Fig. 86).

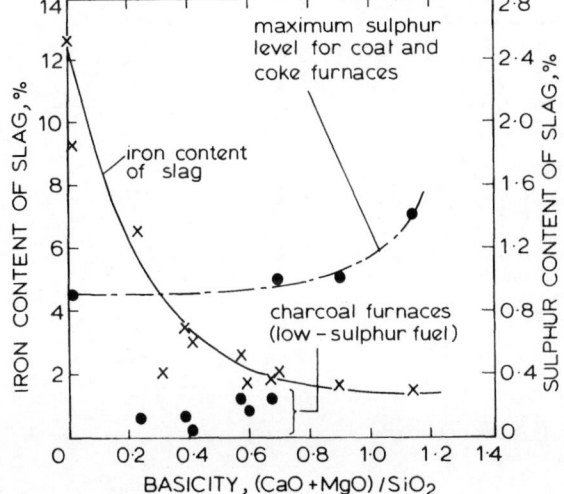

86 *Effect of basicity on the iron and sulphur contents of blast furnace slags*

Table 60 Analyses of 17th-19th century blast-furnace slags

Element	Amount, %					
	*c.*1650 Himley, coke or coal cold blast[7]	1700–1792 Charlcot, charcoal, cold blast[7]	1770–1800 Bradley, coal, cold blast[7]	*c.*1850 Tipton, coke, cold blast[7]	*c.*1850 Dowlais, coal, hot blast[7]	1836 Decazeville,[50] coal or coke
SiO_2	47·80	52·5	38·10	39·52	44·60	46·0
CaO	7·60	17·0	41·10	32·52	29·19	25·3
MgO	3·93	4·57	2·52	3·49	1·70	4·6
MnO	0·55	1·86	1·86	2·89	2·62	1·5
FeO	–	–	–	2·02	2·63	2·0
Fe_2O_3	9·30	4·30	–	–	–	–
Al_2O_3	23·99	20·17	11·18	15·11	18·10	17·0
P_2O_5	0·02	0·32	0·05	–	0·10	–
S	0·12	0·01	1·37	0·96	1·06	2·3
Alk.	–	–	–	1·06	–	–
Fe	6·5	3·0	1·45	1·6	2·05	1·6
Basicity	0·24	0·41	1·15	0·91	0·70	0·65

The Conversion of Cast Iron to Malleable Iron

In view of the fact that the bulk of the demand was for wrought iron rather than cast iron, Darby I's use of coke did not by itself have much immediate effect on the total quantity of metal produced. The charcoal-fired finery was still the only satisfactory method of conversion up to about 1770. Fuelled with charcoal, this was not capable of causing any change to the sulphur content of the 'fined' pig, owing to the low basicity of the fayalite slag. Fuelled with coal or coke, however low the sulphur content, most of the available sulphur went into the iron, thus reducing the quality. Of course coal could be used in the chafery, and had been so used for a considerable time. This is because the area of interfacial contact between metal and fuel was now so reduced and no liquid metal phase was present. (*see* Tables 61 and 62).

The use of coke in the blast furnace tended to increase the silicon content so that the early coke irons tended to exceed 1.0% Si and were all grey or graphitic irons. This in turn tended to make the finery process rather slower than with the former low-silicon or white forge pig, and it seems clear that the finery had been coping with the problem in two stages: first by oxidizing the silicon and thus converting the graphite to combined carbon, and then following with a second oxidation, this time of the combined carbon (*see* Table 61).

We now see the division of fining into two processes: the first was carried out in the 'refinery' and the second in the 'finery'. Actually, the idea of the 'refinery' was not confined to Britain. It appears to have been in use in Styria in the 1750s when its use preceded the finery.[21] Soft (presumably graphitic) cast iron was melted with charcoal in a *braten ofen* 2.4m long and 1.8 m wide, with a tuyere on each side. After 14–15 hours the melted metal ran down the inclined hearths to a channel in the middle. During this process it acquired some malleability as it broke with difficulty, so more than the silicon was removed. After this, the conversion was completed in the usual iron-plated finery hearth (*see* Fig. 87).

The Darby partners tried to do the first stage (refin-

Table 61 Composition of metal used in finery, refinery, and puddling processes

Metal	Element %						
	C(graph.)	C(comb.)	Si	Mn	S	P	Slag
Cast iron							
Cold blast charcoal pig[172]	2·43	1·43	0·85	0·05	0·029	0·11	
Cold blast coke pig[171]	3·52	–	1·86	–	0·05	1·72	
Cold blast coke forge pig[171]	2·81	–	0·57	0·13	0·06	0·29	
Refined iron[185]	–	3·15	0·20	–	0·04	0·80	
Composition of wrought iron produced, %							
From puddled coke pig (Yorks)[185]		0·27	0·11	–	0·01	0·06	3·0
Swedish[30]		0·08	0·11	tr.	0·03	0·004	–
Low Moor[30] (Yorks)		0·016	0·122	0·28	0·104	0·106	–

Table 62 Composition of slags from finery and puddling processes

	Amount, %				
	Finery	Refinery	Puddling furnace, Staffs		Chafery
Element	Sparke Forge, charcoal fuel hematite iron [172]	Dowlais, coke [30]	White iron [30]	Grey iron [171]	Sparke Forge, coal fuel [172]
Fe_2O_3	50·4		8·27	23·75	27·6
FeO	33·6	65·5	66·32	39·83	33·10
SiO_2	8·16	25·8	7·71	23·86	21·3
MnO		1·6	1·29	6·17	
Al_2O_3	4·65	3·6	1·63	0·91	2·52
CaO	2·60	0·45	3·91	0·28	5·68
MgO	0·54	1·28	0·34	0·24	1·77
S		0·23	1·78		0·5–5·0
FeS				0·62	
P	0·25	1·37	3·50	6·42	0·12

ing) with coke in about 1719, but this was not a great success and it appears that the whole process continued to be done with charcoal.[22] Many others had been trying to achieve conversion with coke or coal. William Wood mixed powdered ore, coke, and lime together and charged the mixture into a coal-fired reverberatory furnace. In 1761 his son, John Wood, was also granted a patent.[23] In this, he first fined the cast iron in an ordinary refinery with coal until it was nearly malleable (the metal would contain some sulphur), then broke up the metal, mixed it with fluxes, and heated it in crucibles in a reverberatory furnace (*see* Fig. 88). This was one of the many 'stamping and potting' processes in which

granulated cast iron was oxidized in contact with suitable fluxes to convert it to iron or steel. Bergman also tried such techniques in Sweden in 1781.[24] The pots usually broke during this process leaving small piles of malleable iron in the furnace hearth. This process was carried out at Little Astoll Forge which John Wood leased in 1740.[25] In accordance with his patent in 1761, he seems to have refined granulated charcoal pig from Hales furnace in a coal-fired finery, producing a large quantity of fayalite slag with an unusually large sulphur content (0.59–3.22% S). The high figure must be due to the liquation of the sulphide phase.

There is little doubt that the flux was sufficiently

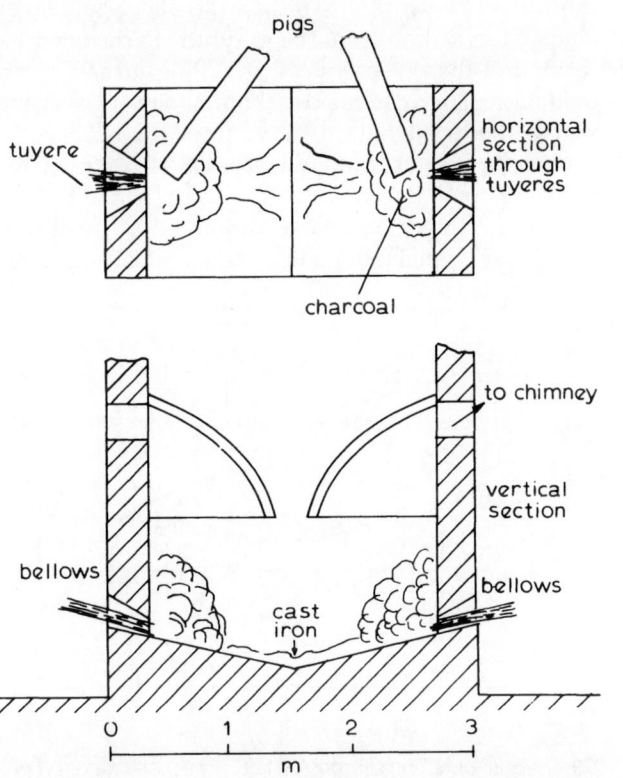

87 Early refinery for the removal of silicon from cast iron by oxidation, thus converting grey cast iron to white cast iron (based on a description of Jars[11])

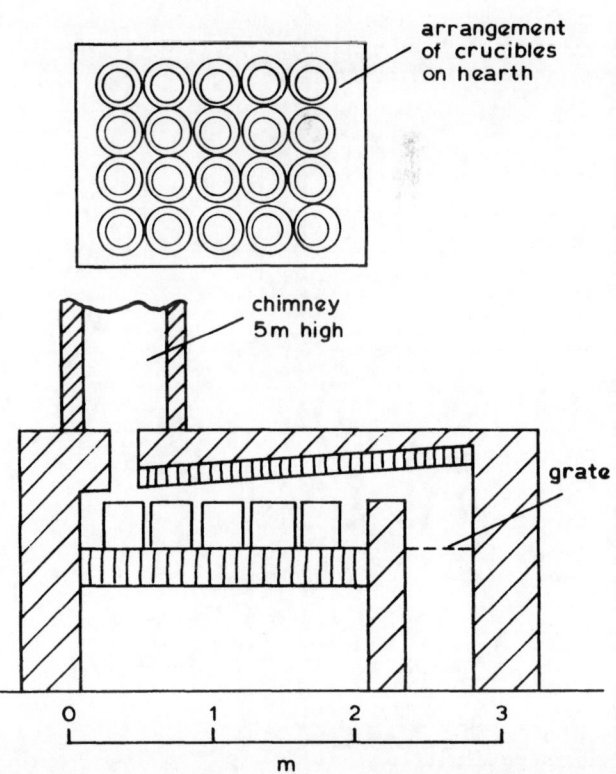

88 Reconstruction of a reverberatory furnace used by John Wood in 1761 for the fining of cast iron by 'potting' (based on Morton and Gould[25])

basic to absorb some sulphur. The slag was found in large masses ('hambones') weighing up to 25 kg. These seem to have been produced by tapping at intervals from the finery hearths.[26] The metal so produced would contain about 0.2–0.3% S: the coal itself was found to contain 1.10%. The fined metal was broken up and placed in clay pots about 30 cm high by 15 cm i.d. and containing about 45 kg of metal Kelp[27] and other alkali fluxes were added, and presumably the lime content would assist the removal of some of the 0.2–0.3% in the form of $CaSO_4$ and Na_2SO_4 or sulphides, in the slag. These crucibles were heated in coal-fired reverberatory furnaces. The second patent of 1763, in the name of John and Charles Wood, indicates that it was sometimes necessary to carry out the refinery operation twice, but this would be carried out in a reverberatory furnace like the second stage. This is the earliest reference to the possibility of complete conversion from cast iron to wrought iron in the reverberatory. De la Houlière visited Wood's works in 1775 and appears to have been impressed by the quality of the bar iron produced by this process.[28]

John Wood died in 1779 and by this time others had been developing slightly different techniques. In 1766, the Cranage brothers 'fined' pieces of partially oxidized cast iron scrap (old bushes) in a reverberatory furnace. At the same time Purnell was using grooved rolls for

making wrought iron bolts (see Fig. 89). Meanwhile, the finery itself was undergoing a change: multiple tuyeres were first used by Cockshutt in 1771 with charcoal, and he found it an advantage to preheat the cast iron charge in a reverberatory furnace. These two changes speeded up the process. In 1775, Jesson and Wright fuelled their finery with low-sulphur coke and fined coke smelted pigs. It seems that this was merely 'refining', i.e. reducing the silicon content and turning the carbon into the combined form.[29] The process was finished by putting the metal in a reverberatory with a hearth 3 m × 0.6 m, capable of holding 20 pots or piles.

It was left to Onions in 1783 to charge remelted and still-molten white cast iron from the air furnace directly into the reverberatory and complete its oxidation there with the aid of a forced blast (see Fig. 90). The puddled ball was worked with a hammer, but the rest of the process was exactly as in Cort's second patent of nine months later (1784). In his first patent of 1783, Cort used a finery, a natural draught reverberatory furnace, and grooved rolls, and emphasized more than anyone else the need for plenty of working to draw out the slag. In his patent of 1784 he dispensed with the finery and the whole process was one of puddling in a reverberatory with a sand bottom, which tended to absorb some of the iron and convert it into slag. This will be referred to as 'dry' puddling.

Gradually, the processes became simplified and, with the increasing availability of coke pig with its high silicon content, and the decline of charcoal pig, we find the general adoption of a two-stage conversion process — the coke-fired refinery, along the lines of Cockshutt's multituyere invention, to reduce the silicon content and change the carbon from the graphitic to the combined form, and the reverberatory puddling furnace for the oxidation of the combined carbon. As a result of Wood's patent the desiliconized metal from the refinery was removed hot and hammered into a cake 2.5–3.8 cm thick, showing some decarburization. Later, in the 19th century, the refinery became a 'running-out fire' and the metal was tapped into plate metal about 2.5–8 cm thick,

89 Grooved rolls for rod rolling designed by J. Purnell in 1766 — an essential component in the development of the puddling process (reproduced from Patent No.854, July 31, 1766)

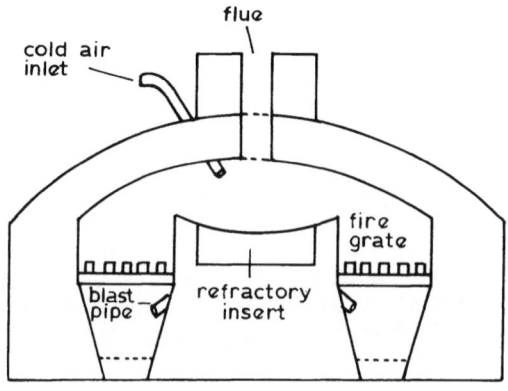

90 Onions' puddling furnace of 1783 in which oxidation of the premelted cast iron was assisted by a cold air blast; the chimney was short and the main part of the combustion air was supplied by air blasts under the grates (after Morton and Mutton[23])

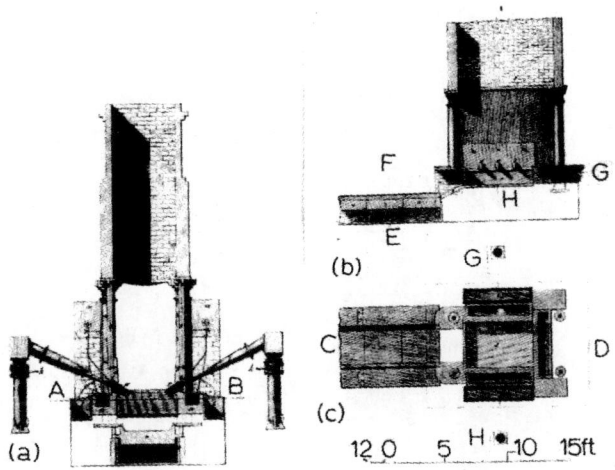

91 Late refinery or 'running out' fire (after Greenwood[30])

30 cm wide and 3.7 m long (*see* Fig. 91). In some cases refined iron was run directly into the puddling furnace, as Onions suggested.[30]

In 1818, Rogers suggested replacing some of the silica in Cort's reverberatory furnace bottoms with iron oxide, to avoid the loss of metal due to its oxidation and combination with silica. The oxygen in the iron oxide contributed to the decarburization of the pig iron.[31]

But Joseph Hall improved on this.[32,33] He worked a charge of iron scale (bosh cinder and slag) and obtained a visible reaction between the carbon in the iron and the oxygen in the iron oxide, producing blue flames of carbon monoxide ('puddler's candles'). This came to be known as pig-boiling or wet puddling, to be contrasted with the previous process of dry puddling. Hall also used roasted puddling cinder (tap slag) for his furnace bottoms instead of the iron oxide and silica sand used by Rogers. This was patented in 1838. A typical puddling furnace of this type is shown in Fig. 92. Another development was the use of the waste heat boiler by Rastrick in 1827.[34] This was a vertical cylindrical boiler through which the flues of the puddling furnace passed.

The process of oxidation in wet puddling was assisted by additions of iron oxide to the slag. This speeded up the process and allowed the working of unrefined grey pig. The first stage in such a case is the oxidation of the silicon to give a highly siliceous fayalite slag. Later, the white cast iron begins to oxidize, assisted by the iron oxide additions, and a reaction between carbon and iron oxide occurs, leading to the evolution of carbon-monoxide, i.e. 'the carbon boil'. Some sulphur and up to 80% of the phosphorus passes into the large volumes of slag produced.

This is perhaps the right place to discuss the technique and reactions involved in puddling in some detail, since from 1830 the process underwent no change.[35] If the charge was mostly white or forge pig the operation started with a higher temperature than if it was grey. This was controlled by a damper on the chimney. The furnace was hot from the previous charge and contained some rich tap slag from this charge. Then, 230 kg of cold metal were introduced through the working door which was lowered and luted. After 15 minutes the pigs were

moved and they became molten after about 30 minutes. The slag, scale, and metal were mixed by rabbling, and after 45 minutes vigorous boiling occurred. Rabbling continued and the bath began to thicken as its melting point rose owing to the loss of carbon. Specks of wrought iron could then be seen and pasty masses were found which were broken up, while the temperature was raised to separate the slag and metal. The masses of metal were rolled over the surface of the bed to the fire bridge, where about six balls, 32–36 kg in weight, were collected. A final superheat was given to promote easy expulsion of slag from the balls under the shingling hammer and allow good welding of the particles. The balls were drawn one by one from the furnace, while a moderately reducing atmosphere was maintained by shutting the damper and keeping the door closed.

The cinder (slag) was tapped out after every second heat. Normally a heat took about $1^{1}/2$ hours and 5–7 heats would be worked per 12 hour shift. With dry puddling not so much slag was formed and the cycle was faster, and 9–10 heats could be worked. But these were usually smaller, about 100–150 kg. Since the slag was usually fayalite with high iron content, a good deal of iron was lost in this way. This usually amounted to 7–18% of the metal charged. Of course, the cinder could be smelted in the blast furnace although it might contribute sulphur and phosphorus to an otherwise pure charge.

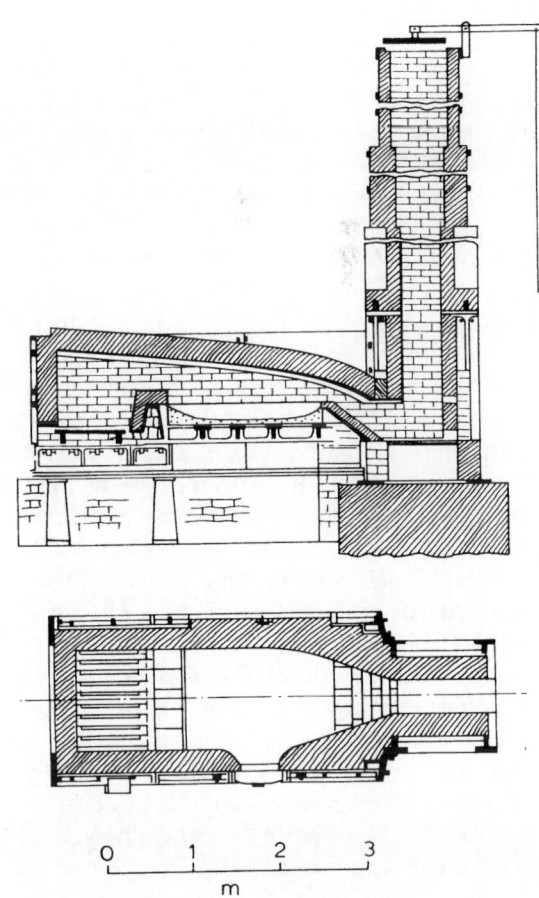

92 Typical Welsh puddling furnace with iron bottom

The Development of The Blast Furnace
AD 1700–1850

In the first section we considered the change from charcoal to coke. While this was to have an important effect on the size of the blast furnace, particularly its height, other changes took place during this period that had quite a considerable effect on the charcoal blast furnace. The fact that low-sulphur coke was shown to be a practical fuel after 1720 did not, of course, mean conversion to coke everywhere. First, not all coals are coking coals, and some coals that do not coke, and swell on heating, are suitable for the blast furnace, as Wilkinson had shown once he could get a strong enough blast. In some places charcoal was still plentiful and in some instances this was because charcoal ironmasters had ensured long term supplies when they built their furnaces. However, it was not easy, in countries with high population densities, to increase iron output by using charcoal. Not only was there a great demand for wood as domestic fuel but there was a strong tendency towards deforestation as a means of increasing agricultural output.

It is interesting to note that in one area of Britain (Yorkshire) charcoal furnaces were put into blast as late as 1761[36] (*see* Fig. 93), yet the Maryport furnace, in that remote area of Cumberland where it might be thought that charcoal was plentiful, was built for coke in 1752.[37] In fact, the charcoal ironmasters were seeking new areas with sufficient fuel as far away as Northern Scotland using North Lancashire hematite, and charcoal furnaces were blown in at Loch Fyne and Bonawe (1753), and at Invergarry (1729).[38,39] New charcoal furnaces were blown-in in Wales in 1720 at Brecon, and in 1755 at Dovey (Dyfi).[40] Another charcoal furnace was blown-in in the Midlands (Melbourne) in 1725, and it is not surprising that it was blown out around 1780.[41] Yet one of the first furnaces in the Midlands to use coal appears to be that at Alderwasley (1764), also in Derbyshire, where coal has been found in the charge.[42] This furnace preceded those built at Morley Park (1780 and 1818) which were designed for coal, and steam blown.

Other European countries sent observers to England in the second half of the 18th century to witness the use of coal and coke in ironmaking and to profit from it. The dates of adoption of coal or coke smelting in other countries (*see* Table 63) indicate to some extent the increasing pressure on charcoal resources. The Swedes and the French were the first to inspect the new techniques of coke-made iron in Britain. The Swedes had no real fuel problem, although a small amount of coal was imported from England. They were mainly interested in the effect that the new coke-fuelled processes might have on their exports to England, and to see whether there were any other new developments that they might adopt. The French sent a number of observers such as Jars and de la Houlière in the 18th century, and Dufrénoy and his colleagues in 1836.

Throughout the 18th century there was a steady increase in furnace size.[43] Table 64 shows the increase in the size of charcoal furnaces starting with the English

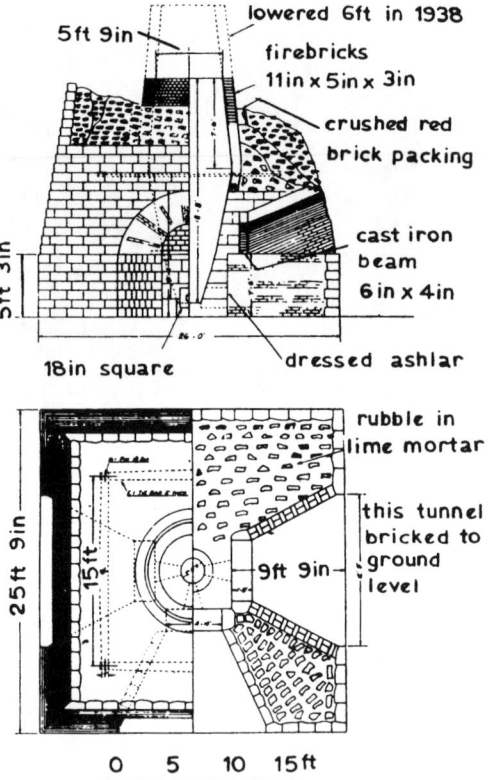

93 Low Mill blast furnace, near Cawthorne, Yorkshire (from Baker[36])

furnaces of 1651–2, and ending with the furnace at Nevyansk in Russia in 1794 with a height of 13.5 m. The height was thought to be limited owing to the poor mechanical properties of the charcoal, but in the case of the tall slender furnaces, much of the load would be transferred to the sides, not to the charcoal. It would seem that the shape and the availability of the air supply had much more to do with the height than the fuel.[44] Many of the early furnaces were below their possible maximum height (*see* Fig. 94).

The earliest furnaces had a minimum number of bottom openings (i.e. two) to avoid weakening the structure, and these were supported by iron lintel beams or stone or brick arches[45] (Low Mill and Duddon). But most of the 18th century furnaces had three or four openings in a square body, suggesting that their designers had envisaged the possibility of having more than one tuyere.[36] By now, the crucible-cum-bosh and shaft were both circular, avoiding the problems of joining a circular

Table 63 Dates for the beginning of the use of coal or coke in the blast furnace

Le Creusot (France)	1785
Glewitz (Silesia)	1796
Königshütte (Silesia)	1800
Seraing (Belgium)	1823
Mülheim (Ruhr)	1849
Vitkovice (Czech)	1836
Donetz (Russia)	1871
Bilbao (Spain)	1880
George Creek, Maryland (USA)	1817
East Pennsylvania (USA)	1835

Table 64 16th-18th century blast-furnace profile data

Location	Date	Height, m	Bosh dia., m	Bosh angle, deg.	Height/Bosh, H/B	Reference
Cannock	*c.*1561	–	–	78	–	Morton[174]
Coed Ithel	1651	6·1	2·2	77	2·9	Tylecote[176]
Sharpley	1652	7·6	1·8	80	4·2	Morton[177]
Rockley	1652	5·2	2·44		2·1	Reference[173]
Dovey	1735	10·4	2·8	60	3·7	Reference[184]
Gunns Mill	1682	6·7	2·1	(40–50)	3·2	Reference[183]
Duddon	1736	8·7	2·7		3·2	Morton[69]
Lamberhurst	1695	7·2	1·6	75	4·5	Swedenborg[175]
Bonawe	1752	9·2	2·44		3·7	–
Maryport*	1752	11·0	3·8	72	2·9	Tylecote *et al.*[37]
Cawthorne	1761	7·6	2·1	80	3·6	Baker[36]
Coalbrookdale*	1777	7·3	2·1	51	1·8	Raistrick[12]
Johangeorgenstadt		6·1	1·5		4·0	Jeniceck[44]
Osek	1750–1800	7·2	1·9	62	3·8	Jeniceck[44]
Strasice	1750–1800	8·0	1·6	63	5·0	Jeniceck[44]
Larvik	1767	7·3	1·8	60	4·0	Jars[11]
Vordernberg	1770	5·5	1·5	81	3·6	Jars[11]
Treybach	1753	6·7	1·1	83	6·3	Jars[11]
Sweden	1770	7·7	2·1	71	3·6	Jars[11]
Le Creusot	1777	10·7	2·9	72	3·7	Reyne[178]
Komarov I	1780	9·0	2·1	64	4·3	Jeniceck[44]
Adamov	1793	8·5	2·3	61	3·7	Jeniceck[44]
Komarov II	1796	11·4	2·7	72	4·2	Jeniceck[44]
Nevyansk	1794	13·5	3·7	53	3·7	Jeniceck[44]
Brymbo†	1798	14·3	3·3		4·3	Davies

*coke, rest charcoal; †part coke; ( ) inferred

bosh to a square shaft. But the crucibles (in those cases where these were separate from the bosh) were often square. By the middle of the century the shaft lining was usually brick, although the bosh and crucible were refractory sandstone. The bosh angles were usually steep—80° at Low Mill,[36] and 59–72° in the Czech furnaces.[44] Horizontal iron tuyeres were inserted into conical holes in the lining made of two or more suitably cut pieces of stone.[41]

The hearth design, consisting of tymp and dam, continued unchanged until the late 19th century, when the closed hearth was introduced. In Central Europe, where the charcoal furnace still had a long life, the main change was the use of improved blowing engines. The fact that the casting of satisfactory iron cylinders was not as widely practiced in Europe as in Britain meant reliance on wood rather than metal. The result of this was the introduction of a three-chamber wooden blower to a design by P. J. Gerstner: the chambers were about 1 m square and powered by a waterwheel operating piston rods.[44] The three chambers gave a more constant blast, usually through only one tuyere, although the Czech furnace at Adamov has four openings. Meanwhile, in Britain and France, the use of coke meant that increase in furnace size was possible whatever the profile of the furnace, and because more efficient blowing apparatus was available.

Wilkinson's cylinder blower was able to provide adequate air providing the power source was sufficient. This called for increased capital resources since, as we see from the experience at Maryport, water power reserves were often not sufficient for the increased demands, and the only answer was a steam engine.[37] The furnace at Maryport was erected in 1752 and designed for coke; its height was 11 m to the charging floor and its capacity was 57 m³. This was an enormous increase in size compared to that of a charcoal furnace, built at about the same time at Low Mill (15.7 m³). Naturally, the leather bellows proved quite unsatisfactory and a new set of iron cylinder bellows were provided by Wilkinson in 1777. The waterwheel ran at 5 rev/min and the shaft contained 8 cams per cylinder, thus giving 40 strokes/min and a blast of about 170 m³/min.

We know a good deal about the materials of construction of blast furnaces of this period. Naturally, some areas were more backward than others and earlier

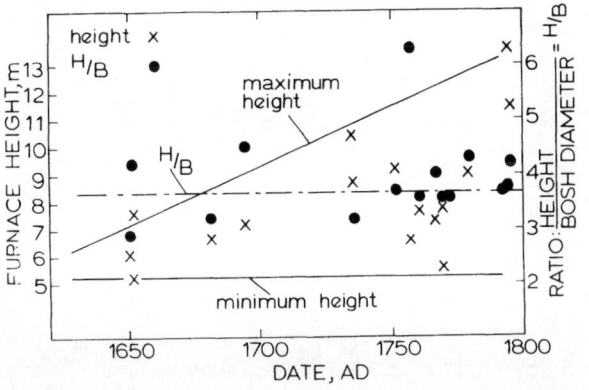

94 *Evolution of the charcoal blast furnace; 1650-1800; note the steady increase in maximum height while the ratio Height/Bosh dia. stays more or less constant at about 3.7*

designs persisted. In the Vordernberg region of Austria we have simple designs in stone consisting of two truncated cones base-to-base.[46] These could be operated either as *Stücköfen* or *Flussöfen*. They were built throughout of the same stone with no special lining; that at Treyback in Carinthia (1758) was of the same type, but slimmer. The more advanced types had separate linings with a space between the stone carcass and the lining which was sealed with sand. The hearths and crucibles were normally made of refractory sandstone but brick began to be used for the stack about 1750 (Maryport, 1752; Low Mill, 1761). The shaft of a Swedish furnace at Soderfors was made from carefully moulded slag blocks.[47] We can presume that the carcass was usually reinforced with iron bars or a wooden frame; the latter can be seen in a representation of a Wealden furnace on a fireback of 1636.[48]

Elaborate precautions were made to drain water from immediately under the furnace and to vent the furnace body to allow easy egress of moisture when blowing in. The drawing of the Larvik[49] furnace shows the former, and the remains of the furnace at Melbourne in Derbyshire[41] (1725–1780) showed an elaborate labyrinth of chambers left in the brickwork between the ashlar masonry and the lining. To some extent the sand fill between the lining and the carcass would assist the venting and avoid catastrophe but, no doubt, accidents had occurred and it was thought better to be over-cautious in this respect.

THE BLAST-FURNACE PROFILE

As far as the internal shape was concerned, one gets the impression that its exact form did not matter very much. No two furnaces were alike unless they were erected in pairs (double furnaces), and still there were often differences. Some furnaces were short and squat with small height/bosh ratios (H/B); others were tall, narrow furnaces.

At Low Mill, the bosh and crucible are all one,[36] like some of the 17th century furnaces, but most furnace builders preferred to start, at any rate, with the three separate parts of crucible, bosh and shaft—the latter always sloping inwards towards the top, although the angle varied enormously. Occasionally we see the oval furnace which was difficult to build, but the smooth lining of this type was finally adopted as standard by the end of the 19th century, and this is essentially the furnace of today with a hearth of the order of 14 m dia., and a height of 30 m.

17–18TH CENTURY FURNACES

There was a steady increase in the maximum height from about 7 m in 1650 to 13.5 m in 1800 (*see* Fig. 94). However, the heights of the smaller furnaces such as those in Austria remained unchanged. It seems that charcoal can withstand the superincumbent weight equivalent to a stockline of 13.5 m, which shows that the limited height of the early furnace was not due to the strength of the charcoal but the problems of blast pressure. The shape, as given by the ratio H/B, remains almost constant over the period at about 3.7 (*see* Fig. 94). However, towards the end of the period the bosh angle underwent a big change from the steep-sided bosh of the late 17th century with angles of about 75°, to the shallow-sided bosh with angles of only 50–60°. This seems to be due to an attempt to enlarge the furnace volume while maintaining the original height and hearth diameter for economic reasons.

19TH CENTURY FURNACES

The coke furnace of the late 18th to early 19th centuries continued the tendency towards increasing height and volume to give a height of something like 14 or 15 m by 1850 (*see* Fig. 95). There was no change in the H/B ratio which remained at about 3.7. Likewise, the bosh angles remain low with a mean value of about 57° (*see* Fig. 96 and Table 65).

There were, of course, one or two exceptions to the general trend. For example, there were the tall narrow furnaces at Swansea and Neath Abbey[50] with heights between 15.5 m and 19 m and H/B ratios of 4 to 5.6. The furnace of Swansea must have been one of the last furnaces with the old steep bosh angle of 73°. Dufrénoy[50] states that the Swansea furnace was thin walled, much more like a cupola than a blast furnace, and the heat losses must have been much greater than normal.

After 1850 there was a reversion to earlier H/B ratios but the size of the furnace went on increasing (*see* Fig. 97). It appears that the advantages of the early steep-sided bosh was discovered anew when the linings of the shallow-angle boshes wore away.[51] Changes took place also at the throat, when it was found by accident that a wider throat had many advantages, including greater

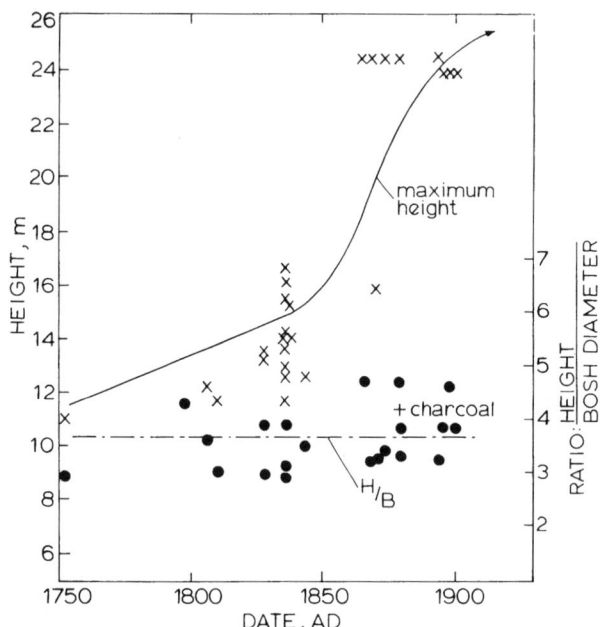

95 Evolution of coke/coal blast furnace, 1750-1900; note continuous increase in height while the height of charcoal blast furnace shows no increase beyond that of Fig.94; Height/Bosh ratio is again constant at about 3.7

Table 65 Early 16th century coke and coal and late charcoal blast-furnace profile data

Location	Date	Height, m	Bosh dia., m	Bosh angle, deg.	Ratio, height/ bosh dia.	References
Whitecliff	1806	12·2	3·35	66	3·6	Reference [182]
Derbyshire	1802/13	11·6	3·8	57	3·0	Farey [62]
Seaton (Cumb.)	1828	11·0	4·4	53	2·5	Curwen MS [179]
Dudley	1828	13·5	3·7	54	3·9	Dufrénoy [50]
Hallfields	1836	11·6	4·0	61	2·9	Morton [50]
Staffs.	1836	14·0	4·5	48	3·1	Dufrénoy [50]
Old Staffs.	1836	11·5	3·7	51	3·1	Dufrénoy
Swansea	1836	15·5	4·0	73	3·9	Dufrénoy
Pontypool	1836	13·6	4·5	58	3·0	Dufrénoy
Birtley	1833	14·0	3·65	56	3·8	Dufrénoy
Monkland	1836	14·0	3·35	61	4·2	Dufrénoy
Glasgow	1828	13·4	4·60	54	2·9	Dufrénoy
Glasgow	1836	12·8	4·0	60	3·2	Dufrénoy
Cyfarthur I	1836	15·4	4·0	63	3·9	Dufrénoy
Cyfarthur II	1836	15·4	5·0	curved	3·1	Dufrénoy
Dowlais	1836	16·1	6·2	60	2·6	Dufrénoy,
Neath Abbey	1836	19·1	3·4	63	5·6	Dufrénoy [50,186]
Lemington	1836	16·6	4·2	55	3·9	Dufrénoy [50]
Bradford I	1836	12·5	4·0	55	3·1	Dufrénoy [50]
Bradford II	1836	14·0	4·0	67	3·5	Dufrénoy [50]
Corbyn's Hall	1839	15·0	3·8	73	4·0	Gibbons [51]
Wylam	1844	12·5	4·0	60	3·1	Jones [180]
Alfreton	1844	12·2	3·4	60	3·6	
*Komarov III	1879	11·3	3·0	59	3·8	Jenicek [44]
*Barum (Norway)	1844	10·0	2·1	75	4·7	

*Charcoal; rest coke or coal

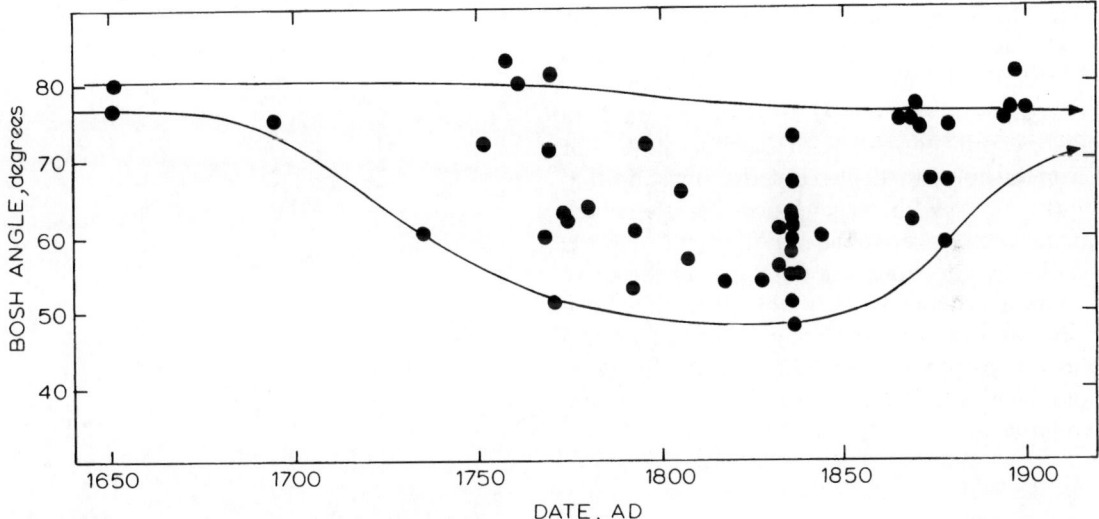

96 *Change in blast furnace bosh angle, AD 1650-1900*

volume. The height of the stockline increased steadily in the 1860s to 20 m by 1930, which resulted in a furnace with an overall height of about 30 m (100 ft) and this has remained fairly constant to the present day. The H/B ratio has stayed at 3.6 and, with the steeper bosh angle of 80°, this has resulted in furnaces with hearth diameters up to 10 m (*see* Table 66).

While those furnaces built between about 1850 and 1870 showed a tendency to return to the steep bosh angles of 70–80° (Fig. 96), it was not easy to get these angles in the old casings without pushing up the bosh so that it was as much as one-third to half-way up the furnace. This led to the bosh running cold and the

uneven descent of a sticky burden, so that bosh tuyeres were brought into use for a time for furnaces making very basic slags.[52] The reluctance to use more hearth tuyeres meant that the designers were tied to small hearths.

The H/B ratio stayed constant in the range 3.0–3.6 as the furnace size increased. Gradually, as the blowing engines improved and high driving rates became normal, the number of tuyeres increased to six by about 1880, to reach 14 in a 6.7 m dia. hearth furnace in 1940–50. Modern theories on furnace shape suggest that the bosh angle should not be less than 75° and that the bosh height (maximum diameter) should be not greater than

Table 66 Blast-furnace profiles after 1950

Date	Place	Height, m (hearth bottom to stock line)	Bosh dia., m	Height/ Bosh	Bosh angle, deg.	References
1860	South Staffs.	15·0	4·6	3·4	73	
1870	Teesside	19·7	6·1	3·0	62–75	180
1900–20	USA	26	6·7	3·8	73	
1930	Typical	20	7·0	3·6	80	55
1944	Lincs.	18–25	5·8–8·2	3·0–3·6	73–77	55
1960–70	Typical	30	9·7	3·2	77	

15% of the total effective height (tuyere to stockline). The tuyeres are normally about 2 m from the bottom of the hearth and there should be 1.5 m between the tuyeres. The stack batter should be about 8 cm/m, i.e. about 1 in 12. Generally, total height and volume do not indicate large output unless the proportions are correct, and for this the hearth diameter is the starting point. At present it is gradually increasing from 9 to 14 m.[53]

The only other change that was to come to the blast furnace was the closing of the forepart. Although the open front with its tymp and dam was difficult and arduous to work, it meant that access to the whole area of the bottom of the hearth was theoretically possible, although practically difficult. Deposits and accumulations of solid material such as bears could be removed, but it is doubtful if they were. The closed forepart was patented in 1867 by Carl Holste on behalf of the inventor, F. Luhrmann of Georgs-Marienhutte near Osnabrück, and was slowly adopted elsewhere.[54] Now it is the standard system used in all furnaces. Essentially, the hearth is circular and two holes ('notches') are made in the brickwork and stopped with suitable clays. For tapping iron the holes are drilled out and any solidified iron is now burnt out with oxygen lances. The slag notch and the metal notch are separated radially by 50° or more. There may be two or more slag notches and these are about 1 m above the iron notch. This all makes for a simpler bottom and easier working and the fear of a solid build-up has proved to be unfounded. However, the taphole clays have to be properly formulated so that they are relatively easy to remove.[55]

Blowing Engines and The Hot Blast

With the exception of China and some primitive plants in Eastern Asia and Africa, the blowing of furnaces had been done mainly with concertina-type bellows and occasionally with wooden box bellows.[41] The leather bellows required considerable care and maintenance and their first cost was by no means low. Considering the superiority of the Chinese piston-type box bellows, and the Japanese *tatara* or hinged box bellows, it is surprising that, in Western Europe, it took so long to replace the concertina type. In Eastern Asia, wooden piston bellows using hollowed-out tree trunks as cylinders had been in use, together with pot bellows, in which skins covered the tops of clay pots; the latter also spread to

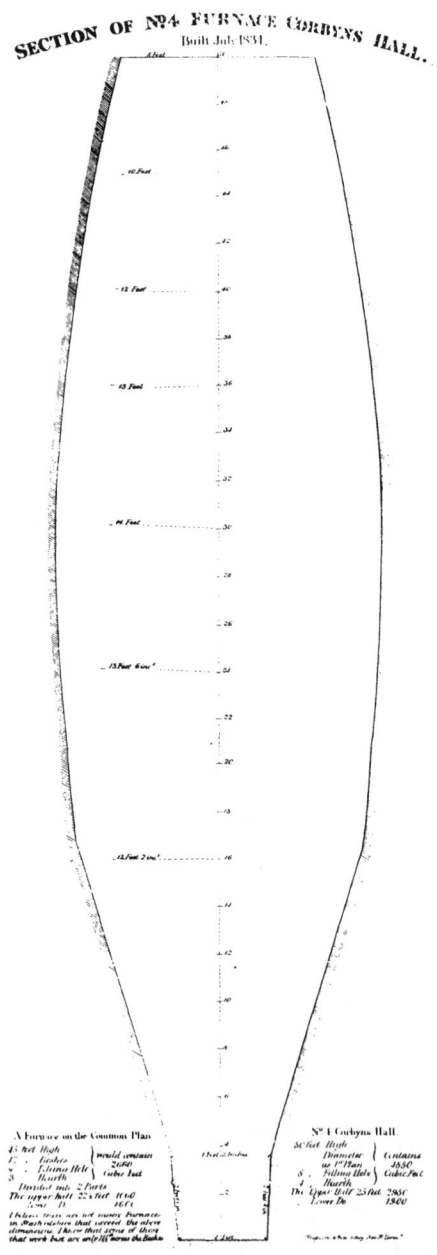

SECTION OF Nº4 FURNACE CORBYNS HALL.
Built July 1831.

97 *Final design of steep-bosh furnace with oval profile as advocated by John Gibbons in 1839 and universally adopted since, but with larger hearths*

Africa.[57] The construction of the concertina type had been described in detail by Agricola[56] and the box bellows by Plot,[58] and no further description is needed here.

It was not until the advent of the steam engine that an alternative to the concertina bellows was considered. It was clear that the old leather bellows were too weak to fully exploit this new means of power, and the production of the steam cylinder itself showed that suitable blowing cylinders could be made from brass or cast iron. To begin with, these cylinders were often powered by waterwheels when adequate water was available, or when capital for a steam engine was lacking.[59,60] In Britain, as we have seen, cylinder blowers were introduced for blast furnaces in 1760. By 1776, Wilkinson had applied the steam engine to such blowers at Broseley in Staffordshire, and at Dowlais in South Wales. The blower at Broseley in 1776 (Willey furnace) was a single-cylinder machine operated by a Boulton and Watt steam engine (*see* Fig. 85). We have a drawing of a more complicated double-acting blower with pressure regulator in William Reynolds' sketch book[61] (No. 41), which is believed to have been designed for the Horsehay works in Coalbrookdale in about 1790.

By the end of the century, air pressures of up to 4 lb/in² (0.27 bar) were available for blast-furnace blowing and this enabled some ironmasters to blow furnaces using raw coal as fuel without hot blast, as appears to have been done at Alderwasley in Derbyshire.[42]

In some furnaces the necessary air was still blown in through a single tuyere, but many furnaces had at least two openings through which tuyeres could be inserted. However, we have no definite evidence that more than one tuyere was actually used until Farey's account of a Derbyshire furnace in 1802–13, in which two were used at opposite sides.[62] All of the furnaces shown in detail by Dufrénoy in 1836 had two or more tuyeres.

The tuyeres were mostly of sheet iron, although copper was used occasionally. Water cooling was not necessary with cold blast and the iron ends were luted into the tuyere arches made in the brick or stonework. The increased and more steady flow of air available from the blowing cylinders with regulators had simplified the tuyere design. Previously, with leather bellows and some earlier types of cylinder bellows, two nozzles had to be inserted into each tuyere.

Up to the end of the 18th century the normal blast pressures would probably have not been more than 1 lb/in² (0.067 bar), but in Staffordshire[54] in 1832, the pressure was doubled to 1.75 lb/in² (0.18 bar), and by about 1870 it was 4 lb/in² (0.27 bar). In 1944 the pressures reached over 20 lb/in² (1.3 bar).[56] The beam blowing engines gave way to direct-acting vertical steam engines in about 1870 at Ayresome,[63] and various methods have been used in the 20th century. In some cases, gas engines working off blast-furnace gas have been used; the present tendency is for electrically-driven turboblowers.

The next big advance was, of course, the hot blast. The history of the hot blast[57] is all the more strange when it is realized that among some ironmasters was the feeling that furnaces ran better in winter owing to cooler

blast. If there was any truth in this it was because, as Neilson suggested in 1825, there was a higher total moisture intake in the summer. Neilson had, in fact, suggested drying the air over quicklime.

There is a certain amount of evidence that people had used heated air in forges before Neilson's patent of 1828: indeed, it is suggested that Wilkinson might have used hot blast in some of his furnaces at Bradley for a short time between 1795 and 1799.[65] This would make sense and might be responsible for the success he achieved there, but there is little evidence for its continued use there or elsewhere, and the Wilkinson firm collapsed in 1836.

J. B. Neilson, the patentee of the hot blast, had had a scientific upbringing and experience in the gas industry. He made experiments on air heating, mainly to increase the volume of air rather than its temperature, but he realized that the benefit was not for the reason he had at first intended and he sought permission to try it out on one of the furnaces at Clyde Ironworks, using a wrought iron box, $1.2 \times 1 \times 0.6$ m, which was heated externally.[66] He raised the temperature to 27°C, and even this amount had the effect of improving the quality of the iron and the fluidity of the slag. On the basis of this work patents were sought and obtained in Scotland and England. Very soon he made an improved apparatus by which he obtained 140°C, and quickly realized that some method of exposing a greater area to the external heat source was necessary. This plan resulted in the use of 'pipe' stoves consisting of a series of 'U'-shaped cast iron pipes through which the blast air was forced (*see* Fig. 98).

There was a certain amount of opposition to continued experiment at Clyde Ironworks, as this had interfered with production, so other ironmasters took up the development of various types of pipe stoves with larger heating areas. Naturally, some of them, for example W. and A. Baird at Gartsherrie, demurred at paying continuing licence fees to Neilson and his sponsors. This

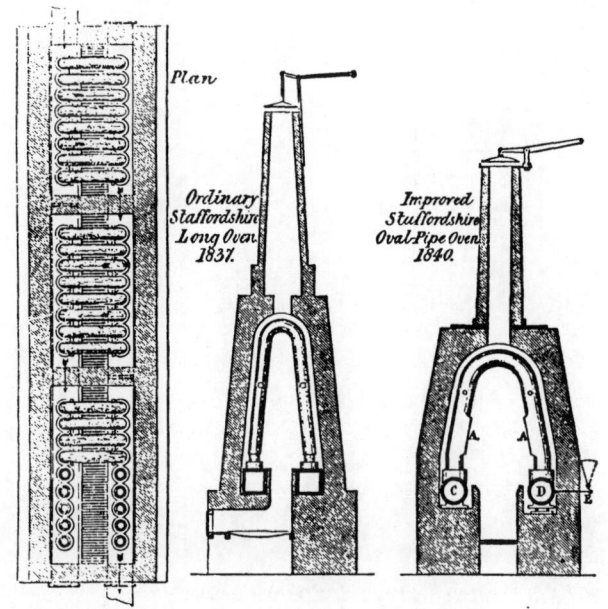

98 Cast iron 'pipe' stoves for blast heating

ended in a famous legal case in which the claims of Neilson were upheld. By 1831, Dixon of Calder Ironworks had raised the temperature to 315°C and started to smelt with coal instead of coke; this allowed the exploitation of the famous black-band ores which, in reality, were mixtures of coal and ironstone.

The problems of blast heating stoves were quite considerable and were never really solved until the invention of the refractory type of regenerator stoves by Cowper and Whitwell (*see* Fig. 99) in the second half of the 19th century.[64] The developed Neilson-type contained a number of cast iron pipes supported within a brick casing (*see* Fig. 98), and differential expansion between metal and brick continually resulted in cracking of the pipes. Nevertheless, the advantages were well worth the cost when iron with high silicon content was required or could be tolerated.

Not only did the hot blast make the use of coal possible and relatively easy, mainly due to the higher temperatures obtainable and the more basic slags that were thus workable, but anthracite could be used and was so used in South Wales and in the USA. Of course, other problems arose: the tuyeres overheated and had to be water-cooled. Two closed water tuyere designs were introduced—the hollow cast iron Staffordshire tuyere which had been used previously in refineries, and the Condie or Scotch tuyere which had a wrought iron tube carrying the water embedded in an iron casing[67] (*see* Fig. 100).

At Clyde Ironworks the blast pressure was 2.5 lb/in² (0.16 bar), and there were two tuyeres of 7.6 cm dia. The charge in 1829 was 260 kg of coke, 160 kg of calcined ironstone, and 45 kg of limestone, while in 1830 it was 260 kg of coke, 280 kg of ironstone, and 70 kg of CaCO₃. In 1833 the charge was 260 kg of coal, 260 kg of ironstone, and 31 kg of CaCO₃. The fuel/ore ratio had thus been reduced from 1.5:1 to 1:1 by the use of coal and hot blast[68] (*see* Table 67).

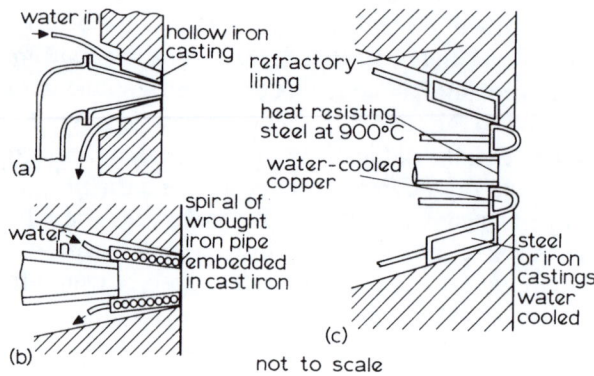

a Neilson's tuyere used at Clyde Ironworks in 1832 based on early types used in forges and refineries; *b* Condie or Scotch tuyere used by Bairds at Gartsherrie; *c* modern blast furnace tuyere capable of operating at 900°C

100 Hot blast water-cooled tuyere systems

Roasting of Ores

During this period it was normal practice to roast the clay (nodular) ironstone to increase its permeability. Roasting would have little effect on the permeability of the hematites but these were often roasted to reduce the sulphur content which was in some cases high (1.0%).[69] This appears to have been the case with the Elban ores used in the Tuscan furnace at Capalbio where we can see today a number of ore calcining kilns.[70] Roasting also assisted the breaking up of the ore to a regular size, which was as small as 12 mm in the case of the less permeable ores. Roasting may be done in heaps on the ground, in stalls, i.e. low walled enclosures, or in kilns. Generally, the more complicated the structure the more economic the process as far as the fuel consumption is concerned. This varies from 100–200 kg/t using wood or coal for heap roasting, to about 30 kg/t in sophisticated kilns[71] (*see* Table 68).

Table 67 Improvements in output and coke rate largely owing to the use of blast heating, 1828-40 (mostly after Corrins[66])

Works	Date	Blast temperature, °C	Coal or coke rate,* fuel/iron	Iron made, t/day
Clyde	1829	Cold	8·05	
	1830	150	5·15	
	1833	315	2·88	
Calder	1828	Cold		5·6
	1831	Hot (about 150)		6·6
	1833	320		9·0

*Including that required to heat the blast

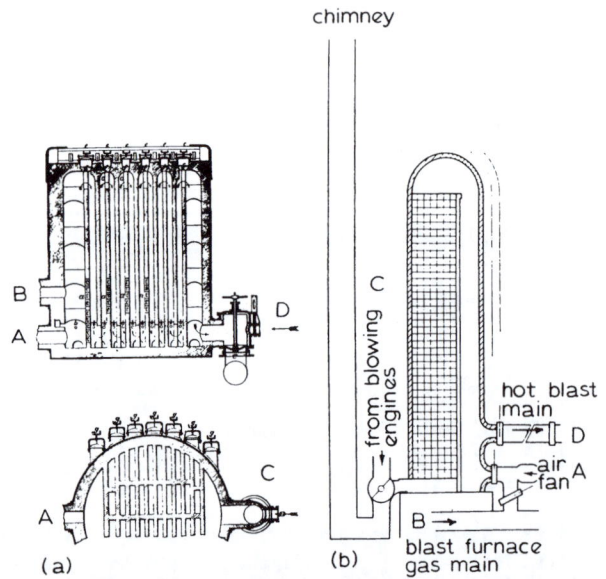

a Whitwell's type; *b* Cowper's type

99 Refractory stoves

Table 68 Calcination of iron ores (after Garillot[71] and Jars[72])

Type	Fuel used	Ore capacity, t	Fuel/Ore ratio, kg/t	Output or duration*	Site
Heap	Wood	–	50–150	15 days to months	
Heap	Coal	–	150–200	15 days to months	
Controlled heap	Charcoal	–	–	7 t/day	Mariazell (Austria)
Stalls	Wood and Charcoal	300–350	–	–	Larvik
	Wood and Charcoal	–	–	–	France
Kilns	Wood	–	–	–	Styria
	Wood	30–35	180	3 days	Allevard (Pyrenees)
	Wood	–	–	–	Tuscan (Elba)
	Wood and charcoal	90	100 kg/coal 5 100 kg/wood } day	22 t/day	Hungary
	Charcoal		50	34(23) t/day	
	Coke			12 t/day	Hungary
	Coal		40	30(20) t/day	Siegerland
	Coal		50	12–15 t	
	Coal		100	–	Clarence (UK)
	Coal		35	100–200 t/day	Gjers (UK)
	Coal		5–7	70 t/day	Sommorostro (Sweden)
	BF gas		–	14–15 t/day	
	BF gas		–	2–3 t/day	Fillager, 1851 (Carinthia)
	BF gas		300 m³ gas	40 t/day	
	Prod. gas		30 kg coal/t	25 t/day	

*Brackets indicate weight of calcined iron ore

Very often, ordinary lime kilns were used; in some cases, the kilns were heated by blast-furnace gas, and in others, after 1851, by producer gas. A rectangular roasting kiln used in Styria[72] was made of stone 3–4 m high, 4.5 m long, and 2 m wide. At the bottom it had an ignition and discharge hole like a lime kiln. The kiln was filled with alternate layers of charcoal and ore, the latter broken to 'nut' size: the layers of charcoal were 30 cm deep and those of ore 60 cm deep (see Fig.101).

At Larvik in Norway in 1767, roasting was done in circular walled enclosures holding 300–350 t of ore. The charge was bedded alternately with layers of ore and mixed wood and charcoal to a height of 2.5–3 m.

Heap roasting with coal was used at Carron in Scotland in 1764. The nodular ore was broken into pieces weighing 3–5 kg and placed on a sloping bed of coal 7 m long, 2.5 m wide, and 15 cm deep. The heap reached a height of 1 m in the centre and was covered by coal dust and ashes. Heap roasting persisted well into the 20th century in Britain.[73] It was used in the Northamptonshire ore field up to 1910. The ore was roasted in the quarry by first laying down a layer of lump coal, then some slack, and finally placing the excavated ore on to this to a height of one metre or so. Now that the blast furnace charge is blended and sintered, the calcining of ironstone is no longer necessary.

Where cold blast was still used in the middle 19th century, blast-furnace gas was available for heating calcining kilns, as at Finspong in Sweden in 1857.[74] Here,

red hematite and magnetite were calcined even if low in sulphur, and then crushed to less than 2.5 cm in size. In the Cleveland (Teesside) area of Britain, blast-furnace gas was not available for this purpose, and kilns were heated with coke. In the original circular kilns at the Clarence works the coke and ore were mixed and the consumption was about 100 kg/t. This was reduced in the later (Gjers) type of kiln[75] which was capable of calcining 200 t/day with a coke consumption of 35 kg/t (see Fig. 102).

LIMESTONE

In the main, limestone was not calcined before being charged to the blast furnace, nor is this done today. However, according to Jars, in a few plants in the 18th century, such as Blankenbourg in Brunswick, the limestone was previously calcined.[76] This seems rather extraordinary as calcined limestone would quickly slake to the hydroxide in a damp climate and become a general nuisance. It would have to be transferred quickly to the furnace after calcination.

Coking Processes

Coke making in some places followed charcoal making practice. At Clifton, in 1765, the circular heaps were 3.7 m dia. and had a foundation of sandstone slabs on which the large material was placed to ensure permeability.[77] The heaps were built to the shape of a cone 1.5 m high, and were then covered with earth and coal dust. At

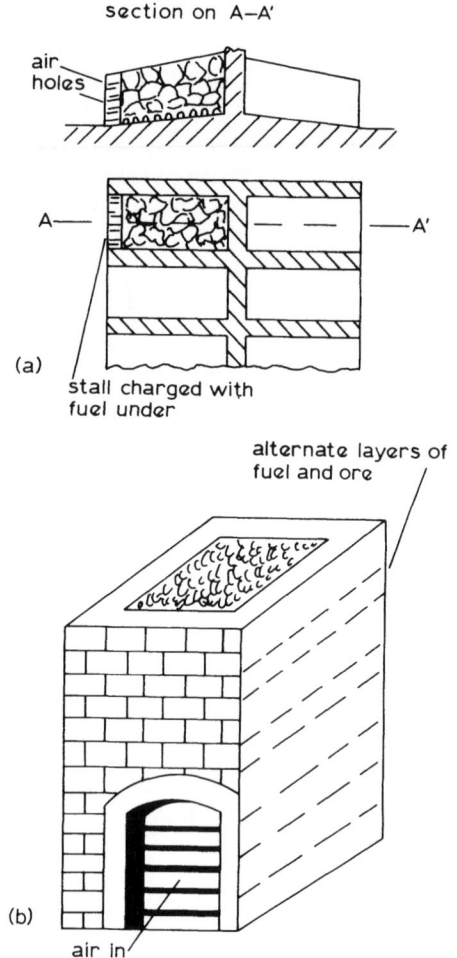

a after Hassenfratz; *b* after Jars

101 Stall-roasting of iron ore (from Garillot [71])

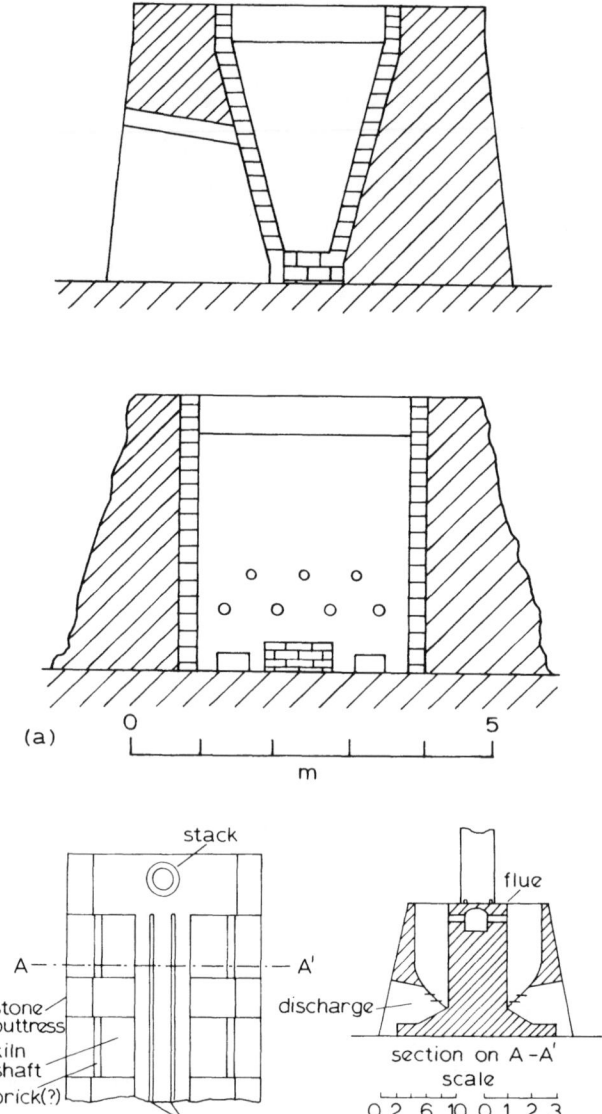

a after Dufrénoy [44]; *b* is based on the remains of a kiln erected by Armstrong at Ridsdale, Northumberland, in 1876

102 Some examples of closed ore roasting kilns

Carron in Scotland, the coking process took 14 hours. A circular bed of clod coal was set alight and sprinkled with coal dust at intervals to control the draught. Coking in heaps was still in use in the Black Country near Dudley, as late as 1830. The floor was earth and there was a central loose-brick chimney (*see* Fig. 103). The piles were about 9 m dia. and 1.5 m high, considerably greater than those at Clifton.

At other works, elongated piles were used as in ore roasting. In South Wales they were 3.7 m wide and 1 m high containing 3 t to the linear metre. Such heaps were replaced in some countries by rectangular kilns to conserve heat and fuel. Horizontal and vertical flues were made in the walls and charging was through openings at the end. In Newcastle, coking was done in ovens as early as 1765 and there were three types, the largest of which was a square masonry structure which made coke for malting. In 1826, similar ovens were being used in Sheffield.[78] These were made of brick, 3 m dia. and 1 m high internally, with external drystone walls to give maximum insulation. These were worked continuously to conserve heat and were filled with small coal through the roof, which was raked level to the springing of the arch through the side (discharge) door. A rather special type was built at Maryport[79] with a sloping charging floor from the back. Even the Welsh furnace at Dolgun

seems to have had at least one coke oven of the domed type. Other 18th century examples exist at Malham Moor and at Norwich.[80]

Coke ovens which were built before 1850 and used up to 1900 were excavated near Mirfield in Yorkshire.[81] These were made of firebrick and were 2.5 m dia. internally and 1.5 m high from the floor to the apex of the roof. Since the door was only 0.6 m high it is unlikely that they would be filled to a height greater than this, as the charge was usually levelled by means of a rake through the door. The charging hole in the roof was about 30 cm dia. (*see* Fig. 104).

At Whinfield in County Durham, we have some of the earliest byproduct recovery beehive ovens[82] which were built in 1861 and lasted until 1958. They were built of brick and are somewhat larger than those at Mirfield, being 3.3 m dia. and 2.8 m high internally, with a door 1.1 m high. The floor sloped to facilitate discharging. At the

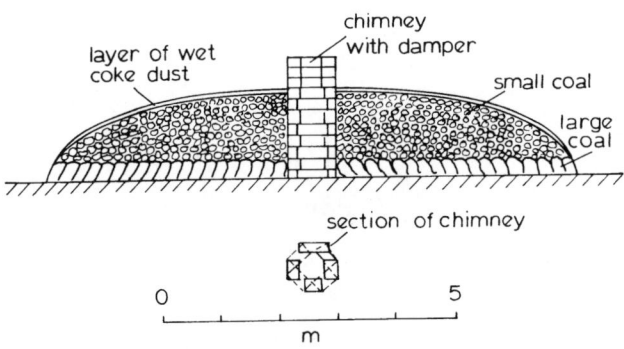

103 *Coking in heaps (after Percy; 'Refractories and Fuels',
1875)*

back was a common horizontal balloon flue which was
connected to each oven by a small inclined flue, and the
tar byproducts were collected in condensers serving a
large number of ovens. As at Mirfield, the circular dome-
shaped ovens were built into long banks with the aid of
stone retaining walls. The space between the brick oven
and the ashlar retaining wall was filled with rubble and
the ovens were heat-retaining to such an extent that a
new charge could be ignited by the heat from the old.
After about 1850, it seems to have been the usual practice
to quench the coke while inside the oven. At Whinfield,
an oven contained 5 t of coal which was charged from
four tubs running on rails above a row of ovens: coking
took 72–96 h.

Iron Foundries

It was the remelting of pig iron from the blast furnace in
a reverberatory or air furnace that was responsible for

the high quality of the British cannon of the mid-eighteenth century.[77] Previously, iron was cast direct from the
blast furnace and the difficulties of proper control of the
cast metal, imposed by the smelting regime itself, was
often responsible for poor quality. Double furnaces were
erected to enable larger cannon to be made, and this
process was more or less universal in Europe until the
end of the 18th century .

However, the work of the Darbys and the Wilkinsons
had revolutionized iron casting, and coke smelting had
made easier the production of soft grey irons instead of
the more brittle white irons. A great deal of the raw pig
iron from the blast furnaces was, however, poor in
quality and contaminated with slag. Remelting in a
reverberatory furnace enabled the slag to float out,
especially as the temperature limitations of the blast
furnace no longer applied and a higher temperature was
available.[83] The melting conditions were slightly oxidizing and some of the carbon was reduced, giving a greyer
and softer iron with high fluidity. Wilkinson seems to
have been aware of the beneficial effect of manganese,
and added manganese dioxide to his blast furnace or
cupola charge,[84] but we do not know whether he knew
that this addition acted favourably on the sulphur content.

By the beginning of the 18th century the reverberatory furnace, fuelled with coal, was common in the lead
and copper industry so it is not surprising that Darby
used it for remelting iron for his pots. For one thing, the
blast furnace had to be taken off blast during casting and
it was not possible to keep going to the hearth and
ladling out a small amount for the odd casting without
the furnace running cold for lack of blast.

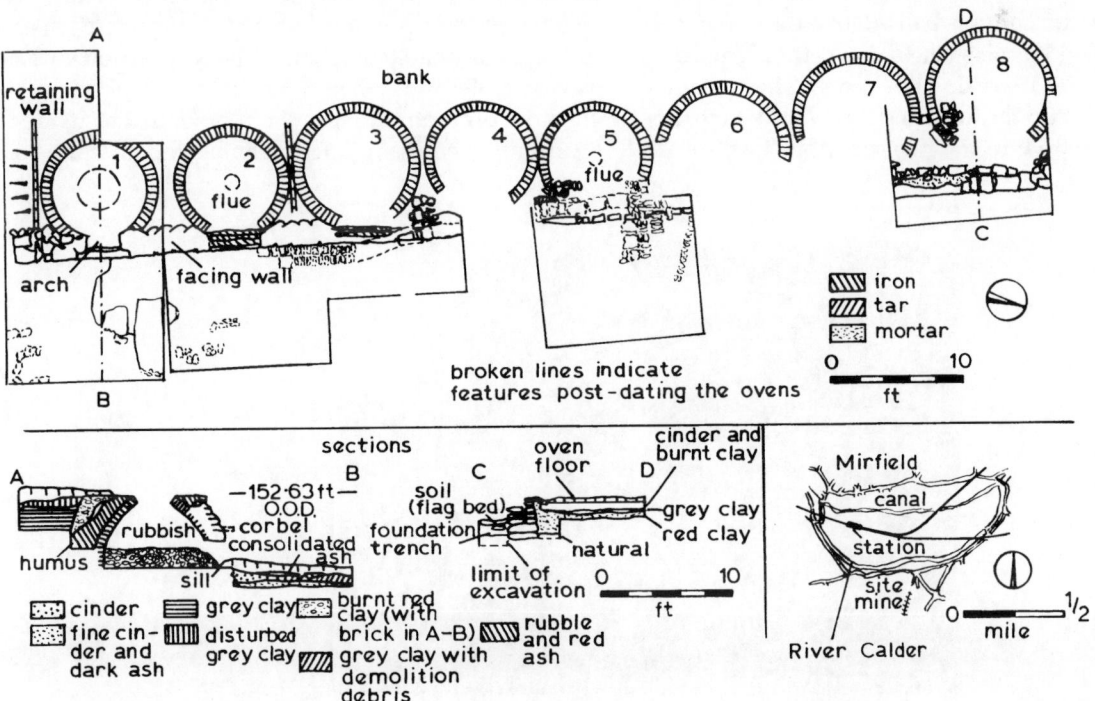

104 *Beehive coke ovens at Mirfield, Yorkshire (after Lyne[81])*

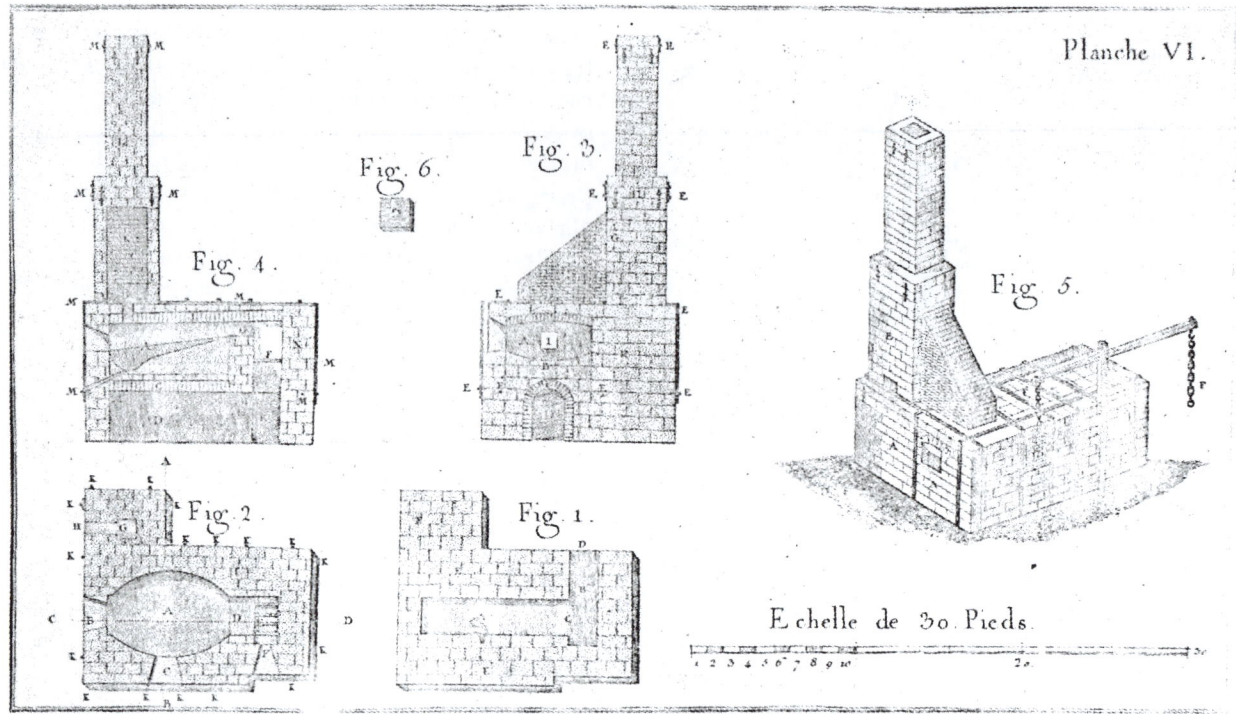

105 An air furnace for the melting of cast iron seen at Newcastle in 1765 (from Jars[85])

We have details of an air furnace at Newcastle[85] which was used for general iron castings in 1765 (*see* Fig. 105). The hearth was made of river or sea sand and fired for 3–4 hours before being charged with 1.6–1.8 t of iron, which took 1½ hours to melt. Some furnaces were capable of taking 3 t of metal, others only 2 t, but it was thought that the smaller furnaces made better cannon.

We see from Réaumur,[86] and from Diderot and D'Alembert's encyclopaedia,[87] that small blast furnaces —charcoal fuelled and hand blown—were in use for remelting iron for cannon balls. How these furnaces came to be called cupolas (the name certainly used by Wilkinson in 1794) we do not know. They bear no resemblance to a cupola dome and this would seem to be a better term to give to the reverberatory furnace. In-deed, in its non-ferrous version the reverberatory was often so called: 'cupulo' is the term used in the minute book of the London (Quaker) Lead Company for their first reverberatory smelting furnaces in 1701, and the use of this term was to continue in the lead trade and gives rise to much confusion among scientific writers.

However, the word *cupola* is the diminutive of the late Latin *cupa*, a cask, and is therefore more aptly used for the small cask shaped forced-draught melting furnaces that were used for melting cast iron. These were segmental and the many segments fitted into each other giving a reasonably tight joint. The bottom portion acted as a crucible, and, when the metal was melted and the charcoal on top removed, the segments that made the shaft could be dismantled, leaving the crucible to be

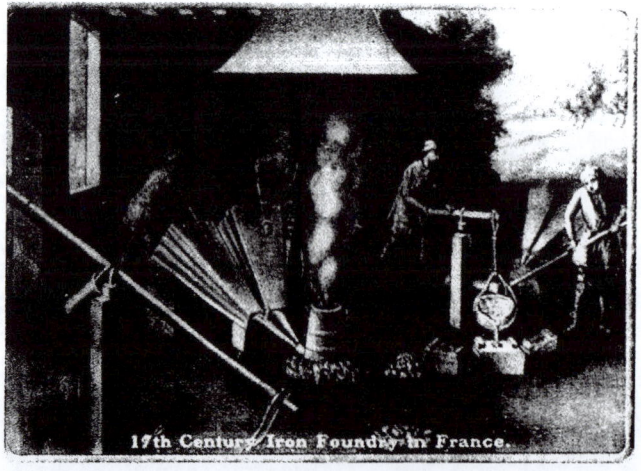

106 Small cupola for melting cast iron for cannon shot, 1722
(after Reaumur[86])

lifted and its contents poured into the moulds. These melting units were portable and powered by hand bellows and were ideal for casting shot[88] (*see* Fig. 106).

One must assume that these furnaces developed like the bell-founding furnace and the small Chinese type of blast furnace, and it would seem that such furnaces were a standard part of mid-eighteenth century chemical and metallurgical laboratory equipment,[89] and as we see were well known to such men as Réaumur who also reinvented malleable cast iron. Wilkinson was using coke fuelled cupolas in 1794, but we are given no details of size; he seems to have been one of the first to use coke and as a result they were often referred to in France as 'fourneaux à la Wilkinson'.[89] By 1836, the cupolas were being blown by air from piston blowers,[90] at Swansea a single blower of 126 cm dia. × 190 cm stroke, at 18 strokes/min, could blow two cupolas, one small blast furnace, and a finery. Cupolas were in use near Bradford and at Lemington near Newcastle.

The reinvention of malleable cast iron by Réaumur was perhaps one of the less dramatic but nevertheless worthwhile developments of the period. Like Dudley, he appreciated the obvious differences between white and grey cast iron but seems to have assumed that 'sulphurs' were the reason for the difference. The fact that he 'annealed' the white iron in a mixture containing charcoal seems to indicate that he was trying to carbonize, i.e. convert iron into steel, rather than decarburize. Fortunately, he used bone ash and iron oxide as the other ingredients and ended with a neutral atmosphere, or one leading to slight decarburization. At first, he was convinced that grey iron contained more impurities than white iron and noticed that the yield of wrought iron in the finery was greater when using white iron rather than grey iron. Later, after noticing that cast iron could be made either white or grey by altering the cooling rate, he realized that the impurities were not the only reason for the difference. There is little doubt that this is one of the first results of controlled scientific experiment in metallurgy.

The other advance on the casting side was the introduction of 'green-sand' moulding. Before this, castings were made in clay or 'loam' which had to be completely or superficially dried, and this was inconvenient. If metal is cast into media containing moisture, such as clay or damp sand, then the permeability of the mould must be very high or else the steam produced, when the hot metal comes into contact with the mould, will affect the metal and render it porous. Before the 18th century, a lot of clay was used in loam mixtures and this detracted from the permeability, so most of the moisture had to be removed by drying or stoving.

By using natural sands in which the clay is as low as 5–10%, and moistening it, it is possible to get sufficient strength and permeability to cast molten iron or bronze directly into the undried sand without difficulty. It is not known who discovered this but it seems to have been used by Darby I for his thin-walled cast iron 'pots' (cauldrons). The alternative dry-sand process uses more clay and moisture to get higher strength but needs drying in a stove and is thus much more expensive than green-sand moulding. There is no evidence that Darby used the dry-sand technique, although he probably used either this or loam for the core (inside) of the cauldron mould. The largest cauldrons, like bells and other large castings, were still made in loam.

At Boulton and Watt's Soho Foundry in 1795, which was mainly turning out parts for steam engines, all three methods of moulding were used but green-sand was still used very little. The air furnace was used mostly for melting metal for the larger items and the cupola for the smaller quantities. We find the same situation elsewhere, and it would seem that the cupolas were too small for using the large scrap available. The smaller foundries ran their cupolas from waterwheel driven 'fans'. In 1837, at Workington in Cumberland, the value of the cupola is given as £10 and it probably consisted of cast iron staves bound together like a barrel with wrought iron bands.[91] This was lined with natural sandstone from local sources, and cemented with fireclay (*see* Fig. 107).

Late Bloomeries

As a means of producing the malleable metal, wrought iron, the bloomery was highly competitive with the indirect method in areas which were well wooded. Thus, we see the use of the bloomery process in the form of the Catalan hearth of the Pyrenees well into the 19th century (1840). Essentially, there was no change in detail from the 15th century process when water power was generally introduced into ironmaking, but blooms now weighed 100–200 kg and the Catalan process tended to use the 'trompe' as a means of providing a blast.[92]

Bloomery iron was still being produced in Sweden in a small way in 1851,[93] but it was in the North American continent that improvements were made. In 1831, about

107 Nineteenth century stave cupola

10% of the wrought iron produced there was by the bloomery process.[94] By 1856, it was still 6% and had expanded greatly as the overall production of wrought iron was over ¹/₂Mt. The bloomeries were iron plated Catalan hearths using heated blast,[95] and the charcoal consumption had been reduced to 2 060 kg/t of raw metal, i.e. giving a 'coke' rate of 2.06, which was better than that obtainable by the indirect process. Furthermore, the production rate had been increased to one 136 kg bloom per 3 hours compared with the Pyreneean Catalan hearth output of one 180 kg bloom in 6 hours.

Mechanical Working of Iron

The forge of the 17th century usually included conversion processes such as finery and refinery furnaces. This tradition was carried on to some extent into the 19th century, and puddling furnaces were naturally placed close to the hammers and the roll trains which they served. In countries which depended on charcoal there were attempts to improve efficiency in various ways, and international competitions were organized to compare the various techniques available.[96] One competitor submitted a design for a completely integrated works which included a blast furnace with cylinder blowers worked from water power. A second and a third set of cylinder blowers blew the fineries, and individual waterwheels provided power for the hammers. All this was to be under one roof, but this was merely the ideal and most works fell far short of this mainly because the water power was not available.

In Britain, however, where reliable supplies of wood and water had always been problems, the separation of the blast furnace and the forge took place early on—in fact, right at the beginning of the blast furnace period—and persisted until towards the end of the 18th century when ironmasters such as Wilkinson decided to go 'to work the forge way' (1777).[97] By this time the use of coal had made this integration possible, and it was merely a matter of learning how to use steam engines to work hammers, slitting mills and rolling mills.

HAMMERS

The waterwheel had provided rotary motion which could be easily harnessed to hammers (*see* Fig. 108) and rolling mills, but the steam engine had grown out of the need for pumping and was mainly designed around the beam. In this form it could be applied to blowing without change. In 1777, Wilkinson had tried a single-acting beam pumping engine which had hammered itself to pieces working a stamp hammer with a head of 27 kg weight.[97] But in 1782, Boulton and Watt supplied a chain-on-beam engine with sun-and-planet motion to work a tilt hammer at 25 strokes/min. This loosened the hammer frame but the engine itself was satisfactory. By 1782, two steam-driven hammers were in operation; one was a tilt hammer, with a head weighing 55 kg (*see* Fig. 109), and the other a lift hammer (belly helve), with a

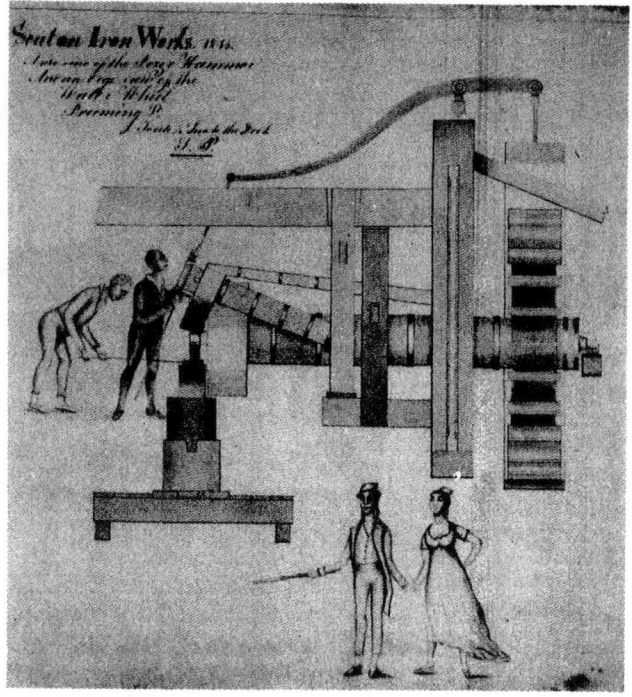

108 *Side lift hammer or belly helve used at Seaton Ironworks in 1816 with water power*

head weight of 300 kg. The Boulton and Watt engine, designed for a belly helve at Horsehay in 1793, used a sun-and-planet gear.

By 1787, two hammers were being worked from one engine of 28 hp—both belly helves. Each hammer would be served by a large brick built coal-fired hearth, blown with an air blast, in which the blooms would be reheated to forging temperature (*see* Fig. 110). This type of hearth was fairly standard and served for ferrous and nonferrous purposes; indeed, it has survived today as smaller versions are used by small cutlers and others in the Sheffield area.

While Wilkinson's engine was capable of giving 300

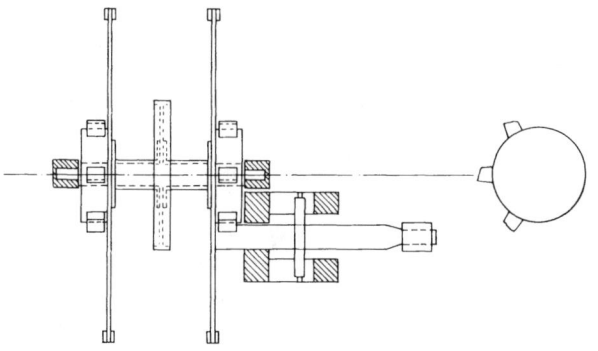

109 *Steam-driven tilt hammer of the type used by Wilkinson at Bradley ironworks in 1788*

blows/min, it was not possible to make use of rates greater than 100 blows/min in practice. With six cams this would mean a speed of 18 rev/min. The first illustration of a nose helve is given by Dufrénoy *(see* Fig. 111); it was made of cast iron and used for shingling puddled balls in Staffordshire.[98] This was really a very simple type of hammer with a minimum number of parts. All three types of hammer could be found in a fully equipped works, each type having its optimum rate of working and weight of head. The smaller tilt hammers (tail helves) were probably the most numerous and were used for beating out bars to sheets for tinplate, and in non-ferrous 'battery works' for 'brass' boiling vessels and pans. Gradually, they were replaced by the rolling mill.

ROLLING AND SLITTING

As we have mentioned, small rolls have been used for some time—mainly for flattening sheet before slitting, and for minting purposes. Generally, it would seem that these applications involved very little reduction in thickness, and therefore the housings did not need to be very strong nor the driving power very great[99] *(see* Fig. 112). In 1792, Wilkinson patented an improved rolling mill, using a chain wrapped around the roll and attached to the beam of a steam engine to give partial (reciprocating) rotation.[100] This was intended for puddled bars, and the rolls were 1.5 m dia., 1.8 m long, and weighing over 8 t each. This was not very satisfactory for rolling but the principle might have been applied to the bending of plates for boilers.

As already mentioned, in 1766 Purnell patented the use of grooved rolls for making bar for ships bolts, and

110 *Forge with hollow fire and water-driven cylinder blower at Seaton Ironworks, Cumberland*

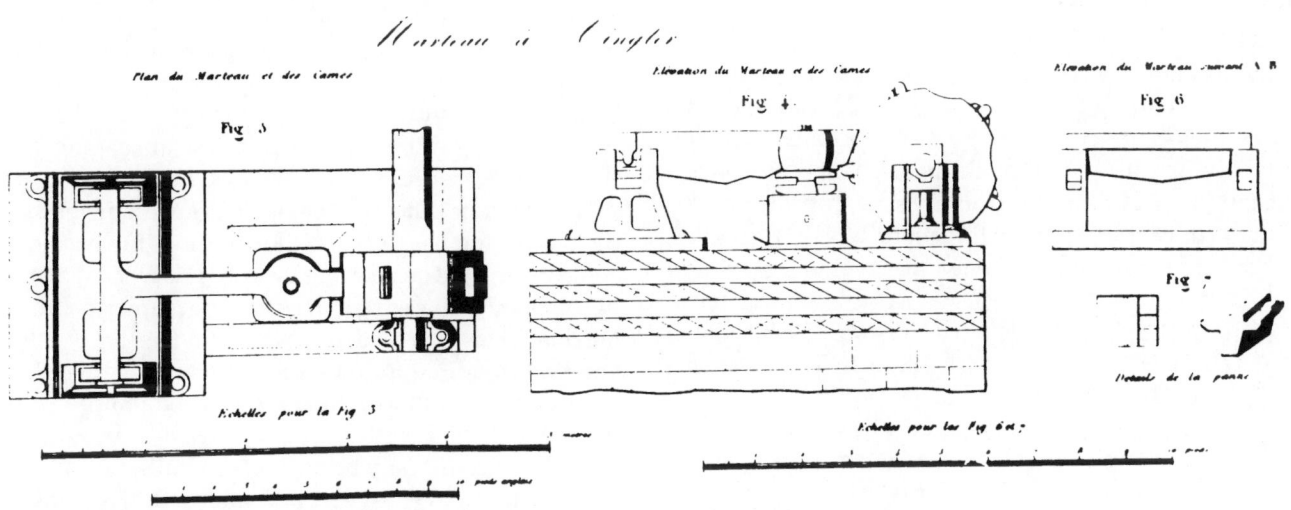

111 *Steam driven nose helve used for shingling wrought iron about 1836 (from Dufrenoy, vol.2, pl. 111)*

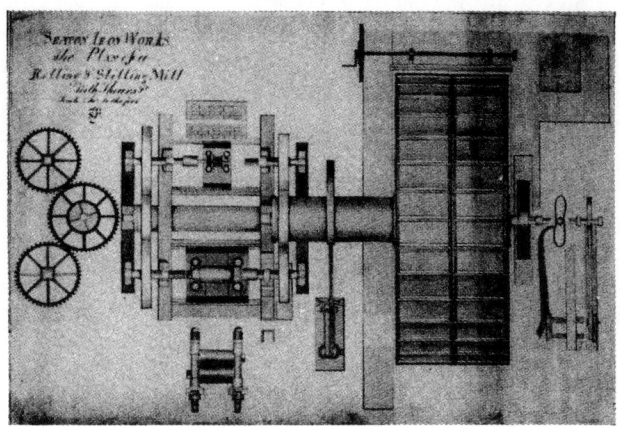

112 *Rolling and slitting mill; Seaton Ironworks*

this idea was later adopted by Cort for working puddled bar.[101] These rolls had three passes and one of the rolls (bottom) was driven via a wobbler from the waterwheel, and the other (top) through a gear from the lower roll. This diverged from the use of independent drives for both bottom and top rolls by separate waterwheels, which was widely applied to slitting mills in the early 18th century. But there is little doubt that Cort returned to the former arrangement in his mill at Funtley[102] which appears to be very similar to the arrangement shown by Diderot.

Dufrénoy gives detailed drawings of rolling mills[50] and their layout for about 1836. Flat and round bar passes followed the nose helve shingling hammer. These were all two-high mills driven from one steam engine through gearing. Two-high sheet mills had rolls 40 cm dia. and 1.4 m long. For small bars, three-high mills were used, the rod being returned through the upper passes: three-high mills are also shown for sheet rolling and there is evidence that this principle was known about by 1825.[103]

BORING AND CUTTING MACHINES

In this period we begin to see the development of more accurate cutting and machine tools. The earlier cannon had their bores cleaned out with the primitive horizontal boring mill shown by Biringuccio, and we have an example of a boring bar from Sussex, probably dated to the 17th century, in which the four bits on the cutting head are made from hardened steel welded to wrought iron backing plates.[104] The use of a core and the problem of drying it adequately led to the casting of solid cannon and the use of more powerful drilling machines to make the bore. There is no doubt that this produced a superior weapon, as the porous region in the centre was now removed completely rather than left, as previously, on the inside of the bore adjacent to the core. The heavier boring was first done by a vertical boring machine in which the gun was suspended over the boring bar which was turned by horses, and this is the type shown by Diderot in 1762. The cutters had numerous hardened steel inserts—essentially no different in principle from those recommended by Biringuccio.

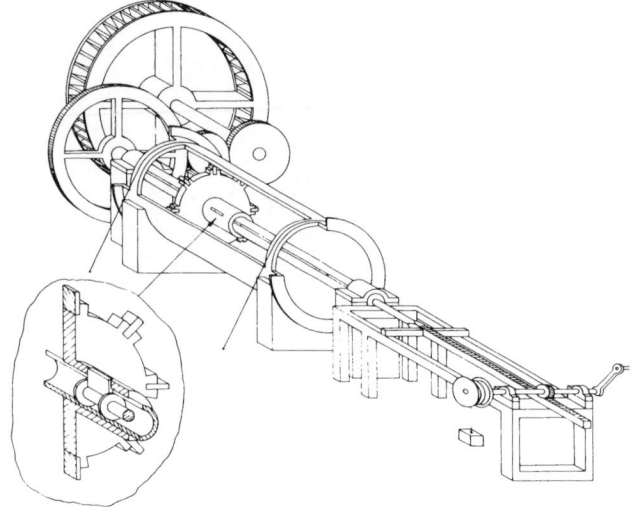

113 *Wilkinson's improved boring machine*

However, the demands of the steam engine and the increasing use of large iron pumping and blowing cylinders required something better than this, and we see a reversion to horizontal boring machines somewhere around the middle of the 18th century.[105]

Bronze guns were being cast solid from about 1715 when the Maritz process of boring was introduced.[105] Here, the gun was rotated and the cutting tool advanced supported on several bearings. Improved accuracy was thus obtained. This process was applied to iron guns by Anthony Bacon in about 1773; the actual work was probably done by John Wilkinson at Broseley, who patented it in 1774.[106] For open ended cylinders, it was possible to use a rotating boring bar with the cylinder fixed, as the boring bar could be supported outside the cylinder at both ends (*see* Fig. 113). The boring bar was hollow (a tube) and carried a slot along its length in which a piece of steel was able to slide; this in turn was connected to a rod which was free to travel horizontally within the tube. The piece of steel, or key, was keyed into the boring disc which slid along the boring bar and contained the cutters on its periphery. The tube was rotated by the waterwheel, later by a steam engine, and the cutters were fed by the controlled motion of the rod within the tube. The tube, being of larger diameter than the old boring bar, was less flexible and, furthermore, could be pivoted on bearings at both ends when machining an open cylinder so that superior accuracy could be obtained. This provided the steam cylinders that Boulton and Watt required for their more efficient engines. The vertical boring mill continued to be used but the work was now rotated on the face plate like a lathe, and the boring tool could only be moved vertically.

In Oslo, in 1760, where old and defective cannon were being melted down, the cannon had to be cut up to get them into the small reverberatory furnaces.[107] This was done with a circular saw consisting of a 30 cm dia. iron disc in which steel teeth were inserted. This was rotated by a waterwheel and the gun was gradually lowered against the cutting disc. This seems to be the earliest reference to the use of a circular saw: the saw normally used for removing gun heads (feeders) was horizontal and hand powered.

The Production of Steel

By 'steel' we mean iron containing appreciable amounts of carbon, and not the so-called steel of today which is mostly low-carbon 'mild' steel—the modern equivalent of wrought iron. At the beginning of our period, steel was being imported by Britain from Sweden and Russia and was used carefully, as its price was about three times that of bar iron. Some of this was produced by cementation and piling, and some direct from the finery. We know that at St. Gallen some of the metal from the *Stücköfen* was separated and used as steel: the Japanese were also separating steel from their direct smelting process—the *tatara*.

In Styria and in the Tyrol, so called 'natural' steel could be made the main product of the finery by using white cast iron, and by increasing the inclination of the tuyere and making certain other changes in the process;[109] this was the main process used in Sweden and Belgium.[110] Another possibility was the direct production of carbon-containing iron in the bloomery, but by this time the bloomery process was a dying craft, extinct in nearly all of Western Europe apart from the Pyrenees and Corsica. It was, however, extremely active in the New England states of North America until late in the 19th century, although there seems to have been no intention of producing steel in this way. Presumably, the slag separation problem militated against this. Carburization of the finished article was also used for certain mass-produced swords and for files,[111,112] but the majority of steel was made by cementation of good quality wrought iron bar.

CEMENTATION STEEL

Naturally, the abundance of coal led to its use in the cementation process which had been formerly carried on with charcoal and wood, at prodigious cost, in Sweden and Russia. In Sweden, the conversion of wrought iron to steel took place in a furnace containing three chests, 2 m long, which were heated for 6–7 days. It needed 100 t of charcoal to convert 450 kg of iron to steel.[113] In order to try and reduce the cost of this material, some coal was imported from England.

There was no difficulty at all in applying coal to the heating of cementation furnaces, and this soon became the only way of making raw steel in Britain in the 18th century. Imported charcoal bar iron was best since it contained rather less slag and phosphorus. The latter tended to slow down the carburization reaction. In Sheffield, there were two types of cementation furnace using Swedish iron.[114] One was a small, single chest type and the other a double chest furnace. The latter furnaces are well known, and at the time of writing at least one exists in Sheffield and one in the Newcastle area (*see* Fig. 114). Those at Sheffield were brick built with an arched chamber 3.7 m across, in which the sand-stone chests

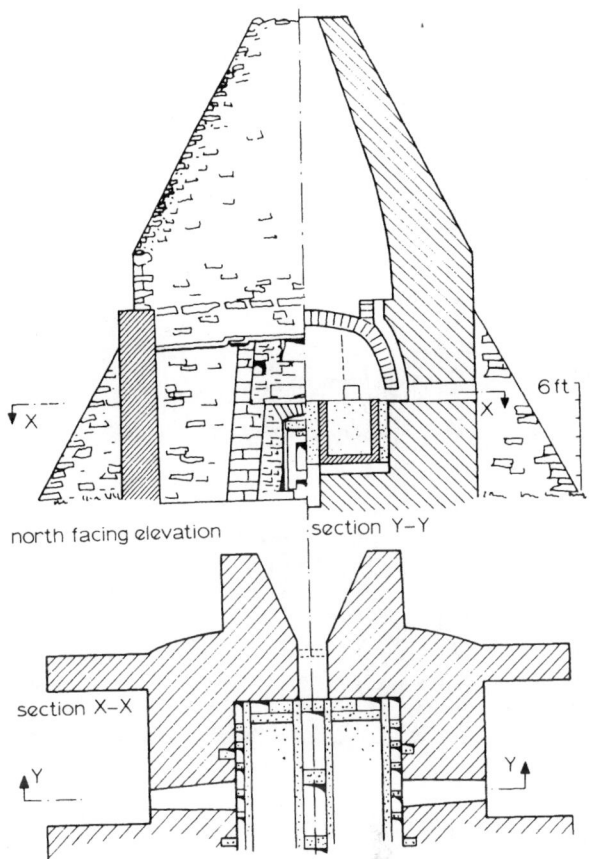

114 *Cementation furnace still standing at Derwentcote, County Durham*

were placed. These held 4–5 t of flat iron bars 7.5 × 1.3 cm in section, embedded in charcoal dust, and they were sealed with a mixture of iron and grinding wheel grit or wheelswarf. The heat from the coal fires burning in the grates below was transmitted to the chests by suitable flues. The sulphur in the atmosphere could have no effect on the iron as it was well protected. The temperature would have to be above 900°C and was maintained for 5 days. The flat steel bars swell somewhat in this process, due to the reaction between the carbon and the residual oxygen in the metal (mainly in the slag), and the resulting material was therefore called 'blister' steel. This was then reheated in forge fires, piled, and worked down to suitably sized bars. The result was not very homogeneous, with 1%C at the surface and very little at the centre and, after forging, the steel had a banded structure, often with residual slag along the welds between the piled layers. No doubt this was why in 1740 Huntsman, the clockmaker, objected to it and preferred his more homogeneous crucible steel.

Huntsman is generally credited with the making of homogeneous cast steel by melting cemented 'blister' bar in a crucible and casting it into ingot moulds prior to forging. But we know that as late as 1765 it was not in general use in the Sheffield area and was confined to steel requiring a good finish such as the best razors, knives, some watch springs, and small clock files. The crucible furnaces were undoubtedly developed from the brassmaking furnaces previously mentioned.[115] But

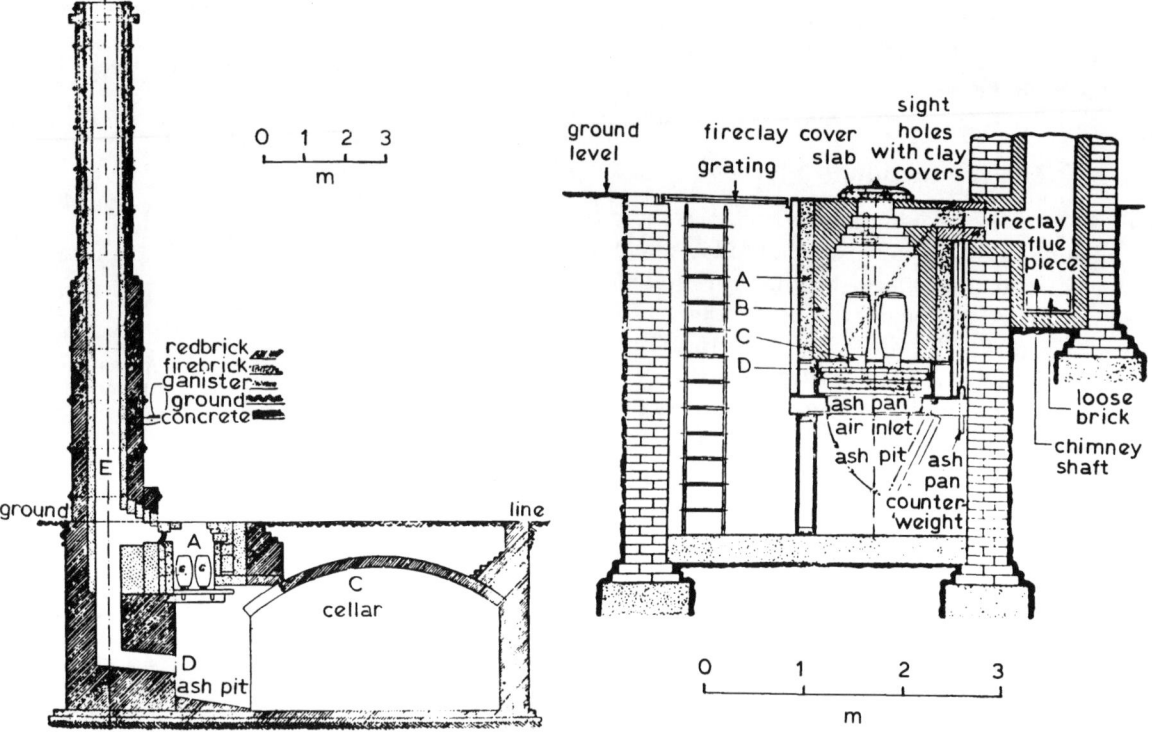

115 Crucible steel melting furnace

most of the steel furnaces contained only a single large crucible, 23–25 cm high and 15–18 cm dia. The 'blister' steel was put in with a mysterious flux (probably powdered glass) and the crucible placed on a 'stool' on the grate. Coke was placed all around and ignited, the air was admitted under the grate bars, and the products of combustion drawn through a side opening above the pot by a tall chimney. The hole was sealed by a lid (see Fig. 115).

The steel took 6 hours to melt and was then poured into cast iron moulds to give ingots weighing about 10 kg. The only difference between this and some forms of the Indian steel (wootz) was the longer melting and cooling time of the latter, which caused the development of a coarse structure upon solidification and allowed more time for diffusion of carbon from both within and outside the crucible. The crucible process became the standard method of making large steel castings throughout Europe and Russia up to the latter half of the 19th century. It served as a method of making certain alloy tool steels well into the 20th century, but has now been replaced by the high frequency and carbon arc furnaces.

PUDDLED STEEL

Considerable amounts of steel were made between 1850 and 1880 by a modification of the puddling process. This was an obvious idea once it became known that the difference between cast iron and wrought iron was that of carbon content, but the puddling process was difficult to control in such a way as to leave a residual carbon content of the same order as that normally associated

with steel, i.e. 0.5–1.2%. Cort had had the idea originally, but it was not until about 1823 that the process had matured sufficiently to be considered for continental use.[116] It was not until 1851 that we learn of its use in Britain, at the Low Moor ironworks near Bradford. Here it is recorded that a total of 1 250 kg of steel were made in an iron puddling furnace in ten heats with a wastage of only 6.75%. Unfortunately, the cost was high and in order to obtain maximum ductility it had to be melted and used in the cast form.

By 1898, puddled steel was having considerable success on the Continent where Krupps had actively taken up its production. As it required considerably more skill than wrought iron the workers were paid a bonus. The carbon content of their product was variable because of the difficulties of control, and the bars were sorted into three grades; A, containing 0.9–0.75%; B, containing 0.75–0.65%; and C, containing less than 0.6%. The bonus

Table 69 Composition of puddled steel
(after Barraclough[117])

		Amount, %	
	Ebbw Vale, 1863		France
Element	Pig iron	Puddled steel	Puddled steel
C	2·68	0·50	1·18
Si	2·21	0·11	0·33
Mn	1·23	0·14	tr.
S	0·125	0·002	nil
P	0·426	0·096	0·02

was only paid for grades A and B. The phosphorus content of the pig iron had to be low (but not as low as that for Bessemer steel), and a good deal of the metal produced was used as melting stock for steel castings.

It is clear that puddled steel was of high quality as far as the dissolved impurities were concerned *(see* Table 69), but there is no doubt that it had a slag content similar to wrought iron. It was used in the puddled state for carriage springs but most of it was remelted to homogenize the structure. It was thus more in competition with orthodox crucible steels rather than wrought iron, and when the puddling furnace ceased to be necessary for making malleable metal with the introduction of the new 'mild' steel, it gradually died out.

Copper

THE EXTRACTION OF COPPER

On the continent of Europe the premier process of copper extraction was still the German–Swedish process in which sulphide ores were roasted and reduced along the lines specified by Agricola. At Mansfeld, the process comprised seven operations of roasting and smelting. First, a low grade matte was separated from the slagged gangue and then this was enriched by concentration smelting, intended to bring about the oxidation and slagging of the iron. The enriched matte was then 'dead' roasted to oxide, which was reduction smelted to black copper and refined.[118]

In the matting process the fused material was tapped through a taphole in the front of the furnace and the slag and metal separated in the forehearth: the taphole normally remained closed. Slag accretions were removed by breaking away the front of the furnace. The result was a lowgrade matte which was roasted in stalls for eight days, and then broken up and reroasted six or seven times until a dead-roast calcine was obtained. The second half of the 16th century saw a slight change when the first operation became a pyritic smelt in the shaft furnace to slag and matte with some metal (bullion). This was the first attempt to make use of some of the heat produced by oxidation of the sulphur in the sulphide ore, and also to economize on fuel.

When the silver content of the copper was sufficient to repay the cost of desilverizing, the black copper was desilverized by smelting with lead, and plano-convex ingots of copper-lead alloy were tapped from the furnace. When solid, the lead is almost insoluble in the solid copper so that the metallic lead which contained nearly all the silver from the copper (and the lead ore) could be liquated by reheating the ingots at a temperature much below that of the melting point of copper (1 084°C). When the silver content was low, as in the Harz, the copper was refined in so-called 'spleiss' furnaces, which were in use at Rio Tinto, Spain, around 1840. There were then used for refining black copper; they were of German design and the details are shown in Fig. 116. The hearth was about 1.7m diameter and was provided with bellows and a tuyere to speed up the oxidation process essential

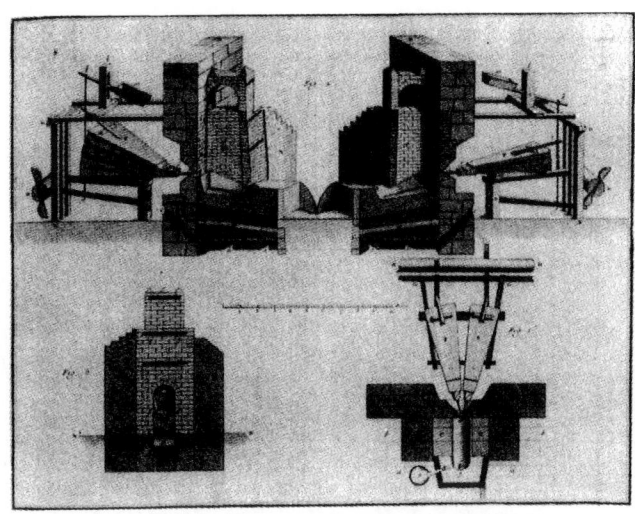

116 *Saxon furnace for copper smelting (after Diderot [87])*

to refining; the fuel was wood and the flue gases escaped through ports in the semi-spherical roof. [119]

The situation had not changed very much by the middle of the 18th century, and Diderot's encyclopedia of 1762 shows in great detail the furnaces used. As an example of the main type of furnace used for extraction, Diderot takes the Freiberg (Saxony) furnace which, by then had reached a height of 2.5 m. His illustration is detailed and very useful as it supplies information on aspects which are lacking in Agricola's description. However, it is clear that the process and the furnace have not changed in principle *(see* Fig. 116). The tuyere was inclined at an angle of 1 or 2° and the hearth was sloping, and liquid slag and matte were tapped from the hearth into a sump or settler.

The matte was roasted in stalls with some vents in the back wall *(see* Fig. 117). A shaft furnace very similar to the first was used for the first fusion (i.e. reduction smelting) of the oxidized matte. For silver recovery, lead ore was added at some stage and separated from the solid copper in liquation furnaces *(furneaux de division)* *(see* Fig. 118). These had remained unchanged since Agricola's time and, in view of the low temperatures required, were fuelled with wood. The cupellation furnaces used for the extraction of silver from the liquated lead were of the normal type with bellows. Similar circular furnaces, often double, were used without bellows for refining the black copper. It is at this stage that we see the use in Continental practice of the standard type of reverberatory with a tall chimney for draught. According to Diderot, it was used in the Vosges at Geromagny, and it is described in Schlüter's treatise of 1738,[120] *(see* Fig. 119). It is not surprising that the reverberatory was first used in Continental practice for refining, as this was impossible in the restricted hearth of the shaft furnace, and slow when done in a smith's hearth[121] *(see* Fig. 120). The very much wider hearth of the reverberatory, and the independent heating from the separate firebox, would make the operations of oxidation and poling much easier and more efficient.

Copper was also being smelted in various parts of the

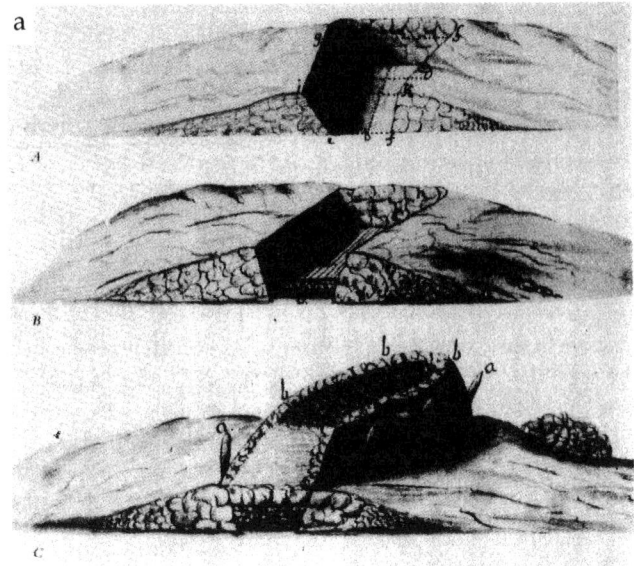

117 Roasting stalls for (a) copper ores and (b) matte (from Lindroth)

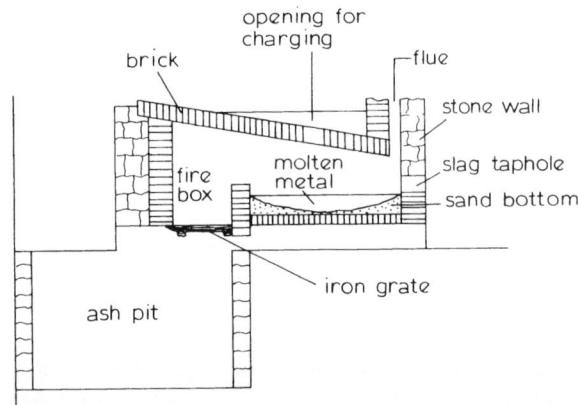

119 Copper reverberatory furnace depicted by Schlüter and many others

118 Liquation furnace for extracting silver-rich lead from cakes of smelted copper (from Erckerl [187])

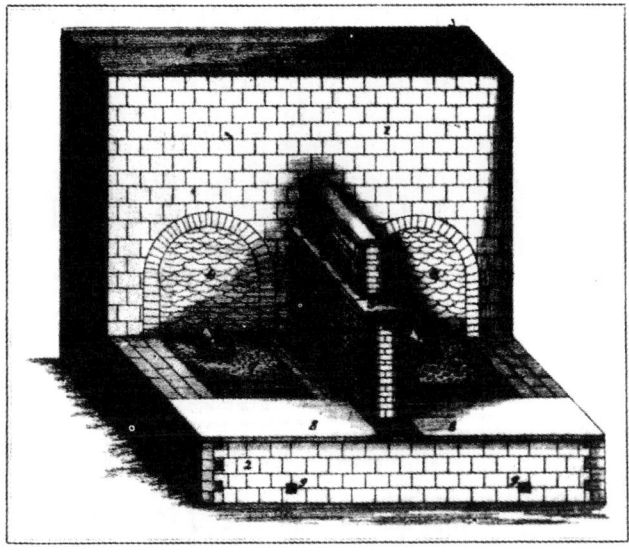

120 Refining hearth for copper (after Schluter [120])

Spanish Empire, and there is no doubt that this was making significant contributions to European needs. In 1786, a Spanish ship laden with over 600 t of copper bars and pigs, the latter weighing about 100 kg, was wrecked off the coast of Portugal. This was part of a large import intended to make up for the interruption of British and Swedish supplies due to war. The Seville archives show that, in the period 1761–1775, Spain received 3450 t of copper from Peru. No doubt most of this came from Chilean mines.[122] We have a description of the bellows-blown blast furnace used to produce this type of ingot.[123]

In Spain itself, various attempts were made at Rio Tinto to work the deposits abandoned by the Romans. In 1839, blast furnaces of the type shown by Diderot and others are shown in Fig. 116—typical of the German tradition—were being used. As mentioned above the refining was done in a speiss furnace which was the same type as used at Massa Marittima in Tuscany in the 15th century[124]. In this furnace the metal was tapped into two rosetting pots, in which the surface of the metal was solidified by the splashing-on of water, and the solid metal levered out. This was the refining system practised on the copper in a hoard found off Helgoland and dated to the 12th–14th centuries.[125]

These processes were heavy users of fuel and in this respect there had been little improvement during the 200 years of their use. In Britain, however, the shortage of timber and the ready availability of coal in some areas had favoured the use of the reverberatory instead of the

blast furnace. The problem of fuel worsened as the available ores became leaner, but on the Continent, as long as smelting was confined to the heavily forested mountain regions, the fuel problem was not pressing.

The English or, more correctly, the Welsh process of copper smelting was based upon the use of the reverberatory furnace which had been pioneered by the end of the 17th century for lead smelting. Of course, the principle was not new; the melting furnaces used for bronze bells were essentially of this type. But the eaand the metal were mixed and the metal melted under the domed furnace roof. The furnaces used for lead smelting with coal were much more sophisticated and better control over oxidizing and reducing conditions was possible.

The Welsh process was introduced around 1700 and at first needed ten operations,[126,128] but by 1830 the Swansea smelters had reduced this to six or seven (*see* Fig. 121). The basic difference was the extensive use of the roast-reaction or double decomposition process, in which use was made of the sulphur as a reductant. This was brought about by preroasting or the use of oxide ores. To a large extent it was based upon the importation to Swansea of the highest grade ores available throughout the world, and these often reached as much as 60% Cu. By blending oxide and sulphide ores in the range 8–60% Cu (*see* Table 70), maximum use could be made of the double decomposition reaction with the minimum of fuel. As far as the furnace conditions were concerned, all processes were oxidizing.

First, the pyritic ores (high in iron sulphide) were calcined to leave the sulphur in the form of Cu_2S and FeS in equimolecular proportions. This was designed to give a matte (coarse metal) in the next stage, with a composition of about 35% Cu, 35% Fe and 30% S. For calcination, a reverberatory was used with a capacity of 3–4 t at a temperature of 800°C for 12–24 hours. The product was quenched in water to break it up. In the next (fusion) stage of the process, the calcined ore was mixed with oxide ore of low iron content. For this fusion a smaller reverberatory (ore furnace) was used with a silica hearth, and temperatures of the order of 1 150°C were obtained. The principal reaction was; $Cu_2O + FeS \rightarrow Cu_2S + FeO$.

The FeO was removed as slag with the aid of silica from the hearth and this produced coarse metal (matte), (*see* Table 71), and black fayalite slag (*see* Table 72). The matte was granulated and, in the next stage, the granulated matte was calcined to remove some of the excess iron. For this purpose it was mixed with sulphide low in iron and calcined to reduce the sulphur from 30 to 15%. Charges of 2 t were treated in 24 hours at 800°C.

The next stage was called 'running for metal' and the aim was to obtain 100% 'white metal', i.e. Cu_2S. In this process the calcined matte from the last stage was mixed with rich oxide ores in the metal furnace. For this, the reverberatory had to be very hot (>1 200°C): the product was white metal containing 75% Cu, 2% Fe, 23% S, and a rather rich slag (containing 5–5% Cu) which was returned to the earlier stages. The flux was sand from the furnace bottom. If the iron content was too high, or the

Table 70 Composition of grey copper ores smelted at Swansea about 1850 (Napier[127])

Element	Source			All ores; mean composition, % (Le Play)
	Cornwall	Devon	Algeria	
Cu	15·5	12·5	20·3	13·5
S	23·7	15·6	14·2	23·1
Fe	41·7	15·0	4·6	19·7
Sb	5·6	4·1	7·5	
As	3·1	0·8	5·0	0·9
Zn			1·1	
SiO_2	8·5	51·6	47·3	38·5
Al_2O_3				2·4
CaO				0·3
MgO				0·4
O_2 etc.				1·2

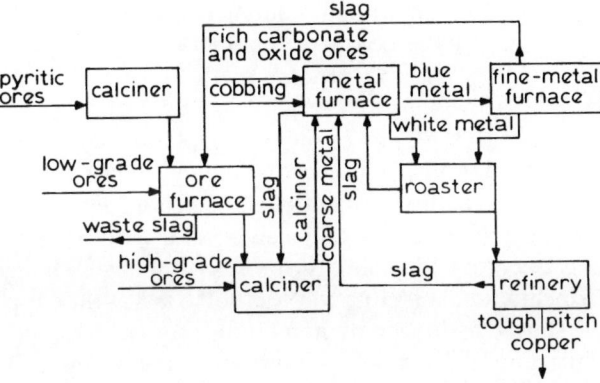

121 *Flowsheet for the Welsh process of copper smelting (after Hopkins[126])*

Table 71 Composition of products of Swansea copper smelting (after Napier[127])

Products	Composition, %						
	Cu	Fe	S	SiO_2	O_2	Sn + Sb	Pb
Coarse metal (matte)	31·4	41·3	27·3				
Blue metal (matte)	55−66	9−18	20−23	2−5	1−4	0·7−1·3	
White metal (matte)	78	2	18	2			
Coarse copper	90−96	2·4	0·6	0·7	2·9	0·5	
Blister copper	98−98·5	0·8	0·1		0·6	1·0	
Refined copper	99·8	0·05			0·04	0·04	0·05

Table 72 Composition of Swansea copper slags (after Napier[127])

Element	Amount, %				
	Waste slag from 'ore furnace'	Slag skimmed from blue metal in 'metal furnace'	Stiff slag from white metal in 'metal furnace'	Cu-rich slag skimmed from roaster	
				Start	Finish
SiO_2	60·5	36−40	60·4	37	33
FeO	28·5	54−58	36·1	49	43
Al_2O_3	2·9				
MgO	0·6				
CaO	2·0				
CaF_2	2·1				
Cu	0·5	1·5−2·0	3·5	8	15 (CuO)
Fe	0·9				
S	0·6			3	2
$SnO_2 + Sb_2O_5$	1·4	0·7−1·3			

rich oxide ores not available, 'blue metal' high in iron was produced and this had to be given an additional oxidizing fusion in a special 'fine metal' furnace.

The next stage was the first to produce metallic copper and was termed 'roasting' by the Welsh smelters. Here the maximum use was made of the double decomposition reaction; $2Cu_2O + Cu_2S \rightarrow 6Cu + 2SO_2$. The pigs of white metal were stacked in the furnace and oxidized, and then gradually melted over an 8 hour period while sufficient oxidation took place to complete the reaction. The slag was skimmed from the surface and the metallic copper cast into pigs of 'blister' copper. These were refined in another reverberatory over a period of 24 hours in an oxidizing environment involving 'flapping', slagging, and poling back to 'tough pitch' copper with low oxygen content. This part of the process is still used today, virtually unchanged.

We therefore see a development along both lines in the 18th century; the development of the Agricolan shaft furnace process, assisted no doubt by the Swedish example, towards the greater height of 2.5 m as shown by Diderot, and the increasing use of the reverberatory in the later stages of the process; and in Britain, the substitution by the reverberatory of the old blast furnace because of the possibility of using coal in the former; this was cheap and easily available in some parts of Britain. It was beginning to dawn on the users of the Swedish–German process in Slovakia that their techniques left something to be desired and, while Jars praises the high level of technique at Banska Stiavnica as compared to other works he had seen, it is clear that the processes

were backward and the furnaces old fashioned.[129] It seems that this was due to inefficient state control, the conservatism of the workers, and a general ignorance of chemistry at this time. This was soon to be rectified by the founding of the Mining Academy and other technical institutions in the region.

This situation was relatively unchanged until about 1850. By this time the blast furnace process had developed at Mansfeld to the extent that only five operations were needed.[118] In Saxony, the Pilz furnace was introduced in 1866; this had a free-standing pentagonal hearth with a sump and water-cooled tuyeres. Water-jacketed furnaces were introduced by Hering at Braubach on the Rhine in 1866. But gradually the two processes merged, with the reverberatory being increasingly used for matting and refining while the recovery of the black copper was left to the blast furnaces. Outside South Wales this was the situation until the coming of the Bessemer-type converter which took over the oxidation of the matte, leaving the matting process to the reverberatory as before.

In South Wales no changes took place,[126] largely due to the monopoly position of the smelters, who could rely on imports of ore from undeveloped countries at prices fixed by the smelters themselves and, with the aid of cheap coal, could make large profits without any changes in technique. This situation was not to last as the suppliers started to do their own smelting, often with the aid of emigrant Welsh smelters who were able to make improvements and combine the best of available techniques in their new locations. With this competition, copper smelting in South Wales very soon became extinct.

The Welsh process, in spite of its use of high grade ores, still needed about 20 t of coal to produce 1 t of copper in 1850. The leaner Anglesey ores required 30–40 t of coal in 1786. The fact that this showed no improvement on earlier times probably reflected the rather higher purity now demanded by the consumer. In spite of, or because of, this South Wales became the largest smelting area in the world during the period 1800–1875, and Britain, mainly because of its imperialist position, became the largest producer of copper, taking first place from Sweden in about 1720. The fall of the Swedish and German and the growth of British output is shown by the figures in Table 73.

A good deal of copper sheet was used for sheathing wooden ships, to reduce damage by teredo worms and to avoid marine growths and the continued dry docking and scraping that they involved.[130] In 1761 the sheet was attached by iron bolts and naturally, after a time, these corroded away and the copper plates fell off. Muntz metal (60 Cu–40 Zn) was tried, but copper sheets and cold drawn copper bolts gave the best results. A very large proportion of the copper produced was made into brass until the rise of the electrical industry in the middle of the 19th century.

The bulk of the copper produced was refined in the traditional way by poling and adding lead.[131] But the purity could be improved, for the making of Muntz metal (60-40 brass) for example, by the use of the 'Best Select' process. In this, only part of the white metal (Cu matte) was oxidized to copper in which it was found that most of the impurities were concentrated. The remaining purified matte was 'selected' and smelted separately to give a much purer copper.[127] This was the best that the industry could do until the arrival of the electrolytic process invented by Elkington in 1865.

BRASS AND ZINC

During this period we begin to see striking improvements in the production of brass, and the large scale production in Western Europe of metallic zinc. The antipuritan reaction following the restoration of Charles II in Britain seems to have stimulated the brass industry owing to the increasing demand for trinkets. Even before this, brass was being made by melting copper and zinc in a reverberatory furnace of Schlüter's design[132] (*see* Fig. 119), although Diderot shows the natural draught

crucible furnaces being used for this purpose.[133] Certainly, the latter process was preferred and was much used in Britain during the 18th century. The molten brass from the crucibles was poured into horizontal stone moulds and the resulting plates hammered into sheet or beaten into holloware by the battery process[134] (*see* Fig. 122).

But there are indications that zinc was being more often added to copper in the metallic form. As has been mentioned, zinc and lead ores occur together and a good deal of the lead ores are contaminated with zinc and, as we see from the analyses of the brass plates (*see* Table 44), the reverse is often true. Diderot[133] shows a lead–zinc furnace from the Harz mountains. It is of the usual Agricolan pattern but with a subtle difference in that near the front wall, which is about 1.5 m high, a space is maintained between the charge and the hearth or sump which traps the vapourized zinc in the products of combustion on their way to the flue (*see* Fig. 123). This space is what Diderot calls the *assiette du zinc,* and below is the usual sump for the receipt of molten metal. It is not clear whether the zinc metal actually falls back in the molten state on to the lead, which will presumably occupy most of the sump, or merely condenses in the solid state in the *assiette.* Most probably it is the latter, as

Table 73 Copper-approximate annual production

| Date | Amount, t | | |
	British	Swedish	German
1570	100	200	1 500
1712	1 000	1 200	700
1800	7 000	500	400
1850	22 000	750	1 300

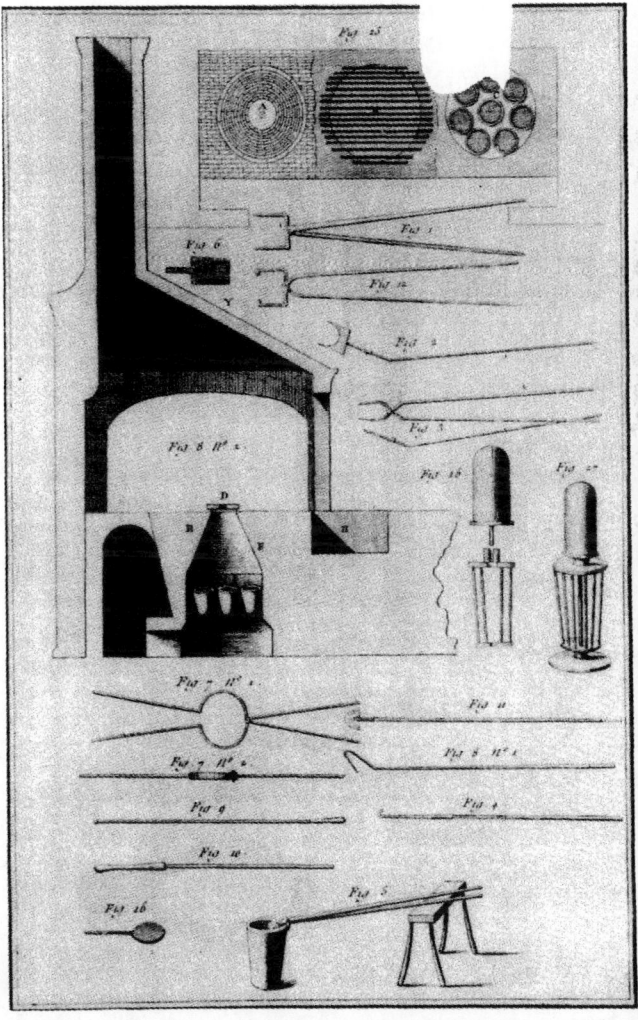

122 Crucible furnace installation for brassmaking (from Diderot [87])

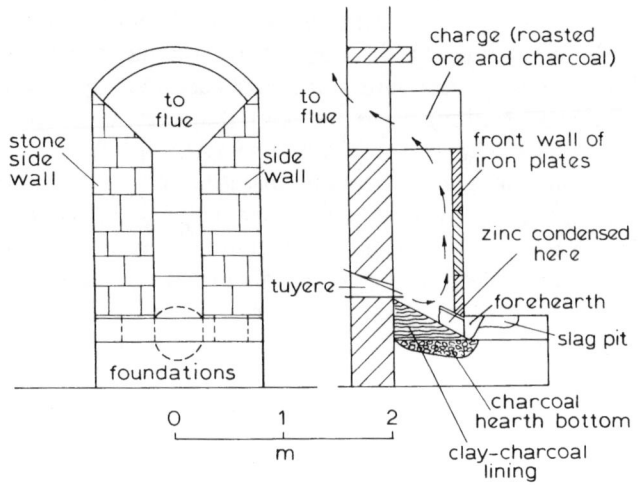

123 Shaft furnace from the Harz mountains for smelting lead-zinc ores (after Diderot [87])

in the tubular clay condenser of a mid-nineteenth century horizontal zinc retort. The lead is ladled out of the sump into cast iron moulds.

This is probably the first indication we have in the West of any attempt to recover metallic zinc, although it was occasionally found in crevices in the lining of early furnaces. The credit for the first intentional production of metallic zinc in the West is usually given to William Champion, who developed a vertical retort method in 1738.[135] His father, Nehemiah Champion, was a Quaker partner in the Bristol Brass Wire Company which was founded in 1702. William Champion built a works at Warmley near Bristol to exploit this process, which was patented in 1740, and a further patent was granted to his brother John in 1788 for the use of the more plentiful zinc blende (ZnS) after calcining to zinc oxide in a separate coal-fired reverberatory.

The problem of zinc metal production is that zinc oxide is not reduced by carbon below 1 000°C, and that zinc metal boils at 907°C, so that it is reduced in the vapour phase and must be condensed to liquid metal before it comes into contact with enough air to oxidize it back to zinc. The essence of William Champion's method was to heat the charge in a lidded crucible which had a hole in the base. This was placed over an iron tube that led into a cool chamber below, in which was placed another crucible containing water and in which the zinc vapour condensed (*see* Fig. 124).

The hot crucibles were charged with alternate layers of coke and calcined blende, and the lids were then luted on. The distillation process took about 70 hours, during which time 400 kg of metallic zinc was produced from six crucibles arranged in a circle. Although this was a great metallurgical advance, the process was laborious and thermally inefficient. As late as 1851, when the process was used at Swansea, it required 24 t of coal to produce 1 t of zinc from easily reducible ores.[135,136]

Meanwhile, experiments were taking place in Belgium—another centre of active interest in zinc smelting. In 1807, the Abbé Dony established a works at Liège and used horizontal retorts in horizontal rows, which gave a more economic heating arrangement as compared with Champion's radial distribution (*see* Fig. 125). The clay retorts were charged with coke and ground calcined ore, and the zinc was collected in clay tubes fitted to the ends of the retorts after charging. Similar arrangements were developed in Silesia at about the same time. As late as 1953, 50% of the world's zinc was still made in horizontal retorts of Dony's type. Improvements involving gas and oil heating and the use of regenerators had been made. In the 1930s, the vertical retort furnace was developed by the New Jersey Zinc Company and later, in the 1950s, the invention of the lead splash condenser made the blast furnace a practical and economic proposition for the recovery of zinc (and lead).

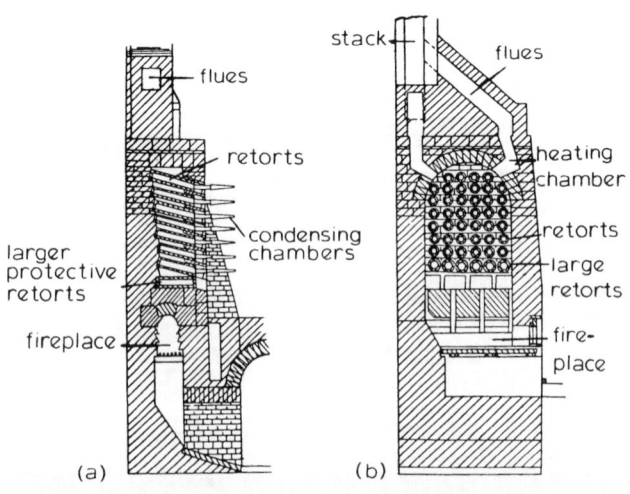

a section along retorts; *b* section across retorts

125 Dony's Belgian zinc-smelting retort furnace (from Cocks and Walters[136])

124 Champion's zinc smelting furnace (from Dufrenoy[50])

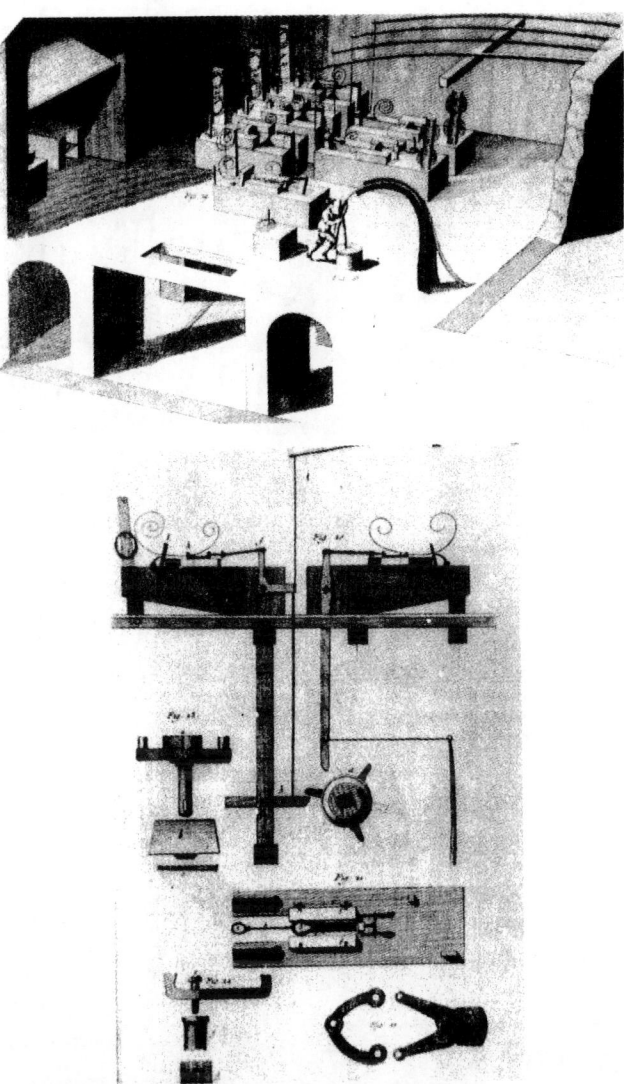

126 *Wire drawing machine (from Diderot [87])*

tory for rods and coarser material, but when applied to thinner wires the productivity was low. Diderot[133] shows the application of water driven automatic machines— five operated by one wheel (*see* Fig. 126). Power is transmitted by cams or bell cracks to the pincers which pull the wire intermittently through the die from a coil behind. The intermittent application of power from the wheel pulls the pincers forward while the return stroke —arranged through the usual spring beam - releases the pincers and pushes them back towards the die. The forward stroke causes the pincers to grip the wire and pull it through another length (about 30 cm). Apparently, one man could control five of these machines which he fed with strips of brass cut from sheet. This sort of non-continuous machine would leave 'stop marks' on the wire, but presumably these were smoothed out by continuous drawing at a later stage.

Much of this wire would be used for pins.[138] From about 1550 until well into the 19th century the heads usually consisted of wire spirals later set by small drop hammers,[139] (*see* Fig. 127). Occasionally, the heads were made by soldering on discs of brass sheet and rounding these to give a semi-circular form. The pins were tinned by an electrochemical process which involved boiling them in an aqueous solution of argol (potassium bitartrate), containing small 0.5 mm thick plates of tin. The tin dissolved slowly in the solution and was plated out on to the pins in contact with the tin plates.[140]

By adding nickel to brass, or by adding zinc to nickel-copper alloys, a white corrosion-resistant alloy may be produced which was known in Europe as 'German Silver', in China as Paktong, and in Malaysia as 'white brass'. These alloys contain 3–15% Ni and 22–26% zinc. Naturally, local nickeliferous ores were originally used such as kupfernickel (NiAs) in the Erzgebirge and garnierites in New Caledonia, Sulawesi and the Philippines.[141]

Lead Smelting and Refining

By the beginning of the 18th century the coal-fired reverberatory was under test for all non-ferrous smelting in England. It had proved itself in the Flintshire works of the London (Quaker) Lead Company, and had the great advantage of using local coal and did not need water power for bellows.[142,143] Where coal was available the process superseded the 'bole hill' and the slag hearth.

In Derbyshire and the North of England the old ore hearth was widely used in 1729[144,145] using water power with peat and wood (white coal) as fuel, although some stone coal was used in Cardiganshire. On the Continent, on the other hand, Agricola's shaft furnaces were still in use for metallurgy generally, the only difference between the furnaces used being the tapping arrangements.[146] The form used for lead is shown by Diderot and D'Alembert in their encyclopedia of 1762, and they say that it is the same as that used for copper.[133] The same situation existed in the Slovakian ore field at Banska Stiavnica.[148] Jars visited these works in 1758 and praised the level of technique as compared with some of the

Most of the zinc produced by Champion's process must have been used for brass, although the calamine process continued to be used up to about 1850. A certain amount would be used for brazing metal (50 Zn–50 Cu) and, after 1836, when the French scientist Sorel patented the process of zinc plating by dipping iron into molten zinc (now known as galvanizing), an increasing amount went into this process.[137] It is interesting to note that Sorel was aware of the electrochemical nature of corrosion and the sacrificial role of the zinc coating on the iron. This, of course is the reason for the application of the term 'galvanizing'; in no case was zinc applied to the iron electrolytically, nor is this normally done today. The idea of stiffening iron sheet by corrugating had been introduced by R. Walter of Rotherham in 1828, and so by 1836 the scene was set for that ubiquitous material known as galvanized corrugated iron sheet or, more briefly, as 'corrugated iron'.

A good deal of brass was made into wire: the increasing use of wire in the 18th century inspired the development of automatic wire drawing machines. The old sling seats and crank operated manual wire-drawing process, shown by Biringuccio and others, was satisfac-

127 Pin making processes (from Diderot[97])

works he had seen elsewhere, but thought that, generally, the processes were backward and the furnaces
dated.[147] No doubt he had heard of the English techniques
which he was to see in about 1764.

We have detailed descriptions of the techniques used
at Banska Stiavnica in 1757 for the smelting of lead and
silver ores, and gold.[149] The furnaces used for lead
smelting were very like those described by Agricola, but
the front walls had been changed so that iron plates were
used at the higher levels. The tuyere was in the back wall
and inclined downwards, and the furnace height was
about 1.8–2.1 m. In some ways, we can see the transition
from the blast furnace to the lower ore hearth, but the
forehearth and slag basin were still like those described
by Agricola, and it is clear that the slag was intended to
leave the furnace in the liquid state.

The liquation furnace was still in use in the Harz
mountains for melting out the lead bullion from the
copper–lead alloy for noble metal recovery. The Hungarian 'Brillen' (brillen = goggle) furnace in use at Mansfeld
appears to have had movable sheet iron covers to allow
charging and to control fumes. Some of these furnaces
were charged by steps from one side; others must have
been filled from the front, and dust chambers were
incorporated in some. The silver content of the lead
produced seems to have been about 310 g/t.

THE NORTH OF ENGLAND ORE HEARTH

Although Dr Burcot (Burkhardt Kranich) had done his
best to introduce the Agricolan furnace to Elizabethan
England—and there is no doubt that it was a great
success as compared to the footblasts and the bole hills
—the English preference for the low hearth (bloomery)
furnace persisted where wood and peat were available.
The best accounts of this process come from the
Cumberland—Northumberland ore field where the
royalties forfeited by the Earl of Derwentwater after the
rebellion of 1715 were being worked on behalf of the
Governors of Greenwich Hospital.[150,151] The furnaces in
use in about 1780 show a close similarity to the modern
ore hearth, which was in use in Newcastle until 1960.
The 18th century ore hearth consisted of a cast iron back
and sides with a cast iron 'workstone' or apron in front,
and a cast iron sump or 'sumpter pot' for the lead to
drain into from the workstone. The tuyere was inserted
through the pipestone at the back and supplied with air
from water-driven bellows (see Fig. 128).

The hearth was worked by first charging fuel and,
when the fuel was hot enough, the unroasted sulphide
ore was put on top and so exposed to oxidizing conditions. The hearth 'pan' was kept full of lead from the last
charge and its buoyancy kept all but the excess lead
floating. Slags were moved forward to the workstone,

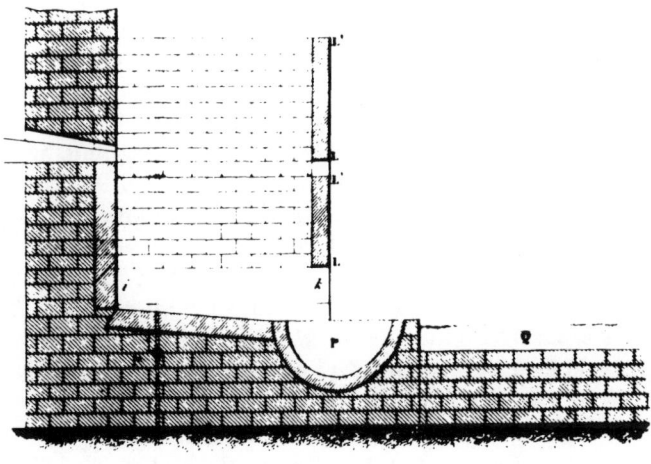

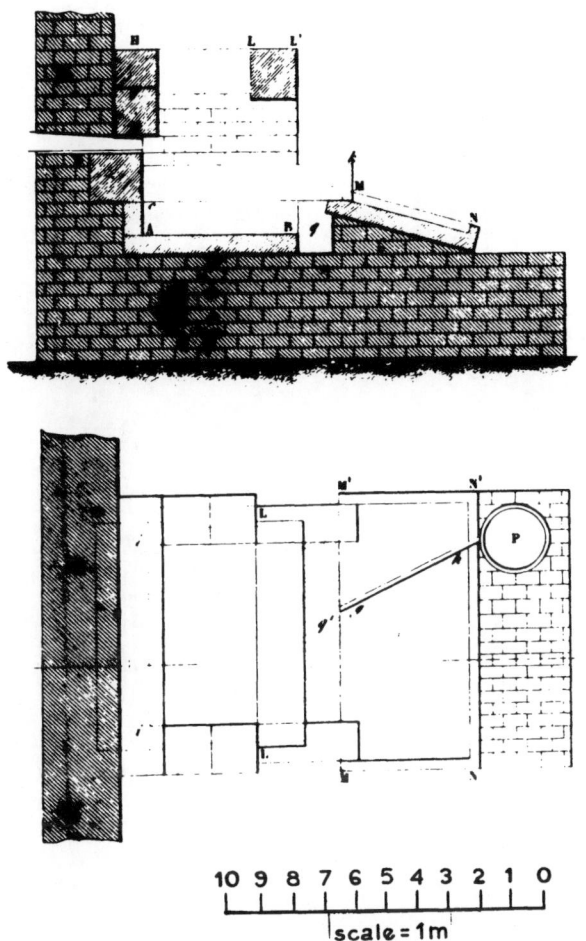

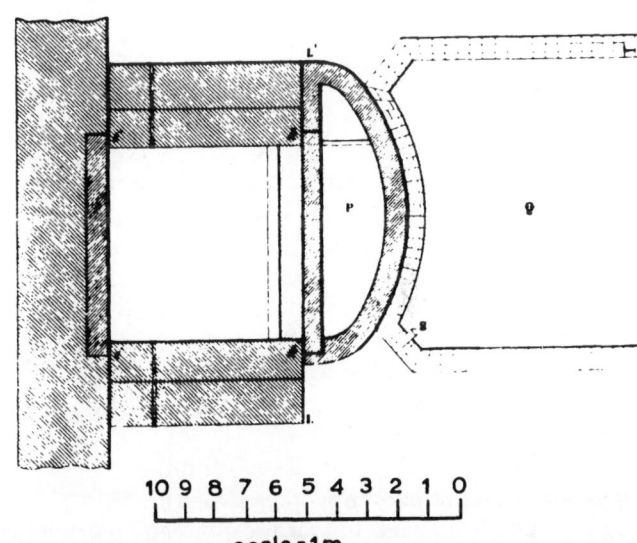

128 *Ore hearth for lead smelting (from Dufrenoy [50])*

drained of their lead, and discarded. The additional lead overflowed from the pan through a groove cut in the workstone and ran into the sump below. No fluid slag was withdrawn from this part of the process. In some areas it was the custom to roast the ore before smelting it in the ore hearth. 'White' coal or dried wood was the principal fuel although peat was popular.

THE SLAG HEARTH

The slag hearth is, at first sight, little different from the ore hearth (*see* Fig. 129), but it had no workstone as the slag and lead sank to the bottom of the hearth as liquid phases and no solid material was removed. Furthermore, the 'shaft' or working part of the furnace was not as short as the ore hearth and this furnace had close affinities to the Agricolan blast furnace.

As the working temperatures were a good deal higher than in the ore hearth (1 200°C) there was a danger of greater losses by volatilization, and the shaft of the furnace was therefore made taller and the dust collected by long flues. The slag was run from the furnace and, if it was not sufficiently free flowing, additions of iron oxides or bloomery slag could be made to raise the iron content. Separation of lead from the slags was brought about by having two pots in front of the hearth; the first received the lead, and the slag overflow went into the second from which it was removed when cold.

129 *Slag hearth for the recovery of lead from ore hearth slags (from Dufrenoy [50])*

THE FLINTSHIRE REVERBERATORY FURNACE

The really big development was in the use of the reverberatory furnace and coal. We have a detailed description of the operations in the Flintshire works of the London Lead Company at Gadlys, which had reverberatories working in 1708.[142,143] By then, there were four smelting furnaces, two slag hearths, and four refining (cupellation) furnaces. In the Flintshire process, the ore was not previously calcined[153] and was oxidized in the furnace as it was in the ore hearth. The oxidation could not be accurately controlled to give up all the lead by the double decomposition reaction

$$PbS + 2PbO \rightarrow 3Pb + SO_2$$

and the charge was therefore over oxidized and reduced back by adding coal later in the process. The slags were usually withdrawn in the solid form and worked in the slag hearth, but in some cases they were free running and could be tapped. The furnace hearth

sloped to one side so that the lead could continuously towards a collecting pot, and the solid slags could
be withdrawn through a door above. Slaked lime was
added as a flux in the reducing stage.

In 1859 the charge consisted of 1.07 t of good Flintshire
ore which yielded 0.74 t of lead,[153] and of this, 91% was
in the form of metal and 9% in the slag and fume. This
would need between 0.61 and 0.81 t of coal. Sixteen
charges of ore produced 11.5 t of lead and 2 t of slag
containing 55% Pb, so the loss of lead in the slag was at
least as high as in the ore hearth. This would be reworked in the slag hearth with iron flux and coal. By
1860, improved reverberatory furnaces were used in
which the temperatures were higher and the conditions
such that comparatively lead-free (black) slags could be
tapped with the lead. Only calcined lead ore was charged
to these furnaces (*see* Fig. 130).

19TH CENTURY TECHNIQUES

Up to the end of the 18th century it was normal British
practice to work the ore hearth with galena concentrates,
although on the Continent stall roasting was often used
before smelting.[152] Reverberatory roasting furnaces were
introduced to Alston Moor in 1810, and roasting preceded
smelting in the ore hearth.[154] The reverberatory was in
general use in Derbyshire at this time for roasting and
smelting but the slags were still worked up in a slag
hearth. The slags from the reverberatory contained a
great deal of lead in the form of lead sulphate (34%)
which was recovered in the slag hearth by the use of
higher temperatures (*see* Table 74). At Whitfield in
Northumberland, lime was thrown on the reverberatory slags to 'protect the workers from the heat'. This
would then be carried forward to the slag hearth and
incorporated into the slag, as we see from the analysis of
that from Matlock (*see* Table 74).

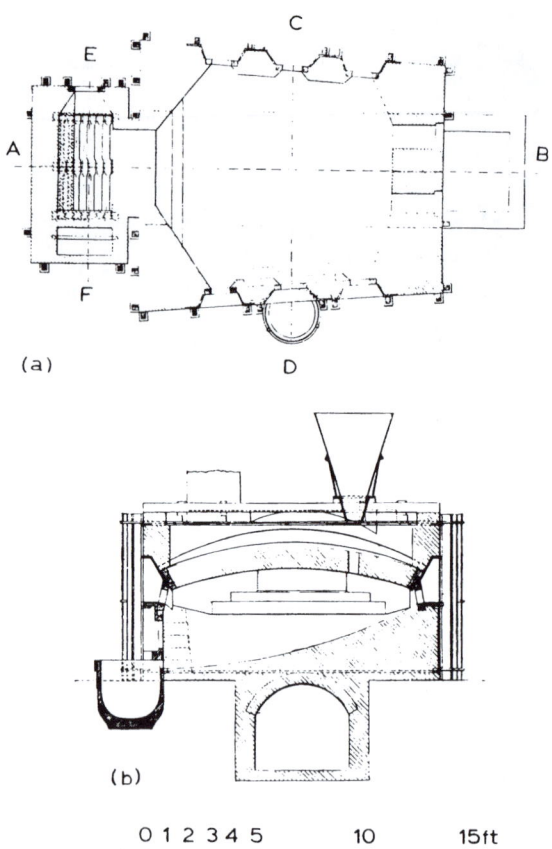

a horizontal section; b vertical section on line CD: the
fire-brick bearers near the fire bridge are shown as left
by the builder, their edges soon became worn off in
the course of working

130 *Flintshire reverberatory or flowing furnace for the reduction of oxidised lead ores (from Percy[153])*

Table 74 Analyses of English lead smelting slags and residues made in about 1836 for Dufrenoy[154] by P. Berthier

Element	Amount, %					
	Ore hearth, Alston	Ore hearth, Grassington	Flue dust, Alston	Reverberatory, Redruth	Reverberatory, Matlock	Slag hearth, Matlock
SiO$_2$	28·5		5·6	35·0		13·0
FeO	25·0	3·0	3·4	22·5		14·5
CaO	24·0	4·5		19·0	17·8	18·5
ZnO	10·6		13·8	6·0	4·5 (includes FeO)	2·5
Al$_2$O$_3$	7·0		(included in SiO$_2$)	3·5		2·0
PbO	3·0	34·0	10·2	12·0	15·9	1·0
MgO	tr.					
PbSO$_4$			65·6			
PbS			1·4			
CaF$_2$		1·5			16·0	13·4
BaO		33·5			16·4	30·0
SO$_4$		23·5			27·8	7·0
Colour	Black	Grey—yellow (too little CaF$_2$)		Black (magnetic)	Grey—yellow (sufficient CaF$_2$)	

It is interesting to note the difference between the slags from two ore hearths, those at Alston and Grassington. The slag from Alston is quite clean (3.0% PbO), while the Grassington slag is mainly lead and barium sulphates. To judge the relative efficiency of the process we need to know more about slag volume, but it would appear that the high barytes content of the Grassington ore was making lead separation difficult.

In 1831 Pattinson, who was to become a famous name in the lead industry, published his report on the techniques used in the North of England.[155] From this we see that apart from the use of the reverberatory for roasting there was little change from 18th century techniques. However, it is clear that the slag hearth shaft was getting taller—by 1836 it was 1.07 m compared with the 0.56 m of the ore hearth. This is the first step towards the water-jacketed blast furnace of modern times. The cast iron sump in front of the taphole was divided into two parts. The first part was filled with cinders and the lead could sink through them and flow under the partition to collect in the other half, from which it could be ladled out. The black slag was allowed to overflow from the first half of the pan into a trough of running water and became granulated. In this way it was easy to recover any metallic lead that it might contain. Around 1850, the Spanish slag hearth was introduced to the Mendips to rework the Roman and Medieval slags.[156] This suffered from the usual problem of slag hearths, that of loss of lead as dust and fume. The furnaces had the bottom filled with 'cinder', so that the lead could filter through it while the more viscous slag could run off the top of the cinder bed.

Pattinson[155] gives some figures for the working capacity and output of the various types of furnaces. He found that the use of roasted ore in the ore hearth gave nearly 50% greater yield of lead compared with the ore in its raw state. As far as Britain was concerned this was only economic as long as coal was used in the roasting stage. While the principal fuel of the ore hearth was peat, a little coal was sometimes used. At this time the concentrate used was still very rich (70% Pb), and an ore hearth could process 70 bushels of ore (about 15 t) during one week of eight shifts, yielding something like 10 t of lead. It would need about eight small cartloads of peat and 0.6–0.9 t of coal.

The slag hearth used coke only, and needed 24 bushels (about 200 kg) of coke per tonne of lead which it could make in a 14–16 hour shift. The slag hearth produced about one-thirteenth of the total lead output. No additional flux was added, but some additional FeO and Al_2O_3 would come from the coke ash. The slag hearth was used to work up the fume and other 'waste', just as the blast furnace does in the 20th century.

SILVER RECOVERY AND REFINING

In this period in Europe the normal silver content of the lead ores varied from about 60 to 750 g/t of lead and, when in the range 250–750 g/t, silver was always worth recovering. This was carried out by cupellation in which

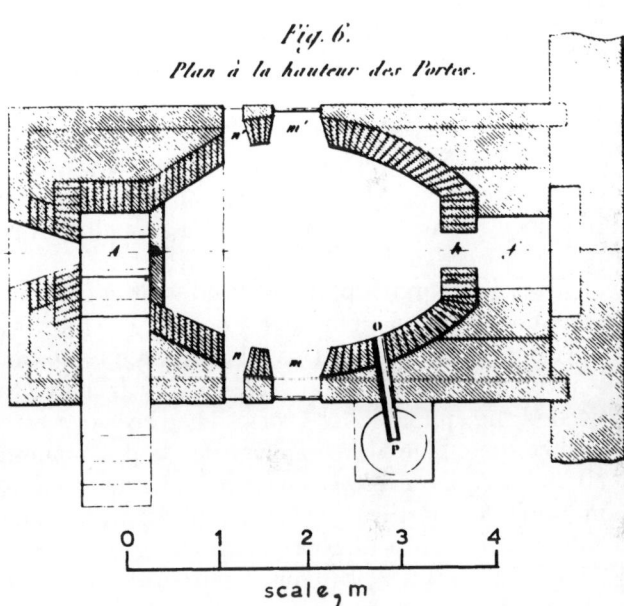

Fig. 6.
Plan à la hauteur des Portes.

131 Reverberatory furnace for the reduction of litharge to lead used at Alston (from Dufrenoy [50])

the lead was oxidized to litharge (PbO) in bone-ash cupels (or 'tests') and the precious metals left as a molten sessile drop in the middle of the cupel. The litharge was originally reduced back to lead in the ore hearth. By 1780, some works used the ore hearth and some the reverberatory with coal as a reducing agent.[150] By 1831, it was normal practice to do it the latter way; the reducing furnace was rather like the Flintshire furnace with a sloping hearth towards the taphole (*see* Fig. 131). One furnace could reduce 24 t/week consuming 180 kg of coal per tonne of lead produced. The small quantity of solid slag was reworked in the slag hearth.

The cupel or test was elliptical with one end more pointed than the other (*see* Fig. 132). This formed the hearth of a coal-fired reverberatory furnace but was

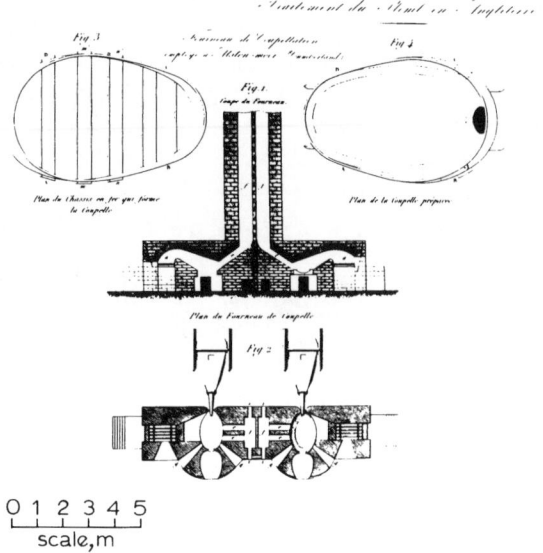

132 Double cupellation furnace for the extraction of silver from lead made at Alston (from Dufrenoy[50])

separate and entirely movable. At the back of the furnace a hole was left, through which the bellows nozzle could be inserted. In 1780, solid lead pigs were placed on the hearth, heated by means of the coal in the reverberatory firebox to 1 000–1 100°C, and oxidized with air from the bellows.[150] The charging of cold pig-lead was not good for the bone-ash hearth, and by 1830 it was normal practice to charge molten lead from a small iron pot furnace placed at the side of the reverberatory.[155]

The lead charge rapidly oxidized to litharge, which was allowed to run off through a slot (gate) in the test at the front of the furnace. This solidified into lumps and was collected for reduction. Molten lead was charged as needed and, in 1831, after 4 t of lead had been processed to litharge to give 51 kg of silver-rich lead (containing about 2% Ag), the test was removed and the metal cast into pigs. The test was replaced and recharged until a workable quantity (3 t) of silver-rich lead had been amassed. This was worked on a fresh cupel with a much more concave bottom to give a cake of silver weighing 50–75 kg. The last litharge to come off this test was separated from the rest as it contained some silver; likewise the tests themselves. These were worked up at the end of the year. The other tests were resmelted in the slag hearth with black slag to give an impure hard lead.

The refining furnaces processed about 4 t of lead in 16–18 hours, or 24 t/week. About 150 kg of coal was required for 1 t of lead. It would seem that about 5% of the silver was lost in the cupel, litharge, or fume, but this was recovered on resmelting. Before the 19th century no refining (apart from silver recovery) took place, mainly because of the comparatively high purity of the lead produced by the low-temperature ore hearth and reverberatory processes. The first big change was brought about by H. L. Pattinson himself and is known as the 'Pattinson' process and, although he does not mention it in his paper of 1831,[155] it must have been in his mind as his patent is dated 1833. He was then working in the Blaydon lead works owned by the Beaumont family; these works were visited by Dufrénoy who records the process[154] which became so well known until it was replaced by the Parkes process. Naturally, Pattinson had been concerned about the high cost of recovering silver by cupellation and reducing back the litharge to lead. In fact, this process could only be made to pay when the average silver content of the lead exceeded 250 g/t, and the silver contents at this time were steadily declining.

Pattinson first intended to distil the lead but later felt that some use might be made of the difference in density of silver and lead.[157] However, he accidentally dropped a crucible containing partially solidified lead upon the floor, and, upon assaying the originally solid lead and then the freshly solidified lead, he found that the latter contained a lot more silver than the former. This was the basis of the process. The work lead (or bullion) was run into cast iron kettles and the temperature slowly reduced until partial solidification set in.[152] The solid lead was removed from the top of the kettle with a perforated ladle, and the silver-rich metal left behind was cupelled. Of course, use was made of the counter-current principle, the enriched lead going one way and the depleted lead going the other. Dufrénoy shows a five-kettle refinery but Percy shows eight (*see* Fig. 133). There were seven actual refining pots and the original lead, containing 250 g/t, was put in the middle pot. The depleted lead gradually made its way to No. 7 pot and the enriched to No. 1. Further lead was added to No. 4 and the process repeated in a rather complicated way until it ended up with enriched lead containing 5 000–5 300 g/t at one end, and depleted lead with only 12–15 g/t at the other. All pots were coal fired so that melting and cooling could be independently controlled.

The Parkes process was patented in 1850 and involves the formation of an Ag–Zn compound which floats to the surface and can be skimmed off. The zinc can be removed by volatilization under vacuum, and the silver-rich residues cupelled. Of the other changes that have taken place, one was the mechanization of the ore hearth—the Newnam hearth—which was still in use in Newcastle as late as 1960: the slag hearth has become the water-jacketed blast furnace, and the roasting of the ore is usually done on a Dwight-Lloyd sintering machine.

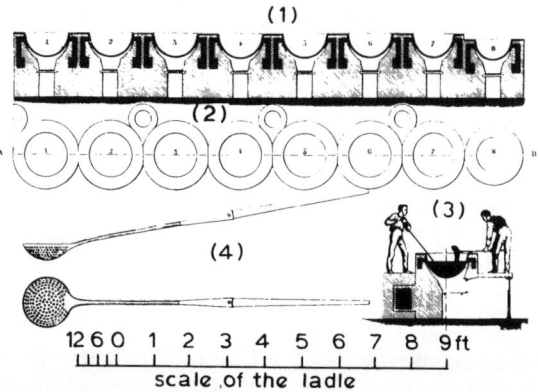

133 Layout of kettles used in the Pattinson process for the enrichment of silver in lead; the scale is not given to 1, 2, 3 of this woodcut (after Percy[153])

Base metal impurities are now removed by oxidation and chemical processes, involving reactions between specific elements and additions to the charge such as sodium hydroxide.

STRUCTURAL REMAINS

In Britain, at least, there are considerable remains of the old processes still visible today.[158] Many of the old smelt mills have been converted to some other use and others exist as ruins. The long flues winding their way miles across the countryside and terminating in chimneys on suitable hill tops can still be seen.[159] The chimney of the Langley Mill still stands; this system was built between 1795 and 1806, presumably because of public complaint of poisoning of the land surrounding the mill. But a well-built chimney and flue was a bag house in itself and, when properly managed, could pay for itself many times over. The best, such as that which served the Rookhope (County Durham) smelter, had small reservoirs at different levels on the hillside and the water from these could be diverted into the flue to wash down the deposits to settling chambers at the smelter. A single year's profit from one of these systems amounted to 70% of the original cost.

In Yorkshire we can still see the layout of some typical mills.[158] Most of them had ore and slag hearths blown by waterwheels, and adjoining roasting houses. Some had reverberatory furnaces for roasting. The flues are much in evidence everywhere and the remains of the wet, labyrinthine Stokoe condensers, introduced around 1850, can often be seen. In the Yorkshire mills there is no evidence of desilverization processes. This is partly because the Yorkshire ores were very low in silver, but mainly because the mills were set up where they were primarily to take advantage of local fuel and water. Their object was to recover as much as possible, as cheaply as possible, and for this the ore hearth was supreme. This function has been the main purpose of the ore hearth in modern plants. Further work on the lead required coal fuel and this was best left to more suitably located refineries.

Tin and Its Uses

The main development of this period was the application of the reverberatory furnace to smelting but, as in other non-ferrous metals, this was a gradual development. The fuel required to produce 1 t of tin was very much less than in copper smelting, because with tin only one process was required, i.e. the reduction smelting of tin oxide (cassiterite). As the purity of the alluvial ore was high no roasting was required but, later, with the increasing use of mined vein ore, roasting was introduced to get rid of some of the impurities such as arsenic.

There is little doubt that in 1698, when Celia Fiennes made her visit to Cornwall, smelting in blast furnaces (blowing houses) was the principal method.[160] She saw the ore and fuel being charged together. The ore had been first dry stamped in stamping mills 'as fine as the finest sand', and the tin from the forehearth was ladled into moulds. The fuel in this case was wood, peat, and charcoal, all obtained locally; but, by the beginning of the 18th century, various patentees were claiming the use of coal in non-ferrous smelting by the reverberatory process. Welsh pit-coal was being used for reverberatory smelting of tin ore at Newham, Cornwall in 1703–4 at the rate of 1.78 t/day. By 1711 tin was being smelted in considerable quantities at Calenick—most of it being done with coal in reverberatories.[161] There is no doubt that this resulted in a good deal of contamination, and was one of the causes of the many complaints of the London pewterers about the poor quality of the pewter.

The blast furnace, used in Cornwall in 1728 and shown in Fig. 134, had changed little since the time of Agricola. It was known as the 'Castle' because of its solid stone structure.[162] It took 12 hours and 18–24 sixty gallon pecks of charcoal (about 1 150 kg) to smelt 8–12 cwt (500 kg) of tin ore, which is equivalent to a fuel/metal ratio of 2.5:1—vastly different from the 20:1 of copper.

The Continental preference for the blast furnace is again apparent in Diderot in 1760 where the only tin smelting furnace shown is a blast furnace. It is probably very like the Cornish model, but Diderot shows only an external front view.[87] However, it is clear from this that his furnace contains a labyrinth of some sort above the shaft to trap the dust before the products of combustion are exhausted by a small chimney to the atmosphere.

Vein tin was being mined in larger quantities by the middle of the 18th century, as pumping became economically possible with the development of the steam engine. In alluvial deposits, arsenic and other impurities suffered selective weathering, but with vein tin this was not possible and it is clear that an increasing amount of impurities were being incorporated into the tin as sold. This was another reason for the complaints of the London pewterers. This problem was overcome in two ways. The first was the calcining of the mineral to oxidize the arsenic—mainly as arseno-pyrite. Diderot shows calciners with long deposition chambers like those in lead flues in his plate depicting cobalt and arsenic production. Details of Cornish arsenic calciners have been given by Earl.[163] These allowed some degree of recovery of the arsenic as oxide.

The early dressers and tin smelters could not get rid of pyrite (FeS_2) and arseno-pyrite (FeAsS) by washing, because their specific gravities were too near that of cassiterite (5–5.2 and 5.9–6.2, respectively, compared with 6.8–7.0 for SnO_2). This meant that the concentrate had to be roasted in order to oxidize these minerals to iron oxide with the evolution of arsenious oxide and sulphur dioxide. The iron could be removed from the cassiterite after roasting by the normal washing processes. For tin roasting, a rectangular calcining kiln was used in which the wood fuel was put in at a lower level (below the hearth) and the ore exposed only to the combustion gases in the floor above. In the roof above there was a charging door, and rabbling could be carried out from a front opening which also served as an opening for the exhaust gases and discharge of the calcine. The arsenic was recovered from long horizontal flues as arsenious oxide which was reduced and condensed in special furnaces.[164]

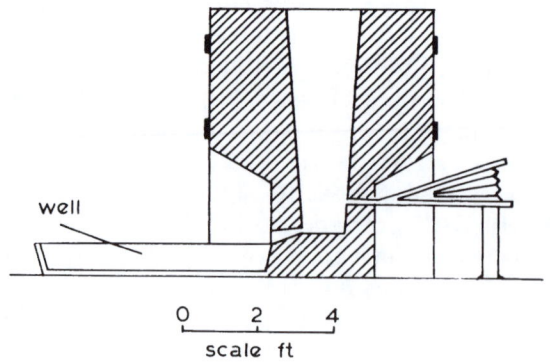

134 Blast furnace for tin-smelting in Cornwall (after Pryce[162])

By about 1829, the Brunton calciner was introduced.[165] The remains of this plant can still be seen in the ruins of Cornish smelters. It consisted of a revolving bed about 3.6 m dia. having two fireboxes set at an angle, and it was used extensively in Cornwall until its supersedence by flotation in the 1920s.

The element cobalt was first produced (unknowingly) by Brandt in 1742 and identified as a metal by Bergman in 1780. However, it had always been a common impurity in non-ferrous metal extraction,[166] and dressed tin ore from the Cornish mine at Dolcoath used to contain 0.5% Co. Cobalt could be recovered from the roasted arseno-pyrite as during calcination the CoAs oxidized to a stable arsenate.[166] It was left behind as a speiss after smelting, and some Cornish material had the following composition in per cent,

Fe	53.0	W	3.5
As	18.0	S	2.50
Co	4.4	Sn	16.25

The tin content was probably not in the form of a compound but mechanically entrapped. These compounds were fairly common in early metallurgy and could have been used as hardeners for copper. They are not dissimilar from the black niello used as a dark background to contrast with gold and silver in art metalwork.

The main impurity in smelted tin was 'hardhead', the intermetallic compound FeSn. This was usually removed by 'tossing', similar to flapping in copper refining, in which the impurities are selectively oxidized when the metal is poured slowly through air.

Once wood and peat ceased to be freely available, the advantage that the blast furnace had over the reverberatory no longer existed. This advantage was, of course, that of being able to produce a purer tin which would sell at a higher price. The use of less pure ores and coal as fuel meant some sort of refining operation, and the smelters became more complex. Cornish tin had been smelted in a reverberatory as early as 1704 and possibly earlier. In 1778 the design of the reverberatory was exactly the same as that used for copper at Hayle— the site of one of the few Cornish copper smelters. We have a plan and sectional elevation of such a furnace[162] (*see* Fig. 119), and we can assume that the design of the furnace was virtually unchanged since 1738 at least, when this furnace was in use near Bristol for copper and brass.[149]

The use of coal in the reverberatory, where it would need to be mixed with the ore, gave rise to the poor quality disliked by the pewterers. Gradually, techniques improved and smelting became concentrated in fewer smelters. In 1794 the Calenick (Truro) smelter had ten furnaces.[167] Slag from this site showed a composition reflecting the lower purity of the ore being smelted by such furnaces. It is probable that iron in some form was being added as flux.

While copper went through many operations during smelting, which were carried out in several reverberatory furnaces of slightly differing size and design, tin could be converted from ore to metal in one operation using only one furnace. This made the smelter a lot simpler and the fuel requirements were so much less that there was no economic reason to move the smelter to the coalfields when the reverberatory started to displace the blowing house. The latter, however, had its advantages, especially for smelting the pure alluvial tin ore, and the last blowing house did not cease work until about 1840.

In Britain, the number of smelters grew as output increased in the early years of the 19th century: during this time a good deal of the tin ore was imported. Later, however, a decline set in as other countries, particularly those in the Malay Straits, built their own smelters. Gradually, the Cornish smelters closed down leaving only one smelter in Britain, near Liverpool, conveniently within reach of seaborne ore and fuel.

The use of tin for tinplated iron gradually became a major industry as containers of tinplate began to displace the more fragile glass and pottery. However, in Germany, the country of its birth, there was a considerable decline in the industry, so much so that at Wunsiedel in 1785 there was only one enterprise left, and this was making spoons of tinned iron which were said to be intermediate in the social order between the silverware of the rich and the wooden spoons of the poor.[168]

In the 17th and 18th centuries various European countries attempted to become self-sufficient in the production of tinplate and France had a works at Beaumont in Franche-Comté in 1665, and one in 1695 at Chenecey which was capable of producing 800 barrels/ year using English tin.[169] But the industry declined slowly here too, as the expansion of the British tinplate industry gradually began to absorb all the European markets. The English tinplate industry is generally as-

135 Tin plating sheet iron (from Diderot[87])

sumed to have started with the visit of Andrew Yarranton, the British industrialist, to Saxony in 1665. He saw an outlet for the output of his ironworks in south west England, together with Cornish tin. It is doubtful whether this visit had any immediate effect but, by the 1720s, John Hanbury was making tinplate at Pontypool in Wales assisted, no doubt, by an import duty levied in 1706.[170]

Apart from the use of the rolling mill and the rolling of thin sheets by doubling and rolling in eights, the technique had not changed much from that described in Chapter 8, p.105. But in some cases three tinning pots were used, i.e. the soaking pot in which the cleaned plates were held for one hour, the wash pot in which the plates were redipped to ensure a more even coating, and a 'list' pot in which the lower edges were dipped in order to remove the excess tin. A separate grease pot, where the plates stood for an hour, was introduced by Mosley in 1745. All this became known as the 'Welsh process' (*see* Fig. 135).

Gradually, mineral acids (vitriol and muriatic acid) came to displace the ferments for pickling. In 1818, charcoal iron was still preferred because of its lower slag content. The plates were pickled and heated in a scaling furnace, and then cold rolled between 75 cm dia. rolls to break the scale. The plates were then pickled again in fermented bran, then in sulphuric acid in a lead-lined tank at blood heat. The plates were stored in cold water before being dipped for $1^{1}/_{2}$ hours in an iron pot filled with low-purity tin, covered with a 15 cm deep layer of tallow. The plates were greased and re-dipped in the wash pot which held $^{1}/_{2}$ t of high-purity, grain tin. One-third of this tin was replaced every 60–70 boxes of 225 sheets, and that removed was put into the first tin pot. The sheets were individually brushed and given a final dip before putting in the grease pot which removed the surplus tin and evened out the coating. The plates were then cooled, and the thick deposit at the bottom edge taken off in the 'list' pot which was only 0.65 cm deep.

The tinplate of 1824 consisted of sheet iron 0.47 mm thickness, containing a 0.013 mm thick layer of 99.92% Sn on each side. This corresponds to 75 kg of iron sheet per box of 225 sheets, coated with tin weighing 3.9 kg. In 1800, there were 11 works in south west Britain and a healthy export trade, because of the availability of cheap coal-produced wrought iron and local Cornish tin. Later, the process became increasingly mechanized with sulphuric acid being used for pickling, and palm oil instead of tallow. Zinc chloride was used as a flux and the coating thickness halved. The industry expanded to such an extent that by 1882 the annual production was $7^{1}/_{2}$ million boxes each of 47 kg, $4^{1}/_{2}$ million of which were exported. Although these were all made of home-produced iron and steel, by this time most of the tin was being imported from the Malay Straits. After this, owing to the imposition of tariffs by the importing countries, the industry declined somewhat, and its post 1914–18 future was largely decided by the progress of the continuous sheet steel industry.

References

1   G. HAMMERSLEY: 'The charcoal iron industry and its fuel; 1540 to 1750', *Econ. Hist. Rev.*, 1973, **26**, 593–613

2   G. WEBSTER: *Ant. J.*, 1955, **35**, 199

3   T. DAFF: *Bull. HMG*, 1972, **6**, 1

4   DUD DUDLEY: 'Metallum Martis', 1665, London

5   E. N. SIMONS: *Metallurgia*, 1956, **55**, 21

6   R. A. MOTT: *ibid.*, 1957, **56**, 296

7   G. R. MORTON and M. D. G. WANKLYN: *J. West Mid Reg Studies.*, 1967, **1**, (1), 48

8   In chemical terms, slags consist of acids such as SiO_2 (sand) and bases such as CaO (lime) and MgO: slags proportionately high in the bases are more capable of absorbing sulphur and keeping it out of the pig iron than the more acid slags with lower lime content. Moreover, the bases replace iron in the slags and there fore give a higher yield. On the other hand, the more acid slags with lower lime and higher iron have lower free-running temperatures and are easier to remove from the furnace in a liquid state. The 'basicity' as given here is the result of dividing the sum of the CaO and MgO contents by the SiO_2 content

9   R. A. MOTT: *TNS*, 1957–59, **31**, 49

10   R. W. BUNSEN and LORD LYON PLAYFAIR: 'The gases evolved from iron furnaces with reference to the theory of the smelting of iron', 1845, Cambridge, British Association

11   G. JARS: 'Voyages Métallurgiques', 1774-81, 3 vols., Lyon

12   A. RAISTRICK: 'Dynasty of iron founders; the Darbys of Coalbrookdale', 1953, London

13   G. R. MORTON and A. F. MOSELEY: *West Mid. Reg. Studies*, 1970, Special Publication No. 2

14   W. H. SANSOM: *Metallurgia*, 1962, **65**,165 Backbarrow accounts 1713–19 University of Newcastle Library

15   R. A. MOTT: *TNS*, 1957–59, 31, 271

16   R. A. LEWIS: Thesis, 1949, Birmingham University; *see also*, N. MUTTON: *Bull. HMG*, 1966, **1**, (6),

17   A. RAISTRICK: *op. cit.*, 241

18   R. A. MOTT: *TNS*, 1959–60, **32**, 43, Pl.V

19   R. A. MOTT: *TNS*, 1957–59, **31**, 49

20   G. R. MORTON and W. A. SMITH: *J. Iron Steel Inst.*, 1966, **204**, 661

21   G. JARS: *op. cit.*, vol. 1, 46

22   R. A. MOTT: *op. cit.*, 81

23   G. R. MORTON and N. MUTTON: *J. Iron Steel Inst.*, 1967, **205**, 722

24   C. S. SMITH (ed.): 'Sources for the history of the science of steel; 1532–1786', 1968, London, MIT Press jointly with Soc. for History of Technology

25   G. R. MORTON and J. GOULD: *J. Iron Steel Inst.*, 1967, **205**, 237

26   T. H. TURNER: *ibid.*, 1912, **85**, (1), 203

27   Kelp is the ash of seaweed and is rich in sodium

28   W. H. CHALONER: *E. A. News*, 1948–49, **27**, 194, 213

29   G. R. MORTON and N. MUTTON: *op. cit.*, 725

30   W. H. GREENWOOD: 'A manual of metallurgy', 2 ed., c.1870, London, Collins

31   W. K. V. GALE: 'Notes on the Black Country iron trade', *TNS*, 1943–45, 24, 13–26

32   Patent No. 7778, 1838

33   TREVOR DAFF: 'The early English iron patents; 1600-1850', *Bull. HMG*, 1972, **6**, (1), 1–18

34   W. K. V. GALE: 'The British Iron and Steel Industry', 1967, Newton Abbot, DAVID AND CHARLES, 66

35   E. GREGORY: 'Metallurgy', 20, 1931, London, Blackie

36   H. G. BAKER: *TNS*, 1943–45, **24**, 113

37   R. F. TYLECOTE *et al.*: *J. Iron Steel Inst.*, 1965, **203**, 867; *see also* T. R. SLATER: *Indust. Arch.*, 1973, **10**, (3), 318

38   A. FELL: 'The early iron industry of Furness and District', 1908, 2nd ed., 1968, Ulverston

39   J. H. LEWIS: 'The charcoal-fired blast furnaces of Scotland; a review', *PSAS*, 1984, **114**, 433-479

40   JAMES DINN: 'Dyfi furnace excavations; 1982-87', *Post-Med. Arch.*, 1988, 22, 111-142

41   Private communication from W. H. Bailey who excavated this site in 1963; there is a short report on it in *Bull. HMG*, 1964, **1**, (3), 3

42   P. RIDEN: 'The ironworks at Alderwasley and Morley Park, Derby', Arch. J. 1988, **108**, 77-107

43   L. KRULIS: *RHS*, 1967, **8**, (4), 245

44   L. JENICEK: 'Metal founding through the ages in Czechoslovak territory', 1963, Prague

45   G. R. MORTON: *J. Iron Steel Inst.*, 1962, **200**, 444

46   G. JARS: *op. cit.*, vol. 1, 29

47   G. JARS: *op. cit.*, vol. 1, 128

48   E. STRAKER: 'Wealden Iron', 1930, London, 2 ed. 1969, 343

49   G. JARS: *op. cit.*, 160

50   M. DUFRENOY: 'Voyage Métallurgique en Angleterre, 2 volumes and atlas of plates', 1837, Paris

51   J. GIBBONS: 'Practical remarks on the construction of the Staffordshire blast furnace', 1839, Corbyn's Hall, Staffordshire

52   ANON.: 'Bosh tuyeres at Ditton Brook, Warrington', *JISI*, 1873, (1), 316-317

53   R. T. KINGDON: 'One hundred years of blast furnace development', *JHMS*, 1987, **21**, (2), 63-76

54   W. K. V. GALE: 'The Black Country iron industry', 1966, London, The Iron and Steel Institute

55   G. D. ELLIOT: 'Ironmaking at the Appleby Frodingham works of the United Steel Co.'s Ltd', ISI Spec. Rep. No. 30, London, 1944

56   R. F. TYLECOTE: 'From pot bellows to tuyeres', Levant, 1982, **13**, 107-118

57   G. AGRICOLA: 'De Re Metallica', 1912, London, (2 ed., 1950, New York)

58   R. PLOT: 'The natural history of Staffordshire', 1686, Oxford

59   W. BROWN: *E. A. News*, 1958, **37**, (429), 61; (430), 88

60   G. R. MORTON: *J. Iron Steel Inst.*, 1962, **200**, 444

61   R. A. MOTT: *op. cit.*, 271, plate 27

62   See article by John and Joseph Farey in 'Pantalogia', 1802- 13, vol. IV

63   J. GJERS: *J. Iron Steel Inst.*, 1870–71, **2**, 202

64   J. GARILLOT: *RHS*, 1966, **7**, (3), 163

65   G. R. MORTON: *J. Iron Steel Inst.*, 1967, **205**, 443

66   R. D. CORRINS: *Ind. Arch.*, 1970, **7**, (3), 233

67   J. PERCY: 'Metallurgy; iron and steel', 428, 1864

68   J. PERCY: *op. cit.*, 399

69   W. E. OSGERBY: *J. Iron Steel Inst.*, 1962, **200**, 1

70   D. W. CROSSLEY and B. TRINDER: 'The Ferriera at Pescia Fiorentina, Tuscany', Ironbridge, 1983

71   J. GARILLOT: *RHS*, 1966, **8**, (2), 95

72   G. JARS: *op. cit.*, vol. 1, 160, 270

73   H. B. HEWLETT: 'The Quarries', Stanton Ironworks Co., 1935, 27

74   J. PERCY: *op. cit.*, 376

75   J. GJERS: *op. cit.*

76   G. JARS: *op. cit.*, vol. 1, 93

77   G. JARS: *op. cit.*, vol. 1, 235

78   J. PERCY: 'Metallurgy; refractories and fuels', 1875, 431

79   DAVID GALE: personal communication, 1989

80   D. CRANSTONE: 'Early coke ovens; a note', *JHMS*, 1989, **23**, (2), 120-122

81   J. R. M. LYNE: *Bull. HMG*, 1972, **6**, 19

82   B. McCALL: *Ind. Arch.*, 1971, **8**, (1), 52

83   W. H. CHALONER: *op. cit.*

84   G. R. MORTON and W. A. SMITH: *op. cit.*, 665

85   G. JARS: *op. cit.*, vol. 1, 213

86   R. A. F. de REAUMUR: 'L'art de convertir le fer forgé en acier et l'art d'adoucir le fer fondu', 1722, Paris; there is an English translation; 'Reaumur's memoirs on steel and iron', (eds. A. G. Sisco and C. S. Smith), 1956, Chicago, Chicago University Press

87   D. DIDEROT and J. D'ALEMBERT: 'Encyclopedie ou dictionnaire raisonné des Sciences, des arts et des metiers', 1771-80, Paris

88   D. H. WOOD: *E. A. News*, 1949, **28**, 437, 461

89   T. A. WERTIME: 'The coming of the age of steel', 1961, Chicago, 177

90   M. DUFRENOY: *op. cit.*, vol. 1, 319

91   C. McCOMBE: *FTJ*, Oct. 10, 1968, 581

92   J. PERCY: 'Metallurgy; iron and steel', 278

93   J. A. W. BUSCH: *Bull. HMG*, 1972, **6**, 28

94   P. TEMIN: 'Iron and steel in nineteenth century America', 1967, London

95   D. FORBES: *J. Iron Steel Inst.*, 1870–1, **11**, 126

96   I. KRULIS-RANDA: *Blätter für Technikgeschichte, (Wien)*, 1963, **25**, 31

97   G. R. MORTON and W. A. SMITH: *op. cit.*, 671

98   M. DUFRENOY: *op. cit.*, vol. 2, plate III

99   J. W. HALL: *TNS*, 1927–28, **8**, 40

100   H. W. DICKENSON: 'John Wilkinson—ironmaster', 57, 1914, Ulverston

101   G. R. MORTON and N. MUTTON: *op. cit.*, 724

102   M. D. FREEMAN: *Ind. Arch.*, 1971, **8**, (1), 63

103   W. K. V. GALE: (*see* Ref. 34), 83

104   D. S. BUTLER and C. F. TEBBUTT: 'A Wealden cannon boring bar', *Post-Med. Arch.*, 1975, **9**, 38-41

105   M. H. JACKSON and C. de BEER: 'Eighteenth century gunfounding', 1973, Newton Abbot, David and Charles

106   JOHN WILKINSON: Patent No. 1063, 1774

107   G. JARS: *op. cit.*, vol. 1, 169

108   G. JARS: *op. cit.*, vol. 1, 133

109   G. JARS: *op. cit.*, vol. 1, 49

110   D. BROWNLIE and BARON de LAVELEYS: *J. Iron Steel Inst.*, 1930, **121**, 474, (discussion by H. W. Dickenson)

111   G. JARS: *op. cit.*, vol. I, 232

112   Anon: *E. A. News*, 1949, **28**, (327), 390; (329), 433

113   G. JARS: *op. cit.*, vol. 1, 151

114   K. C. BARRACLOUGH: 'Early steelmaking in the Sheffield area', 1968, Steel Times Annual Review

115   K. C. BARRACLOUGH: 'Steelmaking before Bessemer', 2 vols., Metals Soc. London, 1984

116   K. C. BARRACLOUGH: *Bull. HMG*, 1967, **1**, (8), 24

117   K. C. BARRACLOUGH: *J. Iron Steel Inst.*, 1971, **209**, 785, 952

118   K. KIRNBAUER: 'Copper in Nature, Technique, Art etc.', 40, 1966, Hamburg, Norddeutsche Affinerie

119   L. U. SALKIELD: 'A technical history of the Rio Tinto Mines; some notes on the exploitation from pre-Phoenician times to the 1950s', (ed. M. J. Cahalan), *Inst. Min. and Met.*, London, 1987

120   G. A. SCHLUTER: 'Essais de Mines et des Métaux', 2 vols., 1764, Paris, (French translation from the German originally published in Braunschweig in 1738)

121   S. LINDROTH: 'Gruvbrytning och Kopparhantering

vid Stora Kopparberget intill 1800-talets början', 2 vols., 1955, Uppsala

122   JOHN FISHER: 'Commercial relations between Spain and Spanish America in the era of free trade, 1778-1796', Univ. Liverpool Centre for Latin American Stud. Mono. 13, 1985

123   BASIL HALL: 'Journal written on the coasts of Chili, Peru and Mexico in the years 1820, 21 and 22', 1826, Edinburgh, Constable, 10–11

124   N. CUOMO DI CAPRIO and A. STORTI: 'Ordinamenta Super Arte Fossarum Rameriae et Argenteria Civitatis Massae', In: 'The crafts of the Blacksmith', (eds. B. G.   S C O T T and H. F. CLEERE), 1984, Belfast, 149-152

125   B. HANSEL and H. D. SCHULZE: 'Frühe Kupferverhüttung auf Helgoland', Spektrum der Wissenschaft, 1980, Feb., 11-20

126   D. W. HOPKINS: *Bull. HMG*, 1971, **5**, (1), 6

127   J. NAPIER: *Phil. Mag*, 1852, **4**, 45, 192, 262, 345, 453; 1853, **5**, 30, 175, 345, 486

128   J. H. VIVIAN: 'An account of the process of copper smelting as conducted at the Hafod Copper Works, Swansea', Feb. 1823, Annals of Philosophy

129   I. KRULIS: *Z dejn vied s techniky na Slovenska*, 1966, **4**, 77

130   J. R. HARRIS: 'The Copper king', 1964, Liverpool

131   B. C. BLAKE-COLEMAN and R. YORKE: 'Faraday and electrical conductors; 1821-1831', *IEE Proc.* 1981, **128A**, (6), 463-471 (It is clear that, at this time, users of copper wire did not know what impurities were influencing the conductivity)

132   H. R. SCHUBERT: *J. Iron Steel Inst.*, 1959, **193**, 1; this is exactly the same as that shown by Schlüter *(op. cit.)* in his Fig. XX

133   DIDEROT: (*see* Ref. 87)

134   JOAN DAY: 'Bristol Brass; the history of the industry', 1973, Newton Abbot

135   S. W. K. MORGAN: *Chem. and Ind.*, May 16, 1959, 614

136   E. J. COCKS and B. WALTERS: 'A history of the zinc smelting industry in Britain', 1968, London, Harrap

137   H. W. DICKENSON: *TNS*, 1943–4, **24**, 27

138   C. CAPLE and S. E. WARREN: 'Technical observations on the method of production and alloy composition of pins', Proc. 22nd Symp. on Archaeometry, Bradford, 1982, 273-278

139   R. F. TYLECOTE: *Post-Med. Arch.*, 1972, **6**, 183

140   A. THOUVENIN: *Rev. d'hist. des Mines Met.*, 1970, 2, (1), 101

141   G. VAN PRAAGH: 'Tregganu white brass', *JHMS*, 1979, **13**, (2), 95-97

142   M. BEVAN-EVANS: *Flintshire Hist. Soc. Publ.*, 1960, **18**, 75

143   M. BEVAN-EVANS: *ibid.*, 1961, **19**, 32–60

144   J. MARTYN: *Phil Trans. Roy. Soc.*, 1729, **36**, 22

145   A. RAISTRICK and B. JENNINGS: 'A history of lead mining in the Pennines', 1965, London, Longmans

146   G. AGRICOLA: 'De Re Metallica', (translation from 1556 edition by H. C. and L. H. Hoover), 1912, London

147   I. KRULIS: *Z. dejn vied a techniky na Slovenska*, 1966, **4**, 77

148   G. JARS: 'Voyages Metallurgiques', 535 et seq, vol. 2, 1774–81, Lyon

149   C. A. SCHLUTER: 'Grundlicher Unterricht von Hutte-Werken nebst einen vollstandigem Probier-Buch' 1738, Braunschwieg

150   J. MULCASTER: 'An account of the method of smelting lead ore as it is practiced in the northern part of England', MS in Lit. and Phil. Soc. Library, Newcastle; transcribed in *Bull. HMG*, 1971, **5**, (2); a similar MS is in the Central Reference Library, Wigan, Lancashire by the same author which is slightly more extensive and dated to 1806 (*see* editorial note in *Bull. HMG*, 1971, **5**, (2))

151   F. J. MONKHOUSE: *Trans. Inst. Mining and Met.*, 1940, **49**, 701

152   D. DIDEROT and J. D'ALEMBERT: 'Encyclopédie, ou Dictionnaire Raisonné des Sciences, des Arts et des Métiers'

153   J. PERCY: 'The Metallurgy of Lead', 1870, London

154   M. DUFRENOY *et al.:* 'Voyage métallurgique en Angleterre', 2 ed., 1837, 2 vols., Paris

155   H. L. PATTINSON: *Trans. Nat. Hist. Soc. Northumberland, Durham, and Newcastle,* 1831, **2**, 152

156   H. C. SALMON: 'Lead smelting on Mendip', *Min. and Smelting Mag.*, 1864, **16**, 1321-8

157   E. E. AYNSLEY and W. A. CAMPBELL: *Chem. and Ind,* 1958, 1498

158   R. T. CLOUGH: 'The lead smelting mills of the Yorkshire Dales', 1962, Leeds

159   D. G. TUCKER: *Bull. HMG*, 1972, **6**, (2), 1

160   C. MORRIS (ed.): 'The journeys of Celia Fiennes', 1947, London

161   R. F. TYLECOTE: 'Calenick, a Cornish tin smelter; 1702-1881, *JHMS*, 1980, 14, (1), 1-16

162   W. PRYCE: 'Mineralogia Cornubiensis', 1778, London

163   BRYAN EARL: 'Arsenic winning and refining in the West of England', *J. Trevithick Soc.*, 1983, (10), 9-29

164   These can still be seen at Botallack; *see also* D. G. Tucker: *Bull. HMG*, 1972, **6**, (2), 1

165   D. B. BARTON: 'Essays in Cornish mining history', vol. 2, 114, 1970, Truro, Barton

166   R. PEARCE: *J. Roy. Corn. Instn.*, 1871, **4**, 81

167   D. B. BARTON: *op. cit.*, 86

168   A. LUCK: *Stahl u. Eisen*, 1965, **85**, 1743–51

169   F. LAISSUS: *Rev. d'Hist. Mines et Met.*, 1969, **1**, (1), 37

170   W. E. HOARE: *Bull. Inst. Metallurgists*, 1951, **3**, 4

171   J. PERCY: 'Metallurgy; iron and steel', 1864, London

172   G. R. MORTON: *The Metallurgist*, 1963, **2**, (11), 259

173   *Bull. HMG*, 1963, **1**, (2), Table 1

174   G. R. MORTON: 'Iron and Steel', 1966, **39**, 563

175   E. SWEDENBORG: 'De Ferro', 157

176   R. F. TYLECOTE: *J. Iron Steel Inst.*, 1966, **204**, 314

177   G. R. MORTON and M. M. HALLETT: *ibid.*, 1968, **206**, 689

178   A. REYNE: *Rev. Hist. Sid.*, 1965, **6**, 87

179   Curwen Papers, 1792-1828; Carlisle Record Office, D/Cu. 5/96

180   G. JONES: *J. Iron Steel Inst.*, 1908, **78**, 59

181   D. CRANSTONE (ed): 'The Moira furnace: a Napoleonic blast furnace in Leicestershire', NW Leics. District Council, Coalville, 1985

182   I am indebted to Andrew Clarke for his survey of this furnace

183   I am indebted to Warren Marsh of the Gloucestershire Council for Industrial Archaeology for his survey of this furnace

184   *Bull. HMG*, 1965, **1**, (3), 7; 1965, **1**, (4), 2

185   T. TURNER: 'The metallurgy of iron and steel', 1895, London

186   E. JENKINS (ed.): 'Neath and its district'; D. Morgan Rees, 'The iron industry', 149, 1974, Neath

187   L. ERCKER: 'Beschreibung aller fürnemsten mineralischen Ertzt und Berckwercksarten', 1574, Frankfurt (English translation by A. G. Sisco and C. S. Smith, 1951, Chicago, Chicago Univ. Press)

188   R. F. TYLECOTE, E. PHOTOS and BRYAN EARL: 'The composition of tin slags from the south west of England', *World Arch.*, 1989, **20**, (3), 434-445

189   J. PICKIN: 'Excavations at Abbey Tintern furnace, Part I', *JHMS*, 1982, **16**, (1), 1-21: Part II, *JHMS*, 1983, **17**, (1), 4-11

Chapter 10
More recent times; AD *1850-1950*

The Large Scale Production of Steel

The second part of the nineteenth century was remarkable for the contributions of Siemens, Bessemer, Kelly, and Thomas to the large scale production of what is still the world's most important metal — mild steel.[1] The process did not immediately replace crucible steel, which was the main source of metal for tools and armaments, but wrought iron. Mild steel is, in fact, little more (or less) than wrought iron with the slag removed.

Wrought iron had resisted all attempts to mechanize its production, which was a labour and fuel intensive process.[2] Even after the introduction of steel, the mechanical methods of puddling which had been tested had been found wanting.[3] In Britain, by 1860, there were over 3 400 puddling furnaces making something like 16 t/day in batches of 250 kg, 6–7 charges being worked in a 12 hour shift. These were responsible for about half of the world's production of malleable metal.[4]

However, William Siemens had been concerned with fuel efficiency since about 1846 and had attempted to apply the regenerative principle, which had been patented in 1816 by Stirling.[5] By 1856, William Siemens' younger brother, Frederick, had been granted a patent for its application to furnaces in all cases where great heat was required, and the brothers saw its main use in the melting of metals and glass.[6,7] In experiments on iron before 1861, solid fuel was used with fireboxes at either end of the furnace, and with air and gas passing over them. There were problems, such as ash blocking the brickwork 'checkers' as the regenerative chambers were called, and difficulty in obtaining refractories that could stand the conditions and temperatures required to separate metal and slag. The eventual success depended on the parallel invention of the gas producer, by which a gaseous fuel could be made from ordinary coal in a separate unit, which removed the firebox and the fuel ash away from the furnace.

The first use of the producer gas fired regenerative furnace was in a glassworks and effected a 50% fuel saving. By about 1862 there were about 100 gas-fired open hearth furnaces in use, mostly in glassworking, and Siemens reckoned that the average fuel saving was 75% (*see* Fig. 136).

Although some of the Sheffield steelmakers soon adopted the process in place of the crucible process for steel melting, it did not immediately replace the puddling furnace. It was Pierre and Emil Martin in France, who started to melt steel and wrought iron in the open hearth, who really sounded the death knell of the puddling furnace, but in Britain the steelmakers had been involved in the early failure of Bessemer's new steel making process and were reluctant to try yet another.

At the same time as the Siemens brothers were carrying out their experiments, Bessemer in Britain, and Kelly in the USA, had been experimenting with the obvious but difficult process of blowing air into molten cast iron to reduce the carbon content. In experiments in which Bessemer was trying to increase the temperature of a reverberatory furnace to fuse steel by the addition of an air blast (as used by Onions), he had noticed that some solid cast iron pigs had become decarburized by exposure to the air blast before melting. This oxidation is, of course, the principle of puddling, but it appears that Bessemer had no real knowledge of the latter process and therefore had the advantage (if such it was) of looking at things in a new light.[8,9,10] He realized that the metal had to be at a very high temperature before decarburization would occur, and tried blowing air into crucibles containing molten iron while they were being heated in a furnace. The next step was to see whether this could be done in a larger vessel without external heating and thus the first Bessemer converter was born in St. Pancras, London, in 1856. This was a fixed vessel containing 350 kg of cast iron with a blast of 10–15 lb/in^2 (0.8–1.0 bar) pressure (see Fig. 137). It is believed that the violence of the reaction surprised Bessemer, who had not appreciated the exothermic nature of the carbon–

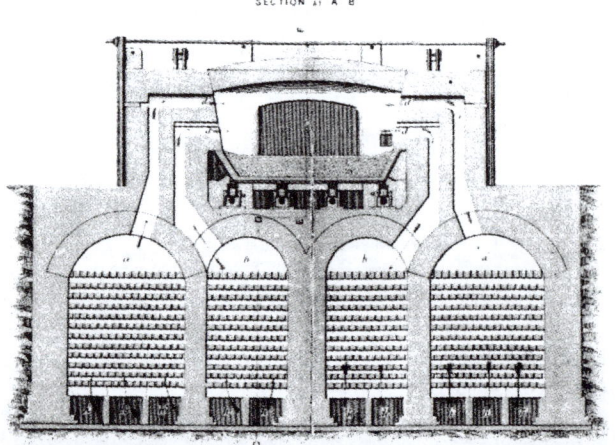

136 *Siemens' open-hearth furnace and regenerators for the melting of steel (from Percy: 'Refractories', 1875)*

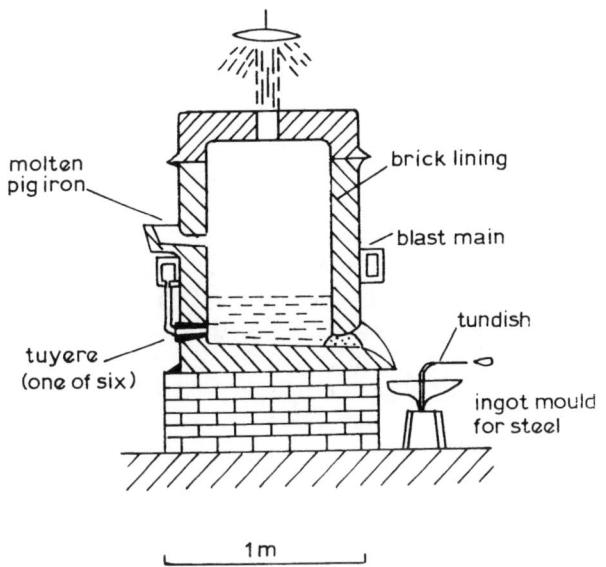

137 *Bessemer's fixed vessel for the conversion of cast iron to steel as used at St. Pancras, London, 1856*

oxygen reaction, nor that of the impurities in the metal. Luckily, after 10 minutes, when all the impurities had been consumed, the flame died down and it was possible to get near enough to turn off the air supply. The metal was tapped into an ingot mould and found to be malleable iron of low carbon content. He was rightly convinced that he had found a method of making a pure wrought iron, and was prevailed upon to make this public at a meeting of the British Association in Cheltenham[10] the following week, on 11th August, 1856.

Very soon he had tilting converters capable of making as much as 5 t/heat (one hour, including time for fettling and teeming, compared with 4-6 hours for 50 kg of crucible steel, and 2 hours for 250 kg of puddled wrought iron). This was revolutionary, and ironmakers and steelmakers from all over the country and abroad rushed to take out licenses (*see* Fig. 138).

But Bessemer, not being a metallurgist, had no idea of the composition of his pig iron and just what impurities were being removed. Furthermore, many of his ingots were either highly porous or over oxidized. These problems were mainly solved by the efforts of a practicing metallurgist, Robert F. Mushet, the youngest son of David Mushet who had contributed so much to the study of the blast furnace in the late 18th and early 19th centuries in Scotland. The Mushets had been working on various crucible steels in the Forest of Dean, [11,12] work which was to culminate in R. F. Mushet's discovery of tungsten steel, known as 'high speed' steel, in 1860. But the Mushet family, like others such as Reynolds in 1799 and J. M. Heath in 1839, had been making additions of manganese to the crucible to improve crucible steels made in the Huntsman tradition. Heath's and Reynolds' addition was manganese dioxide which, it will be remembered, Wilkinson had recommended adding to the blast-furnace charge at Brymbo in 1806. The Mushets, however, had been making crucible steel from Swedish iron, charcoal, and MnO_2 instead of from cemented 'blister' bar, as in the traditional Huntsman process. By

138 *An early tilting Bessemer plant about 1860; this is believed to have been in use at Ebbw Vale, South Wales*

1848, spiegeleisen (mirror white cast iron) was being made in Rhenish Prussia containing 8.5% Mn and 5.25% C, and R. F. Mushet ordered 12 t of this for his experiments, which he probably used for his manganese steels. He obviously appreciated the use of manganese as a deoxidizer, a role it could only fulfil in the reduced form (as Mn metal or Mn_3C) rather than in the oxidized form of manganese dioxide.

With this experience, Mushet was able to appreciate the reasons for Bessemer's difficulties with unsound ingots and was approached by the Ebbw Vale Company which, like many others who had rushed to adopt the Bessemer process, was in serious difficulties. Mushet remelted some of the unsound Bessemer steel and added spiegel.[1] The resulting ingot was 'smooth and piped and had all the appearance of good cast steel'. This was the beginning of the successful development of Bessemer's process and the basis of all good steelmaking up to the present time.

However, there was still one remaining difficulty as far as bulk steelmaking was concerned. The majority of the world's iron ores contain substantial traces of the element phosphorus. During the first part of the 19th century, when the puddling furnace was virtually the only method of making malleable metal, the problem of phosphorus was not appreciated since the copious amounts of mildly basic iron silicate or 'tap cinder' could absorb it. Also, at the relatively low temperature prevailing in the puddling process, phosphorus oxides are stable and are not reduced back into the metal to embrittle it. So, unknowingly, the puddling process was a relatively efficient way of removing phosphorus. As soon as this process was replaced by melting processes (whether Siemens' or Bessemer's) capable of achieving temperatures of the order of 1 600°C, the phosphorus would revert to the metal.

There was, of course, another difficulty. From the earliest times furnace linings had been made of pure sandstone or silica (SiO_2) and such materials were the only ones then capable of standing 1 600°C for any length of time. Such materials are chemically acidic and are slagged, and are therefore quickly consumed if they come into contact with strong bases such as the more alkaline materials of lime and magnesia.

Naturally, Bessemer, like all other furnace builders, had used siliceous materials for his linings, and such linings would not combine with the acidic phosphorus pentoxide (P_2O_5) which would tend to form during the oxidation of phosphorus and remove it from the system as a stable compound. This fact restricted the use of the Bessemer process to non-phosphorus ores, i.e. those containing less than about 0.05% P. As far as Britain was concerned this meant that the only suitable ores were those from Cumberland and North Lancashire, the Forest of Dean, and some of those from South Wales. The coal measure ores, which accounted for most of the production of 19th century Britain, were unsuitable. This was a terrific blow to the Bessemer process, as can be seen by the fact that the output of wrought iron in Britain continued to rise until the end of the 19th century.

The Siemens process was more adaptable than the Bessemer and could make steel by remelting wrought iron, from which the phosphorus has been expelled in the puddling process, or other low-phosphorus metal such as scrap. However, it could not make good steel solely from high-phosphorus pig any more than Bessemer's process could, and it therefore suffered from the same problem.

Meanwhile, new sources of high-manganese low-carbon ferro-alloys were becoming available in France, which encouraged the changeover from wrought iron to steel so that by 1873, in spite of the problem of phosphorus, British Bessemer production was 496 000 t and Siemens 77 500 t; but wrought iron production probably exceeded 3 Mt.

Neither Siemens nor Bessemer was a chemist or metallurgist and it now seems surprising that there was no chemist nor metallurgist available who was capable of solving the problem of phosphorus in steel, and that its solution rested on the determined attempt of one man. This man was Sidney Gilchrist Thomas who, while interested in science, had been forced to earn his living in another capacity.[13] In 1870, he attended an evening course in chemistry at the Birkbeck Institution (now Birkbeck College in the University of London) where he learnt of the problem of phosphorus in the Bessemer converter and the suggestions of Percy, Collyer, and others as to ways to overcome it. Some of these methods had envisaged the neutralization of the acidic P_2O_5 with

lime but had come up against the problem of the consumption of the acidic siliceous lining by the basic slag so produced. All efforts to use lime as a lining had failed, mainly because of the poor performance of limestone, and lime by itself, as refractories.

So, Thomas's problem was mainly a practical rather than a theoretical one. At this time, (1876) Thomas's cousin, P. C. Gilchrist, was chemist to the Cwmavon Steel Works in South Wales and, naturally, Thomas wrote to him to see if he could provide assistance in making experiments. It was at the Blaenavon works during nine months in 1877–8 that the two were able to carry out the experiments that showed that a basic lining and basic fluxes could remove phosphorus in a satisfactory manner from the converter. Large-scale trials with small converters were successful but it took a little time before a basic lining of rammed dolomite and tar enabled the process to operate successfully in the larger sizes of converter at Middlesborough in 1879.[14]

This seems a convenient point to look at the chemical aspects of the Bessemer-Thomas process, known in Britain in its modern form as the BOS process (Basic Oxygen Steel) and now responsible for almost all of the world's mild steel production. The Bessemer converter, in its ultimate (19th century) form, was a tilting egg-shaped vessel in which cold air could be blown through perforations (tuyeres) in the lining in the base of the vessel (see Fig. 138). The vessel was charged with pig iron in the tilted position so that the liquid iron would not flow into the tuyeres and block them. The blast was turned on and the vessel turned into its upright position.

The blast pressure of 15 lb/in² (1 bar) was sufficient to blow air through 0.75 m of molten metal and avoid any blocking of the tuyeres. The oxidation of the impurities in the iron by the oxygen in the air is exothermic and results in a rise of temperature from 1 200°C in the liquid cast iron to 1 600°C in the mild steel, thus keeping the metal molten in contrast to the puddling process. The operation is finished extremely rapidly, which led to difficulties in control in the original process. The order of oxidation is Si, Mn, and carbon in a phosphorus-free iron, and the change in composition of these elements during the blow is shown in Table 75, which gives the initial and final figures for the acid and basic processes.

In the original basic process, the phosphorus content had to exceed 2% to maintain a high enough temperature at the end of the blow to keep the metal and slag molten. In fact, this limitation meant that intermediate phosphorus irons (0.10 or 1.5% P) could best be treated in the Siemens open hearth furnace where the heat was provided externally. But, with the advent of tonnage oxygen, no such limitation existed owing to the higher temperatures attainable in the absence of the cooling effects of nitrogen. This process was developed in Austria by the Linz-Donawitz companies following the original work of Durrer.[16] It uses a vertically positioned lance which is able to blow oxygen on the surface of the metal, thus eliminating the bottom tuyeres. It appears to have triumphed over all other similar processes, and it can be said to be the final and ultimate form of the

Table 75 Changes in composition in the Bessmer-Thomas process (after Aitchinson[15])

Element	Acid Before	Acid After	Basic Before	Basic After
	Amount, %			
Carbon	3·0	0·06	3·35	0·02
Silicon	1·8	0·03	0·448	0·13
Manganese	0·7	0·06	0·85	0·23
Sulphur	0·06	0·063	0·18	0·057
Phosphorus	0·06	0·063	2·01	0·066

Bessemer-Thomas process. The substitution of oxygen for air also eliminated the embrittling effect of nitrogen in the Bessemer process, an effect that was only fully appreciated when welded construction took the place of riveted during the second quarter of this century.

The gradual replacement of wrought iron by Bessemer and Siemens steel took some time, because of the earlier reputation for poor quality that the Bessemer process had gained, and also because of the greater difficulties involved in forge welding it. The welding of wrought iron was facilitated by its slag content and more than 1500 years of experience, whereas steel was slag-free and had to have enough sand added to act as a flux and form a slag. These problems were gradually overcome and Bessemer steel was accepted in the constructional engineering field. The British Board of Trade sanctioned the use of steel for bridges in 1879 and, soon after, the Forth bridge was designed for this new material by Sir John Fowler and Sir William Baker.[17] It was completed in 1890 and 54 000 t of Siemens open hearth steel were used in its construction. The Eiffel tower in Paris, intended as a steel structure, was built in wrought iron in 1899.

The output of wrought iron continued to increase until the end of the century, while steel gradually replaced it on the grounds of cost. Steel gradually took over the tinplate field which had become a major user of thin sheet metal—often made from charcoal iron decarburized in fineries.

In Britain the making of tinplate had become a south-western and, in particular, a Welsh speciality.[18] In the early nineteenth century, the battery process had been entirely replaced by the rolling mill and the process of pack rolling had been developed for the thin sheets needed for this purpose. It is surprising how long the single two high 'hand' mills lasted; in these, the 'packs' were passed back over the top roll, the rolls screwed down further to reduce the gap between them, and the sheets again passed through. When they became too long they were separated and doubled and passed through again until they were reduced to the desired thickness. It was not until about 1930 that we see the introduction of the continuous strip-mill in which, instead of being cut up into short pieces, the plates are welded up into long continuous lengths which are rolled as a ribbon, tinned, and only then cut up into sheets suitable for stamping out tins. This process was extended to the production of automobile body sheet which began to exceed the output of tinplate in the 1930s. Now, of course, this process is used for all thin mild steel sheet applications.

Special Steels and Ferro-Alloys

Much to his chagrin, Bessemer failed to make much impression on the Sheffield steelmakers, whose main product in 1860 was carbon steel made in the crucible furnace. They continued to make and use large quantities of blister bar made from Swedish bar iron in the cementation furnace, as we have seen, although they experimented with such things as 'puddled steel'. Although some steelmakers adopted Siemens' open hearth process, the majority preferred the adaptability of the crucible for the small melts of carbon and alloy steels which were their speciality.

The casting of a 25 t ingot in Sheffield in 1872 needed over 672 crucibles, each containing about 37 kg. The organization of their melting and teeming must have been considerable, let alone the manufacture of the crucibles which could only be used once[19] (*see* Fig. 139).

The Sheffield steel trade was more influenced by the inventions of Mushet, Hadfield and Brearley, and those of the scientists who had discovered the important elements, nickel, cobalt and chromium, rather than those of Bessemer. After the death of his father in 1847, R. F.

139 *The Prince of Wales watches the pouring of an ingot of crucible steel in the Norfolk Works of Thos. Firth and Sons in 1875; the weight of the ingot cast was probably no more than 20 t (courtesy of Firth Brown Ltd, Sheffield)*

Mushet carried on experiments which involved making crucible cast steel from Swedish iron and charcoal (one of the methods of making wootz) and adding manganese. It was this experience that enabled him to come to the aid of Bessemer in 1856. Later in the course of this work (1868–82) he discovered his famous 'self-hardening' steels. These were made out of a mixture of spiegel, cast iron, wolfram (WO_3) and pitch and it is amazing that he obtained a forgeable steel out of this mixture.[20] Some of the early compositions are given in Table 76; in the later alloys, the manganese was replaced by chromium.[11] Such steels were extremely useful as cutting steels and formed the basis of today's 'High Speed Steel', which is capable of cutting in a red hot condition.

The next step depended on the availability of higher manganese alloys than that represented by spiegeleisen. In 1877 an 80% Mn alloy with 6–7% C was made in the blast furnace by the French Terre Noire Co.[21] However, this material was being used mainly for low-manganese steels up to 2.45% Mn, above which point it was felt that the steel was brittle and useless. It was left to Robert A. Hadfield to appreciate the significance of this cheap, high-manganese ferro-alloy and use it to produce his famous high-manganese steel. The higher manganese contents of the alloys of the Terre Noire Co. meant that larger quantities of ferro-manganese could be added to iron without raising its carbon content appreciably. Hadfield sought to produce steels that could be hardened without losing their toughness, and in 1887 patented his 12.5 Mn–1.2 C steel, the latter figure being dictated by the carbon content of the Terre Noire ferro-manganese. This steel was strange in that it was not hardened by quenching and it was non-magnetic. It was tough and could only be hardened by cold working or abrasion.

This was not of much value in the more general engineering field, and even the lower manganese quench-hardening steels produced by the Terre Noire Co. tended to be ignored at this time. It was the invention of the nickel steels in 1889 by James Riley of Glasgow that really shook the engineering and armament world.[22] It was found that the strength could be increased from 460 to 1400 MN/m² , with only a relatively small decrease in ductility, by additions of up to 4.7% nickel.

Others tried additions of chromium, aluminium, vanadium, and titanium and gradually the value of combinations of chromium and nickel became apparent between 1890 and 1914. In 1905, Portevin noticed that steels containing more than 9% Cr were resistant to acid attack, and Strauss of Krupps of Essen was aware that the high chromium–nickel steels were very heat resistant, and patented an austenitic steel composition containing 20% Cr and 5% Ni in 1912–13 for high-temperature applications.[23] However, the high-chromium steels were difficult to work because of the high carbon content of the ferro-chrome then available. It was Harry Brearley of Sheffield who first saw the significance of this work in the cutlery industry, and in 1914 patented the medium-carbon steel with 12–14% Cr, which eventually became the mainstay of the cutlery trade.[24] Later, Strauss' austenitic steels were also recognized for their stainless qualities. Like Hadfield's manganese steel, they were non-magnetic and could not be hardened by normal heat treatment.

After the 1914-18 war, continuous research and development went on in many countries using the numerous new alloys that were commercially obtainable. In fact, very few new discoveries were made during this period and it was a case of the steady improvement of the major types of steel by the use of elements such as cobalt, titanium, molybdenum, niobium and aluminium. In fact, we were to see the complete replacement of iron by nickel in the high temperature field, and the improvement of the nonferrous nickel–chromium alloys which had been used for heating elements since the beginning of the twentieth century. The latest development again concerns nickel, this time on iron with 18% Ni together with small amounts of other alloying elements. Since this steel is an example of the age hardening of martensite it has been given the name of 'maraging steel'. Such steel has the highest combination of strength and ductility yet achieved.

Other developments in the field of special steels and alloys concern magnetic materials for the electrical industry. At the beginning of the 20th century only two materials were known in the magnetic field, (*a*) pure iron (<0.03% C) for magnetically soft uses; and (*b*) hardenable steel—usually containing tungsten—for permanent

Table 76 Composition of some of R.F. Mushet's early alloy steels, compared with modern 'High Speed Steel' (after Osborn[11])

| Date | Element, % | | | | | | |
	C	Si	Mn	W	Cr	V	Mo
1868–69	1·15	0·74	1·15	10·09	nil		
	1·68	0·73	1·21	9·07	nil		
	1·43	0·63	1·21	8·56	nil		
1872–82	2·10	1·11	1·50	6·08	0·50		
	2·31	0·78	1·75	6·72	0·45		
	2·62		1·80	5·68	0·37		
1970 (High Speed Steel)	0·6			18	3–5	1·0	nil
	1·3			14	4	4·0	0·5

magnets.[25] But the intensive research into alloy steels, and the commercial production of a wider range of ferro-alloys that this engendered, brought about the silicon-irons with 1–4% Si for soft magnetic materials with directional properties and a new and wide range of alloys for permanent magnets, some being virtually non-ferrous. Cobalt was the principal component of many of these alloys from about 1920 onwards but later, various amounts of nickel, aluminium and titanium, were added to give such suggestive trade names as Alnico, Titanal, Alcomax and many others.

For a time, low-carbon 1% Si steel was used as a high tensile structural steel, and a small quantity was built into the 'Mauretania' in 1907, resulting in a weight saving of 200–300 t.[25]

The New Metals

NICKEL

The activities of the scientists in the 18th–19th centuries resulted in an enormous increase in the number of metals known, yet comparatively few of these were to take an important part in the modern metallurgical industry. One of the first of these was nickel. This element was discovered by Axel Cronstedt in 1751 and got its name from 'kupfer-nickel', a nickeliferous Saxon copper ore which was called after a local bad spirit because of its refractory nature.[26] However, it remained a scientific curiosity until the isolation of several ounces of the metal by Richter in 1804. An analysis by Fyffe of Edinburgh in 1822 showed that Chinese *paktong* contained a considerable amount of this 'new' element; by 1824 the modern European equivalent of *paktong*, German silver, was first made in Berlin. It was 1865 before substantial quantities of pure nickel were produced by J. Wharton, and 1880 before it reached the 1000 t/a mark when the metal can really be said to have arrived.

New sources had to be tapped to achieve these levels and the first was that of New Caledonia in the Pacific, where a silicate containing 10% Ni occurs near the surface. In 1883, the famous Sudbury (Ontario) deposit was discovered during the cutting of the Canadian Pacific Railway, and this has been the source of most of the world's nickel ever since.

The commercial production of nickel has for a long time hinged on its separation from a copper–nickel matte, the iron having been previously removed by the methods normal to copper smelting. The Orford process was introduced in 1890 and depended on the gravity segregation of nickel from a mixed copper-nickel matte and sodium sulphide fused together with the aid of coke. On solidifying this mixture, two phases separated out, the one on the top was a copper–sodium sulphide while that on the bottom was 99.8% nickel sulphide. This latter could be bessemerized to metallic nickel and refined.

The product of this process contained cobalt, which was itself in increasing demand and it is here that Ludwig Mond plays a part. He was a chemist with a private laboratory in London and had found that nickel combined with carbon monoxide to form a gaseous compound–nickel carbonyl $Ni(CO)_4$. Furthermore, carbonyl formation was selective and reversible, so that although some other metals could be found that underwent the same reaction, each required a different temperature and pressure. He found that he could take up the nickel in impure nickel as carbonyl and decompose it to give pure metallic nickel, leaving behind the impurities in a concentrated and recoverable form. The residue could be treated to recover the copper and cobalt and also the platinum-group metals, platinum, palladium, iridium, ruthenium, rhodium and osmium, as well as gold and silver.

These processes were to remain the basic processes for the recovery of nickel until the 1950s, when the International Nickel Company announced its decision to use an improved process to replace the Orford process. The copper–nickel matte, instead of being smelted with sodium sulphide, is now allowed to solidify slowly so that the matte crystallizes into its components, Cu_2S and Ni_3S_2. The solidified blocks are then broken up and milled to 325 mesh. The crystalline fractions are then separated by various means; some Ni–Cu alloy is removed by magnetic methods, and the rest is separated by flotation. The copper sulphide is melted in an arc furnace and converted to blister copper. The nickel sulphide is oxidized to NiO and either reduced to metallic nickel in a reverberatory furnace or used as the starting material for the Mond carbonyl process.

THE LIGHT METALS

A major achievement of the late 19th century was the discovery of two of the light metals that make up a large proportion of the earth's crust, i.e. aluminium and magnesium. The substance known as alum (impure potassium aluminium sulphate) had for long been one of the most important of early chemicals because of its place as a mordant in dyeing. In 1754, Marggraf showed that it contained two bases, one of which was a lime-like substance, but the other was completely different and one of the substances present in clay.

In about 1807, Humphrey Davy sought for ways of reducing the oxide which by then had gained the name alumina. Having isolated a number of base metals by the electrolysis of their oxides, it was natural that he should apply the same techniques to alumina. But he was not completely successful, although he came near to it when he electrolyzed a mixture of potassium hydroxide and alumina. Many others made attempts, and by 1827 Oerstedt, Wohler and St. Claire Deville had all made minute quantities of the new metal. By 1854, Wohler had made enough to determine some of the physical properties. However, it was Deville's process that proved the most successful; anhydrous aluminium chloride was fused with pure sodium producing aluminium and sodium chloride. The latter removed the alumina that

with another is a useful one and is now used for the production of titanium, the third light metal. Deville finally substituted the mineral cryolite, a double fluoride of aluminium and sodium, for the aluminium chloride which improved the process. A bar of aluminium made by Deville's method was exhibited at the Paris Exhibition of 1855, at which time its production cost was about £130/kg. After Castner had developed his process for producing cheaper sodium in 1886, the price was reduced to something like £5.50/kg.

In this year, however, the situation was completely changed when Hall in the USA, and Heroult in France, announced a new process based upon electrolysis. The idea, which stemmed from Davy's early experiments, had not been entirely dead but, as it depended on cheap electrical energy it awaited the discovery of the dynamo, as battery energy, was too expensive. Not only was electrical energy used to separate the metal from oxygen but its passage also provided the heat necessary to fuse the compounds involved. These were not all that different from Deville's, i.e. molten cryolite in which alumina was dissolved, but the separation did not involve sodium but rather the direct plating out of aluminium on to the cathode of an electrolytic cell (*see* Fig. 140). Both electrodes were carbon and the cathode was a carbon-lined crucible so that the alumina was deposited at the bottom of the cell and the oxygen at the anode at the top, where it reacted to form carbon monoxide and dioxide. This process, the Hall–Heroult process, is still the only commercial process for the production of aluminium. It needs about 25 000 kWh to produce 1 t of aluminium, and is thus extremely dependent on the price of fuel or the availability of cheap hydro-electric power.

The immediate outcome of the Hall-Heroult process was a reduction in price to about £0.44/kg by about the end of the century, when output reached 5 000 t/a. Demand grew slowly because it was dependent on the growing use of physical metallurgical techniques to improve the strength of the metal, as that of the pure metal was very low (about 77 MN/m²). By 1966, the

140 Early electrolytic cell for the extraction of aluminium from bauxite

output of aluminium had reached 8.3 Mt/a, and since its specific gravity is only 2.6 compared with copper at 8.9, this means that it had overtaken copper on a volume basis by a factor of four, and had become the world's second most important metal (Table 77). It is of interest to note that the first aluminium alloy seems to have been produced in China in about AD 300.[27] It was a copper-rich aluminium alloy (aluminium bronze) produced by thermal reduction of copper and aluminium minerals. This process has been tried on an experimental scale in recent times but, as it has not proved possible to use it as an economic method for obtaining metallic aluminium, it is no longer practiced.

Around 1852, Bunsen had obtained a small quantity of magnesium by the electrolysis of molten magnesium chloride, and it was clear that this offered a commercially attractive method once the dynamo had been perfected. It was a simpler process than the electrolysis of the aluminium electrolyte but depended on the prior production or concentration of the magnesium chloride in sea water. For some time it was regarded as a curiosity and did not reach more than 10 t/a by 1900. After all, aluminium seemed to have all the advantages and none of the disadvantages, such as the poor corrosion resist-

Table 77 World consumption of major metals

Metal	World consumption, (in 1 000 t)		Increase, % 1956−66	1966 consumption, kg per capita
	1956	1966		
Aluminum	3 200	7 570	135	2·4
Copper	3 960	6 400	62	2·0
Zinc	2 660	4 240	59	1·32
Lead	2 240	3 300	48	1·04
Nickel	230	460	104	0·14
Tin	177	222	26	0·05
Magnesium	145	159	11	0·05
Total non-ferrous metals	12 600	22 400	77	7·00
Pig iron	201 500	349 000	73	108
Ratio: ferrous / non-ferrous (by weight)	15·9	15·6		15·6

Sources: 'Metal statistics', 1965−68, Frankfurt, Metallgesellschaft AG; and 'Minerals Yearbook', 1956−67, US Bureau Mines, Washington

ance of the early magnesium metal due mainly to the small content of residual magnesium chloride. By 1955, however, world production was 150 000 t and the price had dropped from £0.65/kg in 1900, to £0.11 in 1944.

Furthermore, there were other methods of making magnesium that did not depend directly on electrical energy. The Pidgeon process, developed in 1944, is rather like the horizontal retort process for making zinc but employs a vacuum in the retort, under which condition the direct reduction of dolomite by ferro-silicon becomes a possibility. Dolomite is a double carbonate of calcium and magnesium; the lime is not reduced but forms a slag with the iron and silica. But of course, this process depends indirectly on electrical energy since the ferro-silicon is produced in electric-arc furnaces. While a large amount of the world's magnesium is present in sea water, its concentration is only 0.3%. However, by precipitating this with the aid of dolomite, it is possible to recover the magnesium from both the dolomite and the sea water as MgO, the lime going into the sea water to replace the magnesium. For the electrolytic process the MgO has to be chloridized by reacting it in a reducing atmosphere with chlorine gas, but the MgO can be used direct in the Pidgeon process.

The third light metal, titanium, is of more recent vintage. Its specific gravity is 4.5 and thus is intermediate between the heavy metals at 7 to 9 and the light metals at 1.5 to 2.6. Its oxide was first isolated in the mineral rutile by the French chemist Vauquelin in 1789, but it was 1910 before the first piece of titanium metal was to make its appearance.[28] Again, its growth was slow and by 1954 the world production had reached about 4 000 t/a, but the price of the metal was something like £3.30/kg at which level it still remains, which might suggest that it is at a similar stage to that of aluminium in 1900, and that the take-off point has not yet been reached.

However, the reasons for this are more technological. The only industrially economic process at the moment is the Kroll process, invented in 1940, which reduces titanium tetrachloride with magnesium; a modification of this uses sodium. The tetrachloride, a colourless liquid with a boiling point of 136°C, is made by reacting titanium oxide—much used today as a paint base—with chlorine in the presence of carbon, as in magnesium chloride production. Magnesium metal is melted in an iron pot at about 900°C under an argon gas cover. The liquid tetrachloride is dripped in, vaporizes, and is reduced to titanium metal, (melting point 1 800°C) which collects on the walls of the vessel. The exothermic reaction maintains the temperature and at the end of the reaction a mixture of unreduced magnesium, titanium metal, and magnesium chloride is left. This can be separated by leaching with dilute hydrochloric acid, or by vacuum distillation. The magnesium chloride can be reduced back to magnesium metal in the magnesium electrolytic cell, and thus titanium, to a certain extent, is dependent on magnesium or sodium. But as we have seen, the cost of these metals is far below that of titanium at the present time.

One of the reasons for the high cost of titanium is the problem of fabricating the raw titanium powder resulting from the Kroll process. This has to be melted in consumable electrode furnaces in an atmosphere of argon. In these furnaces the electrodes consist of compressed titanium powder or scrap which are melted in a watercooled copper crucible. The ingots produced have to be rolled with the minimum of air contamination, as both oxygen and nitrogen render the metal brittle. These are only three examples of the new metallurgy. Considering that we now have 100 elements, of which 70 have metallic properties, there are many more applications of the processes recorded above.

Advances in Other Non-Ferrous Metals

COPPER AND ITS ALLOYS

The second half of the 19th century saw the gradual decline of the eminence of South Wales as the main smelting area of the world. Many of the copper mining areas began to build their own smelters for much the same reason as the Cornish miners had in the 18th century, i.e. a belief that the European smelters were taking more than their rightful share of the profits. The rapid industrialization of the USA, Canada and Australia meant that fuel was no longer a problem, and new fuels such as oil were available in many areas. As far as copper was concerned, the principal change arose with the use of the Bessemer process for the rapid oxidation of copper mattes, which had been made by concentrating and separating the sulphides in the blast furnace or reverberatory.

No longer were the old and slow processes of heap or stall roasting the only ones, although they persisted for a long time in Southern Spain,[29] but the preliminary roasting to reduce the sulphur content of sulphide ores to that required to form a rich matte was now done in revolving multiple-hearth roasters of the MacDougall type (*see* Fig. 141).

The principal demand for copper now came from the

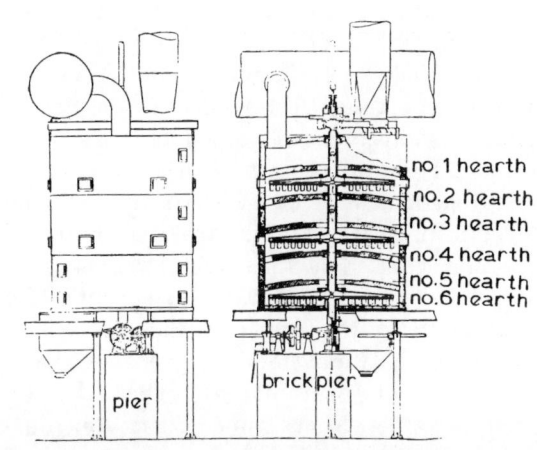

no.1 hearth
no.2 hearth
no.3 hearth
no.4 hearth
no.5 hearth
no.6 hearth
brickpier
pier

141 Improved roaster of the MacDougall type for copper ores

expanding electrical industries, which today consume probably more than half of the world's copper. Its main function here is as a conducting material, and the greater the purity of the copper the greater its conductivity. Since the time of Theophilus, (c. AD 1100) and possibly before, all refining, (mainly the reduction of sulphur and oxygen) had been carried out by the lengthy process of fire refining, involving oxidation and reduction. These processes, when carried out carefully, were capable of producing a copper as pure as 99.25%, but were slow and tedious and consumed a great deal of fuel. The new electrical industry wanted a purity as good or better than this, and electricity itself was employed to this end in the form of electrolytic refining. This entailed using a slab of crude copper from the converter as one of the electrodes (the anode) of an electric cell, and plating the copper from this electrode on to the other electrode (the cathode), which was usually a thin sheet of pure copper. The solution (electrolyte) was an acidified solution of copper sulphate. The exploitation of the process depended on the discovery of the dynamo, and was first successfully put into operation in South Wales in 1869. Only very small changes have taken place in this series of operations on copper up to the present day, and these mainly involve more refined processes of roasting, such as using 'fluidized' beds of finely ground material though which the oxidizing medium, i.e. hot air, is passed.

But the dressing of copper and other minerals was revolutionized by the discovery of flotation.[30] From the earliest times the gangue had been separated from the mineral by grinding and washing, or even winnowing, but these processes depended on a difference in specific gravity between the gangue and the desired mineral. The fact that some of the impurities in the ore were not different in this respect from the desired mineral meant that they were carried forward to the smelter and had to be separated at this stage, or appear as unwanted impurities in the final metal. Flotation takes advantage of the fact that some minerals are not easily wetted by water but are wetted by substances of an oily nature. The ground ore, now a mixture of wanted mineral and unwanted gangue, is agitated in water with the oily material and air so that a bubble attaches itself to the heavier but oily mineral and enables it to float away to the surface, leaving the gangue, wetted by water, in the bottom of the vessel. The process is specific and can separate various types of wanted minerals.

Another group of processes that has been increasingly used is that of metal recovery from solutions. It has been noticed from very early times that if iron is placed in a solution of copper, such as copper sulphate, the iron, being chemically more reactive (basic), enters the solution and replaces the copper which is deposited on the iron. This is what the alchemist noticed when he thought (or wished others into thinking) that he had transmuted iron into copper. Many mine waters contain copper sulphate, and the copper can be recovered by placing scrap iron in them and washing off the fine copper deposited on the iron. Where the ore body is of the sulphide type it must be first partially oxidized to sul-

phate in stalls, either naturally or by heating, and then washed artificially or by rain. This process was for a long time used for copper recovery at the Rio Tinto and Tharsis mines in Southern Spain. It can be extended to oxide ores by using a sulphuric acid solution to dissolve the copper. This group is usually referred to as leaching processes and these have been widely used of late for the extraction of many metals, sometimes employing hot solutions under pressure. They are capable of eliminating both the dressing and smelting processes but produce a finely powdered metal, the consolidation of which can often be costly. However, with the increasing use of metal powders to produce finished components by pressing powders into shaped cavities, known as powder metallurgy, there is clearly an important place for these 'wet' processes.

PLATINUM

As we have seen, owing to its existence as an impure native metal in South and Central America, platinum did not need to be discovered. However, it was the eighth metal to become known in the Western European world and, in spite of a reference to it by J. C. Scaliger in 1557 as a white infusible metal found in the silver mines of Mexico, it only began to be taken seriously by the Europeans about the middle of the 18th century. The impure metal was marketed in Europe in 1741 as 'white gold'. But the impure metal was not easy to work owing to the presence of iron and copper.[3] These elements could be removed to a certain extent by cupelling with lead and bismuth at a very high temperature, after which some forging and hammer welding were possible. A good deal of work went into methods of purification and this continued up to the end of the 19th century. Its melting point of 1 750°C excluded the possibility of casting it in the early days, but it was found that the melting point could be drastically reduced by alloying it with arsenic. After melting, the arsenic could be evaporated by heating to redness in an oxidizing atmosphere. This process could be used to produce platinum powder which could be consolidated by hammering. By 1790, Lavoisier was able to report that he had melted platinum on charcoal with a current of the newly discovered gas, oxygen.

The remarkable acid resistance of platinum meant that its main use was envisaged in chemical apparatus and crucibles, although its great intrinsic value and corrosion resistance suggested other important uses. By the beginning of the 19th century many scientists were still involved with the chemistry of platinum and its malleability. One of these was W. H. Wollaston, who examined the impurities and amongst these identified the element palladium. In about 1801 he was producing platinum powder by precipitation from solution and compressing it in a toggle press to give a hard cake of powder (see Fig. 142), the adhesion of which could be improved by heating to a red heat in a charcoal fire.[32] Alternate heating and forging was used to increase the specific gravity to 21.25–21.5: this must have been one of

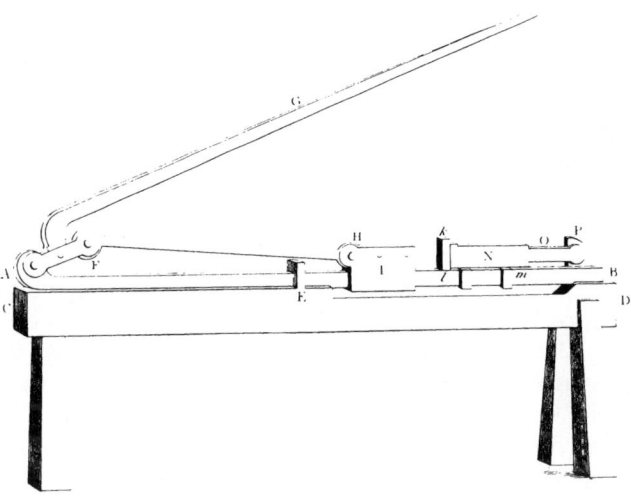

142 *Toggle press used by Wollaston in 1828 for the compacting of platinum powder; 0.67 in = 1 ft*

the first applications of powder metallurgy as we know it today. By the end of 1803 he had used 220 kg of native metal, much of it for large vessels for the concentration of sulphuric acid.

Some of the crude platinum that Wollaston worked on came from a man named Johnson, who was practising in London as an assayer of ores. It is possible that he was one of the members of a family that was to become well known in the platinum industry, and we know that a Percival Johnson was sufficiently interested in the metal to send a communication to the Philosophical Magazine about it in 1812. Percival Johnson specialized in the refining and fabrication of platinum from about 1817 and founded the firm of Johnson, Matthey and Co. in 1851.[31]

Originally, the joints in the platinum sheets forming the acid boilers were made by gold soldering, but by 1861 Johnson and Matthey were fusion welding the metal with a blowpipe. By 1874, the metal could be melted and a cast of some 236 kg of a platinum–iridium alloy was made in an oxygen–coal gas fired furnace in the Conservatoire des Arts et des Metiers in Paris, under the direction of Deville.

OTHER TRADITIONAL NON-FERROUS METALS

The recovery of lead continued along the old lines with the blast furnace and the reverberatory almost completely ousting the ore hearth. But the refining processes were speeded up by the substitution, in 1850, of the Pattinson process by the Parkes process for the recovery of silver.[33,34] This process involves a reaction between the silver dissolved in the lead, and zinc added to produce a silver–zinc compound (Ag_2Zn_3) which floats on the surface of the lead. These 'crusts' can be skimmed off, the zinc removed by heating under vacuum, and the silver recovered in pure form by cupellation, as before. Other impurities such as antimony and tin (from scrap solders) can be removed by oxidation, or selective reactions with reagents such as caustic soda.

A really big development occurred in the recovery of precious metals. The opening up of the USA, Canada, and South Africa produced a 'gold rush' as the placer deposits became known. In South and Central America at least, these had been known to the Indian inhabitants and utilized in a small way but, for some reason, the North American Indians had been quite content with copper, as noted in Chapter 1. It was found that the original deposits, from which the placer deposits were derived, contained particles of gold too fine to be recovered efficiently by the old traditional methods of washing and amalgamation. Mercury was becoming too expensive to use on some of these ores, and this is where the discovery in 1887 of cyanidation by J. S. McArthur and R. and W. Forrest was to prove so important.[29] These men had been working for the Tharsis Copper Company in Southern Spain, which was mining pyrites low in copper but with an important precious metal content. They had found that a dilute solution of an alkaline cyanide would dissolve the precious metals and the gold and silver could be precipitated from the solution with zinc powder. By this means, gold could be profitably recovered from sands containing as little as one part in three million, and from quartz rock and gossan containing one part in 100 000 (0.001%).

Changes in Melting and Casting Technology

Up to the middle of the 19th century only three types of furnace were in existence for metal melting. These were the reverberatory or air furnace fired with coal, the crucible furnace, and the cupola fired with coke. The aluminothermic method, whereby heat is generated by an exothermic reaction between, for example, iron oxide and aluminium powder, had been invented by Goldschmidt in the 1890s and was used on a small scale for reduction, melting, and welding; it is still used to a limited extent today.

The advent of cheap electricity from dynamos allowed Moissan,[35] in 1892, to use an electric-arc furnace for the reduction of lime with coke to produce calcium carbide, which in its turn made the production of acetylene for gas welding an economical proposition. In the electric-arc furnace an arc is struck between two carbon electrodes and the radiant heat used to heat the charge, or else the arc can be struck between an electrode and the charge itself. The latter is the more efficient but is often difficult to control.

In 1898 the Italian, Stassano, obtained patents for the use of an arc furnace for the making of steel direct from iron ore. After intense development, mainly in Norway, this type of furnace (known as the Tysland-Hole furnace) has been used since the late 1940s for the making of pig iron in Northern Norway.

An important development in this field was the Soderberg electrode which is baked in situ by the passage of the current and the heat of the furnace. The carbon–pitch mixture is placed in a thin mild-steel tube, which is fed down into the furnace as the electrode is consumed, and undergoes the baking process. This allows very large electrodes to be used and has been adapted to the Hall-Héroult aluminium furnaces by using an aluminium casing.

The three-phase arc furnace, with an acid or basic lining, has now proved a most economical way of making steel by melting and alloying scrap and is beginning to be used for steelmaking by oxidizing pig iron with an oxygen lance, thus replacing the open hearth furnace altogether.

The inductive effect of electricity has also been exploited; by using high-frequency alternating current, heat may be induced in a charge within a coil by the eddy current effect. The coil has to be water cooled, and the furnace resembles a crucible furnace with the coil built into the lining. Alternatively, low-frequency current may be used in a furnace in which the charge is the short-circuited secondary winding of an iron cored transformer. This type was patented by S. Z. de Ferranti in 1887. More recent developments have shown that it is possible to dispense with the iron core in the case of ferrous metals and still make use of normal 50 Hz alternating current.

The electric furnaces are proving more economical as the gap between the cost of electricity and other fuels narrows. The higher outputs obtainable per unit of capital cost, and the reduced maintenance costs, counter the slight additional cost of electricity.

On the foundry side, sand moulding is still the predominant technique on a tonnage basis, but many other processes have been developed. Sand moulding itself has been automated by the use of pattern-plates, which incorporate a wide range of split wood or metal patterns, of different size and shape, mounted on a board so that no space is wasted in the moulding box. Moulding itself is done by machine, the sand being slung and pressed into place; the finished mould is then placed on a conveyor and automatically filled from a ladle and tundish.

Centrifugal casting, the use of centrifugal force to improve the density of a casting, was introduced in the leading cast-iron pipe foundries after the 1914–18 war.[36] The horizontal spinning technique has always been favoured for pipes, and a leading pipe founder in Britain installed such a unit in 1919. This process is now used for internal combustion engine cylinder liners, and for bronzes for gear blanks.

For smaller components with considerable detail, the lost-wax process has been revived in modern form using disposable plastic patterns, or solidified mercury in the case of the Mercast process. Both these techniques are capable of giving superior surface finish to the castings, so much so that very little final fettling (finishing) is needed. Components intermediate in size may now be made by shell moulding in which resin–sand mixtures, thinly applied to metal patterns, are hardened by infra-red heating. Alternatively, the sand may be mixed with sodium silicate (water glass), and hardened by the passage of carbon dioxide; this method is often used for cores.

Welding

Up to the end of the 19th century the main joining processes were forge welding and riveting. The latter could only be applied to relatively thin plate, and the difficulty of the former when applied to large fabrications often led to failures. Even so, forge welding was used with success in the making of stern frames, anchors, and large structures, and was applied to the making of aluminium vessels before the advent of oxyacetylene fusion welding (*see* Fig. 143).

The idea of high-temperature gas flames using forced draught was not new, and had been employed by Johnson and Matthey for the welding of platinum in 1861. The easy and conveniently portable production of acetylene by the reaction between calcium carbide—itself produced cheaply in electric-arc furnaces—and water, and the high heat of combustion of acetylene with oxygen, produced just the localized heat source required for fusion welding. This process, when applied to thin sheets, rapidly replaced blacksmith's welding and riveting in numerous applications, and was extended to the new metals, aluminium and magnesium.

The principles of the electric arc were first discovered by H. Davy in 1801, but it was another 80 years before carbon arc lamps were to be seen in the streets, and the carbon–metal arc used for the electric welding of lead by De Meritens.[37] In 1887 a Russian, Bernados, took out a patent for the carbon arc welding of steel in which the carbon electrode was positive and the metal the negative electrode. However, this introduced carbon into the melt and made it brittle; this problem was later overcome by using the carbon as the negative electrode and the process became of commercial value in about 1902. The work was melted, and additional metal could be added by means of the melting off of the end of a 'filler rod' introduced into the arc itself. The combustion of the carbon to carbon monoxide protected the welded area from oxidation.

Metal arc welding, in which the arc is struck between a rod of the metal to be welded and the work, was patented by Slavianoff in 1897 and was in use in the UK by 1888.[37] However, this was not very successful, as the

143 *Forge welding a ship's stern frame (courtesy of R.C. Benson)*

molten steel weld pool was embrittled by contamination with oxygen and nitrogen from the atmosphere. This problem was eventually overcome by shielding the arc with a flux coating, a process introduced by the Swede, Oscar Kjellborg, in 1910, and this process is still responsible for the majority of fusion welds. Some flux must be removed from one end, however, before contact can be made between the electrical source and the metal core. By about 1934, this process was being used in civil engineering and naval construction. As the flux is not electrically conducting, in order to make the process continuous and automatic the flux may be applied to the joint area as a powder before the arc reaches it, as in the 'Unionmelt' process. Alternatively, the flux coating may be rendered conducting by incorporating a spiral wire, through which the current can be carried from a contacting sleeve to the metal core as the coated wire electrode passes through the welding head.

The advent of cheaper helium, argon and carbon dioxider has meant that in many cases fluxes can be dispensed with, and shielding, obtained from these inert and therefore non-oxidizing gases, supplied to the work by means of a tube concentric with the welding wire. In some cases it is more convenient to use non-consumable electrodes of tungsten and use a filler rod, as in oxyacetylene welding.

New high-temperature heat sources, such as the electron beam and the laser, have provided yet more ways of welding metals; no doubt there are many more to come. But at the lower temperature end we have seen the revival of forge welding so that it may be carried out cold at room temperature on clean metals with high pressures, or at moderate temperatures—still below the melting point— using somewhat lower pressures. Even friction is now utilized as a heat source to join metals.

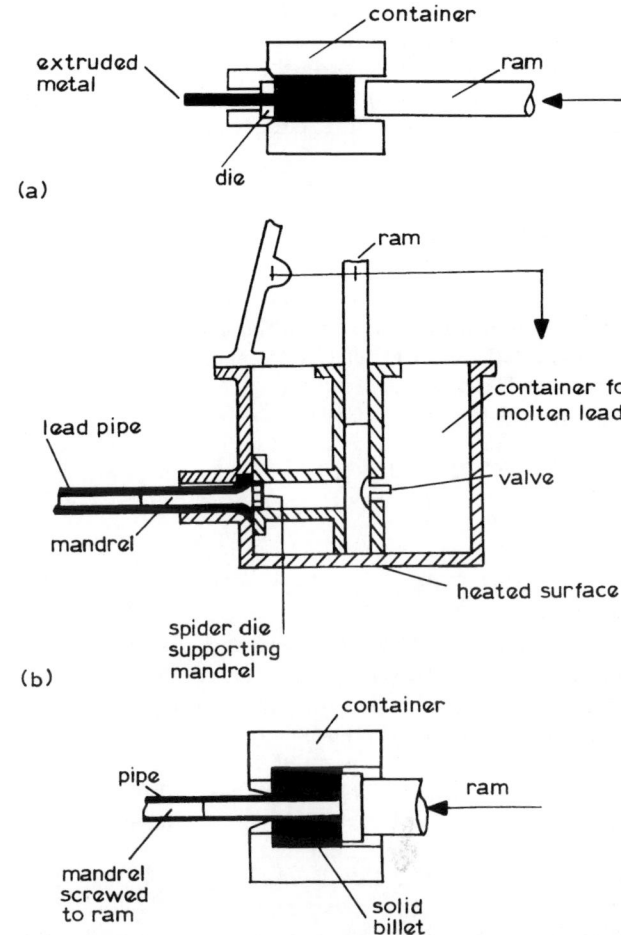

a principle of direct extrusion; *b* extrusion of molten lead for pipe by Joseph Bramah in 1797; *c* Burr's Press for solid lead extrusion (1820) operated vertically—ramdown

144 Details of extrusion processes

Extrusion

A new production process began to be introduced in the early years of the nineteenth century. This was extrusion,[33] the first patent for which had been issued to Joseph Bramah in 1797 (Fig.144). His process was used for production of lead pipe, but used molten lead and was in principle more like pressure die casting than the extrusion of solid metal, which is what we understand by the process today. The extrusion of solid lead, using much higher hydraulic pressures, was first used by Burr in 1820, again for lead pipe for plumbing. In 1897, the same process was being used for the sheathing of electric cables.

The application of extrusion to metals with high melting points depended on the invention of special high-temperature steels, and it was not until 1894 that G. A. Dick was able to extrude some of the copper-base metals. When aluminium became more common, it was soon found that it was an ideal metal for the process, and increasingly complicated shapes were extruded in this metal and its alloys from the 1930s onwards.

The final development was its application to steel, which began to appear in the 1930s. While copper-base metals tend to form an oxide with lubricating properties between the billet and the container, this is not the case with steel. In this case the lubricant has to be applied artificially, and this was done from 1950 onwards with the aid of glass which was applied to the container as a bundle of glass fibres.

References

1   J. C. CARR and W. TAPLIN: 'History of the British steel industry', 1962, Oxford, Blackwell

2   H. D. WARD: *J. Iron Steel Inst.*, 1972, **210**, 396

3   J. A. JONES: *ibid.*, 1872, 278

4   W. K. V. GALE: 'The Black Country iron industry', 1966, London, The Iron and Steel Institute

5   CARR and TAPLIN: *op. cit.*, 31

6   H. O'NEILL: *Metals and Materials*, 1969, **3**, (8), 312

7   M. L. PEARL and J. P. SAVILLE: *J. Iron Steel Inst.*, 1963, **201**, 745

8   A. BIRCH: Nachrichten aus der Eisen-Bibliotek, Schaffhausen, 1963, (28),129;1964, (30),153

9   W. M. LORD: *TNS*, 1945–7, **25**, 163

10   H. BESSEMER: Report to The British Assoc., Cheltenham, 11 Aug., 1856, (*The Times*, 14/8/1856)

11   M. OSBORN: 'The story of the Mushets', 1952, London

12   R. F. MUSHET: Patent No. 2219, 22 Sep., 1856

13   L. G. THOMPSON: 'Sydney Gilchrist Thomas—an invention and its consequences', 1940, London, Faber and Faber

14   S. G. THOMAS and P. C. GILCHRIST: *J. Iron Steel Inst.*, 1879, (1), 120

15   L. AITCHISON: 'A history of metals', Vol. 2, 512, 1960, London

16   H. HELLBRUGGE: *Durrer Festschrift*, 295

17   T. K. DERRY and T. J. WILLIAMS: 'A short history of technology', 746, 1960, Oxford, Clarendon Press

18   W. E. HOARE: *Bull. Inst. Metallurgists*, 1951, **3**, (1), 4

19   K. C. BARRACLOUGH: Personal communication

20   L. AITCHISON: *op. cit.*, 574

21   SIR R. HADFELD: *Chem. and Ind.*, 1925, **44**, 1

22   JAMES RILEY: *J. Iron Steel Inst.*, 1889, (1), 45

23   C. S. SMITH (ed.): 'The Sorby centennial symposium on the history of metallurgy, (1963, Cleveland), 1965, London, Gordon and Breach

24   W. H. HATFELD: 'Cutlery; stainless and otherwise', pp.31, Sheffield; address to the Sheffield Trades Technical Society, 17/12/1919

25   SIR R. A. HADFIELD: Metallurgy; its influence on modern progress', 1925, London, Chapman and Hall

26   R. A. MACKAY: *Bull. Inst. Metallurgists*, 1951, **3**, (3), 15

27   YAN-HANG: *Rev. Aluminium*, 1961, (283), 108

28   M. COOK: *J. Inst. Metals*, 1953–4, **82**, 93

29   S. G. CHECKLAND: 'The mines of Tarshish' 1967, London, Collins; and D. AVERY: 'Not on Queen Victoria's birthday', 1974, London, Collins

30   E. J. PRYOR: 'Mineral processing', 3 ed., 458, 1965, Elsevier

31   D. McDONALD: 'A history of platinum', 1960, London

32   W. H. WOLLASTON: *Phil. Trans. Roy. Soc.*, 1829, **119**, 1; reprinted and edited by J. Gurland in 'Metallurgical Classics', 573, Jul. 1967, ASM

33   J. PERCY: 'The metallurgy of lead', 1870, London

34   J. SMYTHE: 'Lead', 1923, London, Longmans

35   M. LORIA and B. BONI: 'Contribution Italiennes a l'Electrosiderurgie', Vol. VI, 280,1965, Warsaw-Krakow, Actes du XI Congres Int. d'Hist. des Sciences

36   K. R. DANEL: *Mech. Engineering*, 1951, **73**, 644

37   K. WINTERTON: *Welding*, 1962, **30**, 438, 488

38   C. E. PEARSON: 'The extrusion of metals', 1 ,1944, London, Chapman and Hall

Chapter 11
The contributions of the scientists

The earliest metallurgists were empiricists and traditionalists and the lack of chemical knowledge until the later Middle Ages almost entirely precluded the application of science to the art and craft of metallurgy. Metallurgy today is divided into two main components, extraction metallurgy and physical metallurgy, and, while both are amenable to scientific observation, it is in the field of physical metallurgy that we first see signs of the application of science. Roman, and to an even greater extent Anglo-Saxon and Merovingian peoples, were acquainted with the structure of the pattern welded sword.[1] They knew this as a hallmark of quality and they must have known that it resulted from a carefully worked out forging schedule. Theophilus[2] had also made observations of a physical character, particularly malleability tests for quality, and many others were familiar with the processes of heat-treating steel. But lack of knowledge about the reason for the difference in properties of wrought iron, cast iron, and steel acted as an inhibiting factor to their general understanding.

The 17th century was important for new innovations in science. The compound microscope had been invented by Jansen in about 1600, and the Royal Society of Charles II of England was founded in 1662, and the French Academie des Sciences in 1666. Leonardo da Vinci, both a keen observer and an engineer, had died in 1519. Astronomers and dynamicists such as Galileo (1564–1642), and mathematicians such as Napier (1550–1617), Descartes (1596–1650) and Pascal (1623–1662), had made major contributions to science but the properties of the materials on which their observations and experiments depended had not appeared worthy of rare investigation. With exceptions, such as Sir Isaac Newton, even the masters of the Mint—one of any country's main metallurgical establishments—were seemingly interested less in the physical properties of metals, than where they could be obtained and how much profit could be made for their often impecunious patrons.

Metallurgy and Metallography

One of the first to record scientific observations on the properties of metals was Jousse, a Frenchman who published a book on the art of locksmithing in 1627.[3] Like others before him he embraced a wide range of subjects, including a rolling mill for lead cames and a file cutting machine, as well as a detailed description of the operations of hardening and tempering. He seems to have been the first to associate hot shortness in wrought iron with the smell of sulphur. He gives the earliest description of cementation of small pieces of iron packed with alternate layers of charcoal in a crucible. He advised against quenching steel from too high a temperature, which would render it worthless, but he does not comment on its grain size.

Another contribution to the properties of metals was made by Henry Power in 1664 who, with the aid of the microscope, noticed that polished surfaces of metals were full of fissures, cavities, asperities and irregularities.[4]

Réaumur, in 1722, seems to have been the first to discuss the properties of ferrous metal in terms of grain structure.[4,5,6] Again, like Leonardo, he was a man of wide knowledge and a true scientist in the days when it was possible for one man to understand almost all there was to know about science. He actually seems to have reinvented the process of malleablising cast iron, and describes the processes involved in one of his books. It seems that the actual invention can be laid at the doors of the Chinese[7] soon after they began to use cast iron in the last few centuries BC. By the 18th century, there were many hypotheses on the corpuscular nature of matter and it is clear that Réaumur was much influenced by them.[5,6] He observed with the aid of the microscope that a particle of steel consisted of a large number of grains. Although he gave the name 'molecules' to these small entities, he appreciated that these were made of still smaller entities which he claimed he could see. One wonders if he was looking at the structure of pearlite. He seems to have been unaware of the effect of incipient melting on the grain boundaries. Furthermore, he was the first to advocate the use of hardness tests, using both indentation tests of crossed prisms and scratch hardness tests, thus anticipating that of Mohs in 1822.

For the mechanical testing of wires he devised a machine by which a quenched steel wire was flexed until it broke. One end was gripped in a vise, and the effect of bending was measured by the number of turns of a screw thread which applied the load to the wire. This was a machine of somewhat limited application, presumably of more interest to watchmakers for determining the properties of springs than anyone else, but he does refer to the earlier dead load fracture test in which weights are added to a vertical wire, as, used by Leonardo da Vinci, Galileo and Hooke.

The contribution of the French scientists in the 18th century was remarkable. Following the work of Jousse and Réaumur, we have the *Encyclopedie des Arts et des Metiers* edited by Diderot and d'Alembert,[8] to which nearly every scientist of the day seems to have made a contribution.

One experimental scientist of the period was T. Desaguliers[9] who, during the course of work on astronomical models and friction electric machines in 1724, discovered that large lead spheres could be pressure welded merely by twisting flattened and cleaned surfaces under normal load. Very soon after this Van Musschenbroek of Holland developed improved mechanical testing machines.[10] Then, in 1734, we have Swedenborg's treatise on copper and iron *(Opera Philosophica et Mineralia)* which brings together into two volumes a good deal of metallurgical technique of the period 1690–1730. Although this contains very little original thought of value, his descriptions of metallurgical plant and techniques nevertheless fill a time gap between the works of Biringuccio and Agricola, and those of Diderot and Jars.

On the whole, the majority of industrialists of the 18th century were content to rely on tradition and made few contributions to scientific advance. But one of a different calibre was Grignon,[3] who owned an ironworks in Champagne and who contributed a series of memoirs to the Académie des Sciences, which were printed together in 1775; he is also responsible for many other works. Grignon certainly had a real feeling for metallurgy and his enthusiasm was supreme. He was the first to describe accurately the changes in the blast-furnace burden as it descended down the furnace. He noticed that gas was evolved during casting, and ignited around the moulds. If the mould was too impermeable explosions could result and castings could be too porous to give satisfaction. He observed the formation of kish graphite on the surface of cast iron but assumed that it consisted of 'attenuated' cast iron. Its formation took place when the fuel/ore ratio was great, the blast too great, or the iron left too long in the hearth, and he observed, correctly, that its occurrence was wasteful. His description of wrought iron as a bundle of fibres separated from one another by vitrified matter is remarkably accurate and his detailed illustrations show dendrites and masses of polygonal crystals with great accuracy.

He regarded his furnaces as instruments in which art most closely approached nature in the imitation of igneous products and he even took a full-sized blast furnace and cooled it slowly for 15–20 days to obtain large crystals of metal and slag.

One of the first to begin to shed light on the difference between iron and steel was Torbern Bergman,[3] a Swedish chemist (1735–1784) who was for most of his life a contemporary of the chemist Antoine Lavoisier. His main contributions to metallurgy are to be found in his *Dissertatio Chemica de Analysi Ferri*, written in 1781. At the time, Swedish chemists were leading the world in the development of new analytical techniques. Oxygen had been discovered, and Bergman clearly understood its role in combustion but he continued to use the earlier phlogiston theory to explain his results on iron.

At about the same time another Swede, Sven Rinman,[1] who was interested in welded gun barrels, had shown that steel contained something additional to that of iron in a paper published in 1774. He found that, while pure Osmond iron lost 87% of its weight in acid, a similar weight of steel after the same treatment lost only 30% and was covered in an adherent dark ash-grey film. However, he did not identify the film.

Bergman dissolved various ferrous materials in acid, and measured their rate of dissolution and the gases evolved from them (mainly hydrogen). He regarded iron as containing something he called phlogiston which was released by the attack of the acid and went into his inflammable 'air'. Cast iron produced the least gas ($38 in^3$), wrought iron produced between 48 and $51 in^3$, and steel was intermediate with $45–48 in^3$. He was a little uncertain as to whether his phlogiston came from iron, or some impurity that was more capable of giving phlogiston than iron.

He confirmed these results by what we would now call 'dry assays', or fusions in sealed crucibles, in which iron ore was reduced with the carbon in cast iron to produce metal and slag (and unreduced ore). He found, for example, when he fused 200 lb[11] of cast iron with 50 lb of black hematite, that he obtained a regulus weighing $201^{1}/_{2}$ lb covered with black slag: the metal blob at the bottom could be flattened under the hammer to a diameter of 12 mm, showing that the carbon in the cast iron had reduced some of the ore. He carried out over 270 experiments on wet and dry assaying and concluded that ductile iron contained almost no plumbago (carbon), while steel contained a lesser amount, and cast iron was saturated with it.

A lot more came out of Bergman's experiments which show that he had a good appreciation of the principles of thermochemistry, although his experiments were far too crude to give useful results in this sphere. When he collected the residues from his wet assays he noted that those from cast iron had the appearance of plumbago (graphite). When expressed as a percentage they showed that cast iron had 4% C, while English steel had 0.4%, and wrought iron from Ullfors 0.1%. He investigated the differences between the various residues; some of course would contain slag, others sulphur, and some phosphorus, but he was not able to analyse these. Apart from plumbago, his only other analyses are for 'siliceum', the insoluble components in the slag, and manganese. The amounts of the siliceum in the wrought iron and steel (max. 0.9%) are much lower than one would expect considering that the silicon content of the metal itself must have been negligible. If correct, this shows that the iron must have been well wrought and the slag very efficiently extracted.

He also made observations on magnetic properties, noting that iron was not the only magnetic metal; cobalt, nickel and manganese were also magnetic. Furthermore, he noticed that magnetism could be induced in wrought iron, and he also knew of Swedish magnetic ores and that other iron ores could become magnetic on heating under certain conditions. He equated this with the introduction of phlogiston.

The Swedes were not the only analysts in the 18th century; various scientific institutions in Western Europe and Russia were interested in similar work, and a French group, Vandermonde, Berthollet, Monge,[3] and the industrialist Clouet,[12] also made many experiments more or less contemporary with those of Bergman. Their work was read before the Académie des Sciences in 1786. This was the time when metallurgical problems tended to inspire the chemists, including Antoine Lavoisier himself, and it appears that it was in Sweden and France that there was the closest cooperation between industrialists and scientists, rather than in Britain where such contacts were limited to those of Boulton and Watt and the ironmasters of the Black Country and Shropshire. In fact, it was the lack of interest shown by the Scottish ironmasters in his work that lead David Mushet to move southwards in the 1790s.

In spite of the unsettling times in France in 1789 and the years following, France has a splendid record of the interaction of science and industry from the time of Réaumur (1722) onwards. No doubt this was assisted by the realization that French industry was behind that of Britain in the middle of the 18th century and the sending of various observers, including Jars and de la Houliere, to study British methods. The same feelings inspired Swedish visits, but here it was not easy to take advantage of the new techniques since they were based on the utilization of coal. The interest of the French in the new industry is clearly evident in the encyclopedia of Diderot, a highly detailed record of the French industry and crafts in mid-century, an account emulated by no other country.

Let us now return to the advances on the physical side of metallurgy. In 1788 a German, Achard,[13] published in Berlin a book on the properties of metallic alloys in which he gave the results of experiments on over 900 alloys. In the course of his research he discovered the low melting point of the platinum–13% arsenic alloy (597°C) which he used as the basis of the first method of fabricating platinum. In this work he used eleven metals, namely; iron, antimony, copper, lead, tin, bismuth, zinc, arsenic, silver, cobalt and platinum. The only known metals that he ignored were nickel and gold.

For those alloys that looked useful he measured the density and mechanical properties, including tensile strength and hardness. However, there was no theorizing in this book; it was solely a table of properties, in fact, a sort of early *Metals Handbook*, and it seems to have had very little impact on current scientific thought. There is no evidence that it was known or made use of in the metallurgical industry of the time.

Structure, as distinct from properties, began again to interest the scientists of the early 19th century. In 1804, William Thompson[14] identified the duplex alpha–gamma iron structures in the Krasnojarsk meteorite, the intriguing structure that came to be named after Alois von Widmanstatten who was studying the collection of meteorites in Vienna. At the same time, structures were being studied by George Pearson[15] who read a paper before the Royal Society in London on the texture of wootz (1795). This material aroused the interest of many scientists in the 19th century: even David Mushet considered the problem and concluded that it was a mixture of iron and steel in an imperfectly fused state. Of course, its use in the mysterious 'damascened' swords of the period was well known, but nobody quite knew how the structure of the latter stemmed from the former (perhaps we are not quite sure even today). The next to consider the matter was Michael Faraday who, in 1819, published his analyses of a cake of wootz and incorrectly concluded that its properties were due to minute amounts of the earths, 'alumina and silex'.[16] With the help of James Stodart, Faraday[17] later came nearer the truth when he said that the damask structure 'is dependent on the development of a crystalline structure which is drawn out and confused by the hammer'. The two of them went on to try and simulate wootz by mixing various metals and 'earths', such as alumina and silica,[18] but Faraday soon lost interest and went on to consider other things, including the important field of electromagnetism for which he is better known. Even so, Hadfield ascribes the birth of alloy steels to his metallurgical work.[19]

At the same time, similar work was being carried out by Breant and others in France.[20] Again, it was the damascened sword that inspired it; Bréant found that the 'arrangement of the molecules' mattered more than their composition, although he was aware that carbon was important.

Similar Russian experiments on damascening were made by Anosoff in 1841 and described by Belaiew.[21] This work led indirectly to the famous work of Tschernoff[22] on the structure of cast steel ingots. Belaiew was still contributing to the subject of damascening as late as 1950.[23] All this is an example of how interest and work on one aspect of a subject can have important results on another—an example of 'spin off', to use a modern phrase. The conclusions on damascene swords proper were of no importance since by the time that they were published the duplex structured gun barrel, let alone the sword itself, was superseded by improved solid steel barrels or fabricated barrels strengthened by shrinking-on (the Armstrong type). But the experiments that led to these conclusions contributed to our understanding of steel structures generally, and gave rise to improved cast steel ingots.

By the middle of the 19th century one could discern the two branches of metallurgy, the physical one arising out of the experiments described above, aided by the efforts of the geologists and mineralogists (crystallographers) on crystal structure, and the extractive branch, aided by the efforts of the analysts and the rise of chemistry. This is not the place to discuss the rise of chemistry in the 18th and 19th centuries but merely to note that by 1864 it was possible for Percy to publish detailed analyses of ferrous and non-ferrous ores, which we know from more recent work to be of high accuracy. The blast furnace itself was being examined and its gases analysed, as we see from the paper presented to the British Association in 1845 by Bunsen and Playfair.[24] These were all byproducts of the rise of analytical chemistry but the more original contributions to metallurgy were to come from the metallographers, and later the physicists. We must not forget that this was a two-way process since the work on the properties of metals was, in the end, to influence the whole of the mechanical properties of solid materials.

But the 19th century physicist tended to think in terms of 'molecules' and completely ignore the spatial arrangement of these molecules, i.e. the space lattice and the crystalline structure resulting from it. For clarification of this aspect we have to look at the efforts of the metallographers, and for this we need to return to Sheffield where H. C. Sorby, descended from a long line of master cutlers, was inspired to start on a scientific career.[25] By 1849 he was applying microscopic techniques to polished surfaces of rocks. After a visit to Germany he revived an old interest in metallic meteorites and went on to the study of artificial alloys. In 1863, he claims to have discovered an example of the Widmanstätten structure in some Swedish iron. His specimens of polished and etched steels were exhibited at various meetings but do not seem to have aroused much interest among the industrialists. It was found to be possible to take prints from deeply etched specimens using printers' ink, a process used for meteorites in Vienna by Schreiber as early as 1820.

While photography had been invented by 1864, it was not easy to get the conditions of intense illumination and fast plates that would make photomicroscopy possible and produce a permanent record. Sorby succeeded, probably by using an oxy-hydrogen flame, oblique illumination, and wet collodion plates. For visual work he used a vertical illuminator with a mirror. From about 1887, photomicrographs became a normal feature of the new race of metallographers.

Now that they could clearly see the crystalline structure of metals, the main problem was one of identification. In 1827 Lampadius[26] had stated that 'the best cast steel has so fine a fracture that the platelets of iron carbide can be clearly disclosed only with a powerful microscope'. Adolf Martens was the next to make an important contribution, and he followed Schott in Germany who had been content to use a hand lens with great effect. Martens[27] was an engineer with the Prussian State Railways and does not seem to have had as much time as he would have liked to pursue his interest in metallography. Like Sorby, he concentrated on ferrous metals, etching in dilute acid solutions in alcohol for very long periods. He published his first work in 1879, and again, like Sorby, he seems to have been a practical man and shied away from hasty inferences. He seldom drew conclusions nor advanced any theory. But he clearly inspired others, and one to come to Martens' institute in Berlin was E. Heyn, a student of Ledebur.

The latter was responsible for many papers on structure from the beginning of the present century.

By 1885, the value of metallography was appreciated throughout Europe. At the French armament works of Le Creusot, an engineer, Floris Osmond, and his colleagues Barba and Werth, took up the study of metallography and heat treatment.[28] This team were the first to combine the chemical and structural viewpoints with thermal analyses to give a truer understanding of steel. In this way, they were able to demonstrate that there was no loss of carbon during quenching but only a change of state. They concluded that carbon could exist in the combined form as iron carbide and occurred in annealed steel as *carbone de recuit*, and was also dominant in the outer layers of quenched steel (i.e. in martensite) as *carbone de trempe*. They found that the amount of heat evolved in dissolving steel filings in cuprammonium chloride was 4–6% greater if the filings were dissolved in the cold-worked condition than in the annealed condition. The filings gave out a further 5–15% after the steel had been quenched, depending on the carbon content. These 'phases' were to be the source of much controversy. It was noted that the carbide often formed a cementing phase (*ciment* in French) in steel, which was later to be equated to the cement carbon introduced by the cementation process—hence 'cementite'.

However, it was thought that the grain-boundary *ciment* in wrought iron was analogous but lower in carbon content. There was still a feeling that crystals had to be stuck together with something because of their lack of regularity; the idea of their natural cohesion during solidification took some time to develop. So the mechanical properties of steel were thought to depend on the respective properties of the core and the envelope.

Osmond's later work was carried out in the laboratories of Le Chatelier and Troost at the Ecole des Mines and the Sorbonne. He owed much to the development of Le Chatelier's thermocouple which made accurate thermal analysis possible, although some work had been done previously by Roberts-Austen and others using crude temperature measuring methods. Osmond used both the direct and the inverse rate method of plotting cooling curves, and in iron he found three arrests, while in steel the number varied according to the carbon content. He correctly identified these with structural transformations but incorrectly assumed that all quenched steel was beta iron in a continuous mass or quasivitreous state. This was soon to be attacked by Howe,[29] an American, and others. Howe was strongly influenced by mineralogists who had been in the habit of calling their constituents after famous workers in the field, an idea that had also been used for the phases seen in meteorites. After noting the use of descriptive names for the more common constituents, ferrite, cementite, and pearlite, Howe proposed calling an acicular compound (nitride?), reported by Sorby, by the name of 'sorbite'. This convenient idea spread rapidly and was taken up by Osmond in his 1895 paper.[30] Thus martensite, austenite, and troostite were quickly added, while sorbite was reused for a more common constituent than that originally proposed by Howe.

However, it was not until 1901 that we see the nearly correct iron–carbon equilibrium diagram of Roberts-Austen, based on the phase rule and the accepted thermodynamic principles of chemistry. It was another Englishman, J. E. Stead, who applied the principles of crystallography, which changed the thinking of Osmond away from the continuous mass and quasivitreous state to one of crystallinity and the conclusion that gamma iron was cubic.

To a large extent this conversion was due to the appearance of etch pits on metal surfaces, the significance of which was beginning to be appreciated by metallographers.

In spite of the lead given by Osmond, the allotropy of iron and its relationship to the hardening of steel was not universally accepted. In Britain the main proponent was Roberts-Austen, but J. O. Arnold and R. A. Hadfield were not convinced, and controversy reigned far and wide. In 1912, a new tool became available, that of X-ray diffraction, which was demonstrated by Friedrich and Knipping. This enlarged the bounds of metallurgical science, and metal crystals were soon to have their lattice parameters determined by the Braggs in 1913.[31] This is where the physicist returned to the field; metallurgists were soon to realize the implications, but it was 1922 before X-ray diffraction was applied to the study of alloy constitution and the structures of alpha, beta, gamma and delta iron finally resolved.

With the aid of the detailed and painstaking work that went into the accumulation of data on alloy constitution,[32] metallurgy was all set for a smooth advance of its knowledge. The properties of grain boundaries, the effect of impurities, the symptoms of diffusion in the solid state, and the effects of mechanical deformation on crystal structure, were all fertile fields investigated for the first time.

What were the special advances? The first was undoubtedly the age hardening of aluminium alloys discovered accidentally by Wilm of Berlin in about 1906.[33] Previously, aluminium had been mostly used in the cast state and the commonest alloying elements were copper, zinc, and magnesium. The basic incentive for this research was a military one, an interest in the lightness of aluminium, particularly as a material for cartridge cases. Wilm found that certain alloys containing magnesium were unstable after rapid cooling from a high heat-treatment temperature (for aluminium). When tested almost immediately after quenching they were comparatively soft, but after they had stood undisturbed in his laboratory over the weekend they were found to be much harder. This 'age hardening', as it came to be known, was found to be most pronounced on those alloys containing 4% copper, and about 0.5% each of magnesium, manganese, iron and silicon. As compared with the commercial purity aluminium then obtainable, the strength was increased by a factor of nearly two and the elongation reduced by the same factor, but the density was only about 5% greater. This type of alloy

came to be known by its trade mark 'Duralumin', and although it was useless for shell cases, it was used for airframes during the 1914–18 war (e.g. in the airships known as Zeppelins).

The reason for the phenomenon of age hardening was soon found by ordinary metallographic studies to be based on changes in solid solubility. However, the microscope could only show evidence of the production of a second phase —a 'precipitate'—when the process had gone past the optimum and the strength was decreasing with increase of time. It required the efforts of the X-ray diffraction workers to show that submicroscopic aggregates were involved, and subsequent work in the 1950s, using the newly discovered electron microscope was able to confirm this (*see* Fig. 145). No time was lost in applying this discovery to other alloys and we now have copper–beryllium alloys in which the strength is increased from 540 to 1400 MN/m² by quenching from 800°C and age hardening for about 1 hour at 350°C, and high strength (maraging) steels

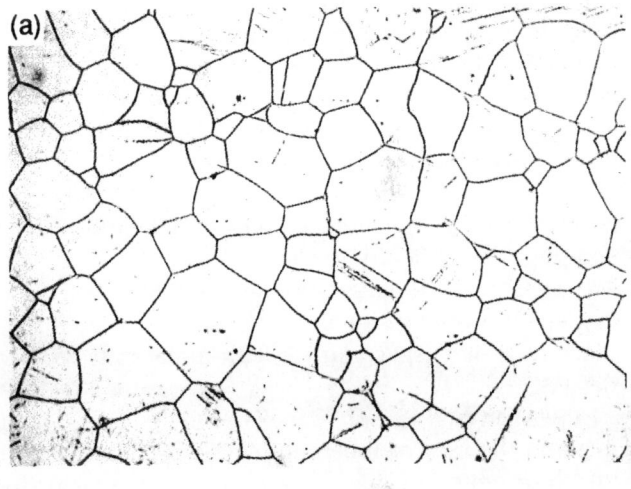

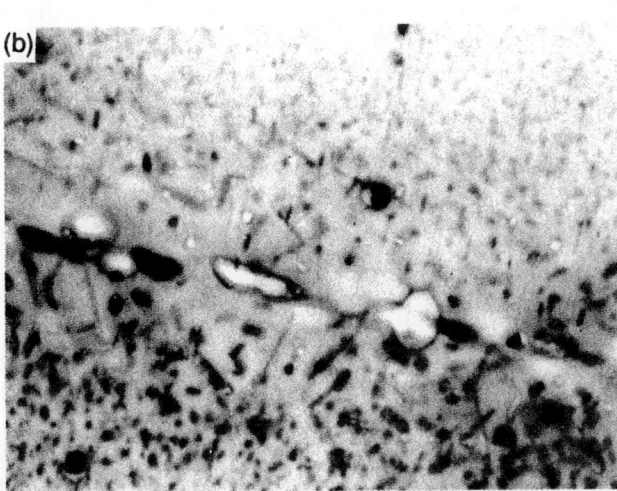

145 Photographs of the same piece of aluminium under (a) the light microscope with a magnification of 52, and under (b) the electron microscope with a magnification of 24,600

containing Ni and Co, in which the quenched low-carbon martensite is increased in strength to give a yield stress of 2 200 MN/m² and yet retain its ductility, by age hardening at 480°C. This principle had been extended to an enormous range of alloys, and the limit has not yet been reached.

The impact of the discovery of age hardening was limited to metallurgy but the next important discovery, that of dislocations in metals, made its mark on the whole of solid state physics. It probably was metallurgy's most important contribution to science in general. That defects were present in metallic materials was clear from the diffusion studies of Roberts-Austen,[34,35] for how else could gold diffuse a distance of 7.65 mm into a solid lead block? The physicists preferred to believe in perfect idealized models which had forced them in the 19th century to ignore all properties of matter related to structure—precisely those with which the metallurgist was mostly concerned. The metallurgists had to tackle these problems in their own way, after the concept of the space lattice was accepted by physicists and metallurgists alike, the puzzling problem of the nature of deformation had to be solved. The metallurgists Carpenter and Elam[36] did the first extensive tests on the properties of metal single crystals in 1921 but, as late as 1930, even Miss Elam found it difficult to believe in the lattice diffusion of zinc in copper and brass.[37]

Even when the idea of point defects, such as vacant lattice sites through which atoms could diffuse, was accepted the difficulty of deformation still remained. The fact of slip in single crystals and polycrystalline aggregates was clear for all to see, but how could it take place in a rigid space lattice with apparently so few defects?

It was an engineer, C. E. Inglis, who first (1913)[38] put forward the idea of inherent defects in solid material which would propagate as cracks. This was taken up by A. A. Griffith and the Griffith crack theory was propounded in 1920, originally for a non-crystalline material such as glass.[39] It was well known that if the surface of glass was being continuously dissolved away it had a higher strength, and this was correctly believed to be due to the presence of fine cracks in the surface. Griffith proposed the initial presence of crack-like defects in a material and calculated the stress to propagate them. This was very much lower than the theoretical breaking stress of a material which the physicists had begun to calculate on the basis of their ideal model, and was of the same order as the stress required to deform a metal by slip. But how were these crack-like defects brought about in metals? It seemed fairly clear that defects as large as cracks were not present in wrought metals initially and that some other mechanism was required to initiate them. Gradually, the idea of the dislocation was born. This was a line defect—effectively an additional line of atoms normal to and terminating on the slip plane, i.e. the surface on which slip was taking place, so giving rise to a line of disregistery where the forces between neighbouring atoms were disturbed. While this was only a theory as far as metals were concerned,

it could be shown on a bubble 'raft', a device invented by W. L. Bragg consisting of a layer of soap bubbles between two glass plates.[40] The bubbles formed a space lattice which was surprisingly regular, but here and there could be seen the point defects representing vacant lattice sites, and others where additional bubbles (atoms) had been incorporated. By moving the plates slightly it was possible to move a defect through a large part of the raft in the way that it was imagined that the line defects were moved. This defect was called a dislocation (see Fig. 146), and calculations on this basis were extended and a comprehensive theory propounded. We had to await the invention of the electron microscope by M. Knoll and E. Ruska in 1931,[41] and its development by Hirsch and others between 1951 and 1956, before dislocations could actually be seen to move through metals and be photographed cinematically. Even here their visibility was not due to actual atomic sized defects, but to the elastic strain field surrounding the dislocation which causes the electrons to be diffracted. It was the development of Muller's field ion microscope, also invented in 1951, which allowed individual point defects to be seen.[42]

Atom Theory

The constitution of matter had been argued since the days of the Greek philosophers. Democritus and others held the view that every substance was made up of an aggregation of very small units, beyond which no further division was possible. Later, this theory was abandoned by Plato and Aristotle who thought that all matter consisted of the same substances and the differences arose out of the varying proportions of these substances, air, water, earth and fire. The final phase of this theory was the phlogiston theory of 1702 when all phenomena observed during the combustion of matter were associated with the evolution of phlogiston, which was an essential component of all combustible matter. The products of this combustion were their elements! Of course, the phlogiston had to have a negative weight, as it was apparent that in many cases the combustion products were heavier than the original material.

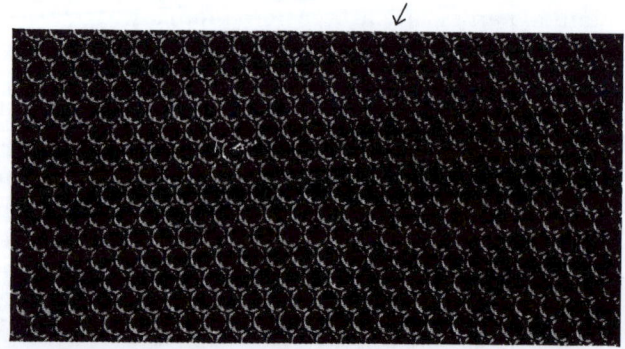

146 Bubble raft used by W.L. Bragg for the demonstration of defects in crystalline materials; the photograph shows a 'dislocation', an extra row terminating in the centre, indicated by the arrow (from Bragg and Nye[40])

It was not until the end of the 18th century that the facts produced by Joseph Priestley, Lavoisier, and many others rendered the phlogiston theory untenable. A new atomic theory was put forward by John Dalton in 1804 which, in fact, was a return to the doctrine of Democritus.[43]

The atom was the indivisible unit of matter, but the atoms of each element (40 were then known) were unique. It was Prout who believed that the atoms of each element were made up from basic substances common to all but, for many reasons, Dalton's theory was accepted up to about 1900.

Gradually, as more and more elements came to be discovered during the 19th century, it became clear that there were certain similarities between them, and the experimental chemists received help from their more theoretical colleagues in arranging them into groups. It seemed that these similarities had some fundamental significance but, before this could be made use of, a reasonably accurate knowledge was required of the atomic weights of the several elements.

By 1860, sufficient data had been obtained, and Newlands published a table which was intended to illustrate his 'law of octaves'. In 1869 Mendeléeff, the Russian scientist, put forward a classification in many ways similar to that of Newlands with all the elements with closely similar properties in one horizontal line and in vertical order of their atomic weights (*see* Fig. 147). He noticed that there was a kind of periodicity in their properties, and hence this table became known as the Periodic Table. The extraordinary thing was that by this means Mendeléeff was able to show that there were gaps and even, in the case of the then unknown element hafnium, to predict an atomic weight of 180 (top, right): actually, it turned out to be 178.5. In the succeeding years the gaps were filled one by one, but some were not filled until 1925 and, later, other heavy elements were added to the end of the list. The modern periodic table (*see* Fig. 148) lists the elements in order of their atomic number instead of Mendeléeff's atomic weight. This number represents the number of electrons that orbit the atoms and are, therefore, whole numbers while the atomic weights are not, owing to the fact that natural elements are composed of various mixtures of atoms of slightly different weight but all having the same chemical properties, i.e. isotopes. Each isotope has an atomic weight which is a whole number multiple of the atomic weight of the normal hydrogen atom.

The modern theory of the atom suggests that it consists of a heavy nucleus and a 'cloud' of orbiting electrons. The nucleus is made of elementary particles, sometimes called nucleons, of which there are two principal sorts, protons and neutrons. The protons are positively charged and have a mass which is about 1 on the atomic scale and which is about 1800 times that of the electron. The second important particle, the neutron, is slightly heavier than the proton but has no electric charge. The total number of protons in a nucleus will determine its electric charge, which is normally balanced by the opposite (negative) charge of the electrons. The sum of the masses of the protons and neutrons will

но въ ней, мнѣ кажется, уже ясно выражается примѣнимость выставляемаго мною начала ко всей совокупности элементовъ, пай которыхъ извѣстенъ съ достовѣрностію. На этотъ разъ я и желалъ преимущественно найдти общую систему элементовъ. Вотъ этотъ опытъ:

	Ti=50	Zr=90	?=180.
	V=51	Nb=94	Ta=182.
	Cr=52	Mo=96	W=186.
	Mn=55	Rh=104,4	Pt=197,4
	Fe=56	Ru=104,4	Ir=198.
	Ni=Co=59	Pl=106s	Os=199.
H=1			
	Cu=63,4	Ag=108	Hg=200.
Be=9,4	Mg=24	Zn=65,2	Cd=112
B=11	Al=27,4	?=68	Ur=116 Au=197?
C=12	Si=28	?=70	Sn=118
N=14	P=31	As=75	Sb=122 Bi=210
O=16	S=32	Se=79,4	Te=128?
F=19	Cl=35,5	Br=80	I=127
Li=7 Na=23	K=39	Rb=85,4	Cs=133 Tl=204
	Ca=40	Sr=87,6	Ba=137 Pb=207.
	?=45	Ce=92	
	?Er=56	La=94	
	?Yt=60	Di=95	
	?In=75,6	Th=118?	

а потому приходится въ разныхъ рядахъ имѣть различное измѣненіе разностей, чего нѣтъ въ главныхъ числахъ предлагаемой таблицы. Или же придется предполагать при составленіи системы очень много недостающихъ членовъ. То и другое мало выгодно. Мнѣ кажется притомъ, наиболѣе естественнымъ составить кубическую систему (предлагаемая есть плоскостная), но и попытки для ея образованія не повели къ надлежащимъ результатамъ. Слѣдующія двѣ попытки могутъ показать то разнообразіе сопоставленій, какое возможно при допущеніи основнаго начала, высказаннаго въ этой статьѣ.

Li	Na	K	Cu	Rb	Ag	Cs	—	Tl
7	23	39	63,4	85,4	108	133		204
Be	Mg	Ca	Zn	Sr	Cd	Ba	—	Pb
B	Al	—	—	—	Ur	—	—	Bi?
C	Si	Ti	—	Zr	Sn	—	—	—
N	P	V	As	Nb	Sb	—	Ta	—
O	S	—	Se	—	Te	—	W	—
F	Cl	—	Br	—	J	—	—	—
19	35,5	58	80	190	127	160	190	220.

147 *The Periodic Table according to Mendeléeff*

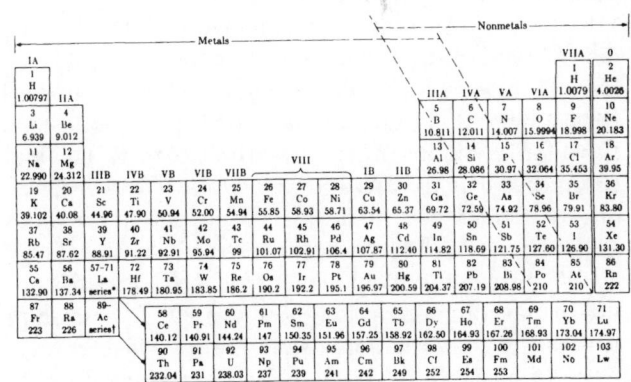

148 *A modern form of The Periodic Table; the atomic number and the atomic weight (carbon = 12.000) are shown for each element (Van Vlack: 'Elements of material science', ed. 2, 1964, Reading, Mass., Addison-Wesley)*

give approximately the mass of the element. The number of protons is called the *atomic number* of the atom, and the total number of particles (protons plus neutrons) the *mass number*. The element iodine has an atomic number

53 and mass number 127, since its nucleus contains 53 protons and 74 neutrons. The *atomic weight* of the element is 126.91, the slight discrepancy being due to the presence of isotopes of slightly different mass number in the natural material (see p.185).

Magnetism

The attractive power of the magnet, either made from iron or the magnetic mineral magnetite, was well known to the Greek philosophers of the sixth century BC, and was described by Lucretius[44] in the first century BC. Chinese literature,[45] covering the period between the third century BC and the sixth century AD, contains many references to the effect of the lodestone (magnetite) on iron, and quantitative measurements were being made by the fifth century AD.

The application of magnetism to the compass, i.e. the realization that the earth itself is a magnet, is more obscure. Ptolemy mentions islands that attracted ships with iron nails. While attraction seems to have been accepted, the idea of magnetic repulsion of like poles, although known to natural philosophers like Pliny, seems to have been difficult to comprehend.

The first description of the magnetic compass is undoubtedly Chinese; we have a mention of a needle magnitized by rubbing with a lodestone and suspended on a silk thread in 1088. Furthermore, the idea of declination, i.e. the failure of a magnetic needle to align itself accurately along the north–south meridian, was also known at this time. However, the water compass and thermoremanence, i.e. the fact that a heated piece of iron quenched with its axis lying along the north–south direction becomes north–south seeking, was known from about AD 1044. By AD 1116 the declination was known to be about 15°E in China.

All this suggests that the magnetic compass was probably used for navigation somewhat earlier than the 11th century in China, and some would date the true discovery to the first century AD (Han period), when we have the mention of what is thought to be a compass consisting of a balanced spoon-shaped piece of lodestone capable of swinging on a polished bronze plate. This had been reconstructed and found to work. There are various other references in the intervening period to spoons pivoting on polished surfaces which are 'south seeking'.

We can safely conclude that, from 1088 onwards, the use of the compass on long voyages by Chinese ships was perfectly normal; its use in Europe was not mentioned until about 1190, which suggests that the knowledge of its use might have come from China.

The use of magnetite, soft iron, and steel for magnetic experiments continued through the centuries but it is only with the electromagnetic work of Faraday that intensive research into magnetic materials began, and a theory of magnetism evolved. Faraday distinguished between three sorts of magnetic materials,[19] for the most part metals: (*a*) diamagnetic materials, in which a rod became magnetized perpendicularly to the direction of the applied magnetic field; (*b*) paramagnetic materials which were magnetized parallel to the field; and (c) most important, the ferromagnetic group, which retained their magnetism when the field was removed. These last became paramagnetic at high temperatures, and it was therefore assumed that the ferro and paramagnetic materials were related and that diamagnetism was completely different. The transition temperature between ferro- and paramagnetism is now known as the 'Curie point'.

It is now accepted that ferromagnetic materials consist of elementary magnets which, when magnetized, all take up the same orientation. The three ferromagnetic metals are iron, cobalt and nickel, and these and the other ferromagnetic materials all have certain characteristics, amongst which are incomplete outer electron shells, low electron cloud densities near the nuclei, and a critical distance between the atoms so that the electron clouds do not overlap too much, nor are too far apart. The elementary magnets are, in effect, spinning electrons revolving in incompletely filled orbits.

By alloying, improvements can be made to the properties of the ferromagnetic materials. Today we need two types of material; magnetically soft, which does not retain its magnetism and which is necessary for electrical transformers, and magnetically hard materials which should have high remanence. This latter requirement used to be best fulfilled by quenched high-carbon steels, but these have been much improved since the end of the 19th century by large additions of tungsten, cobalt and chromium. In the 1930s still better alloys were discovered containing substantial amounts of nickel and aluminium, and directional properties could be conferred by cooling in strong magnetic fields. Improvements also took place in the magnetically soft materials, and now iron containing silicon is used for electrical laminated material and has markedly directional properties.

Radioactivity

One of the most important discoveries for science in general and metallurgy in particular was that of radioactivity by Henri Becquerel in 1896. Induced radiation and the emanation of X-rays had been discovered four years previously by Rontgen, who had obtained his X-rays by bombarding a metal target with high energy electrons. Becquerel felt that some metals or their compounds might be capable of sending out radiation similar to X-rays by themselves, and proved this by placing metallic compounds on photographic plates wrapped in black paper and waiting for several days. He found that some of his plates were blackened as though they had been exposed to X-rays. In such cases, he found that these compounds contained either uranium or thorium, both metals of very high atomic weight.

In 1898, Marie and Pierre Curie found that one sample of an ore known as pitchblende ($2 UO_3.UO_2$), from Joachimsthal in Czechoslovakia, emitted very much more energy than the uranium and thorium content of the sample would have led them to believe. It was obvious that this ore contained yet another radioactive element which was given the name radium. It soon became clear that radium was continuously disintegrating with the emission of energy which slowly decreased in quantity. In 1903, the Curies noticed that the temperature of the radium containing material was a little higher than the surroundings, thus confirming that energy was being released by the disintegrating matter itself. It was later found that the rate of emission at any given time was a constant proportion of the amount of matter remaining in the sample. For radium, this was decreased by 4% in a hundred years.

In the next hundred years this decrease would be 4% of the 96% remaining, and so on. If one plots the rate of emission (such as the degree of blackening of a photographic plate in a given time) against the lapse of time, one gets a smooth curve of the exponential type. The rate of emission from the different radioactive elements varies, and in order to make comparisons, it is normal to talk about a 'half-life', or the length of time that might elapse before the level of emission falls to one half of the original.

The value of this parameter for radium is 1600 years, but that for uranium and thorium is much longer—some thousands of million of years.

Of course, these discoveries had a profound effect on chemical thought. Dalton's theory of the atom was of an indestructible and permanent single particle. Perhaps Prout's theory of an atom built up of a number of similar units—hydrogen atoms—was nearer to the truth. But how could one explain their instability and the fact that the atomic weights of many elements were not whole number multiples of the atomic weight of hydrogen?

This was even more upsetting to the chemists who in the 18th century had finally rid themselves of the old fashioned alchemical ideas of transmutation. The work of the Curies and their contemporaries showed that uranium was changing into thorium, and radium into polonium, and finally into lead. So the atom had to be changed to something containing numbers of 'particles' that could be progressively lost. About 1912, a new atomic theory was put forward which explained why the atomic weights of so many elements were not accurate multiples of that of hydrogen but so close to it. The idea of the isotope was born; this was an atom of slightly different atomic weight which behaved chemically like the majority atoms of an element. An element was therefore made up of a mixture of several isotopes each of a slightly different atomic weight. In the case of uranium, most of the atoms have an atomic weight of 238, but there are some (0.7%) which have an atomic weight of 235, and others of atomic weight 239 all contributing to the overall atomic weight of 238.03 (hydrogen = 1.008).

We may now return to look at the present theory of the atom. Soon after the discovery of the electron by J. J. Thompson in 1897, a particle very much lighter than the nucleus, it was supposed that the atom itself consisted of a heavy nucleus surrounded by very much lighter particles. The hydrogen atom, being the lightest, was likely to have the simplest structure of any atom, that of a nucleus consisting of one proton with one electron orbiting the nucleus. All the other atoms should consist of more electrons and heavier nuclei in proportion to the atomic weight. Today we know more about the nucleus, and suppose it to consist of both protons and neutrons. It is extremely small, taking up a space of no more than $1/10\,000$th of the whole atom. It has a positive charge which is neutralized by the negative charge of the orbiting electrons. In chemical reactions, it is only the electrons that take part, and the characteristics of the elements in this respect are governed by the number and arrangement of the electrons.

The difference between one isotope and another obviously lies within the nucleus. While the nucleus contains both protons and neutrons having equal mass, the latter has no electrical charge. If we take chlorine as an example here, we find that the nucleus of one of its isotopes contains 18 neutrons and 17 protons, while the other isotope has 20 neutrons and 17 protons; both atoms contain 17 electrons orbiting the nucleus.

The neutron was discovered by Chadwick in about 1930. Since this has no charge it can be made to penetrate the negatively charged 'cloud' of electrons in a heavy element and invade the nucleus, which Fermi was able to show in 1934. This upset the delicate balance within the atom and produced a large number of radioactive isotopes of lighter elements, with a total weight less than the original, together with the liberation of an enormous amount of energy. This brings us to Einstein's relationship between mass and energy in which he showed that this disappearance of a small amount of mass is equivalent to the release of an enormous amount of energy. So, Fermi had proved the hypothesis of Einstein and converted matter into energy. This may not seem at first sight to be a new concept to the chemists for, after all, the combustion of carbonaceous fuels seems to be no more than this. However, such reactions affect only the electrons; the nuclei are not involved and the sum total of molecular weights is the same after as before.

This was fundamentally a new concept and, after its application to uranium by Frisch and Leitner, was called nuclear fission. The metal uranium had become a fuel for the production of energy and, as with bronze in earlier times and iron and steel today, it could be used as a tool for peace or as a weapon of war.

In order to use uranium as a source of energy a supply of neutrons must be produced to cause nuclear fission, but it was soon found that, under certain circumstances, the reaction could be made self-sustaining since neutrons are released in the fission process. If enough of the unstable uranium isotope ^{235}U is brought together, so that the neutrons resulting from the breakdown of one nucleus are not lost but slowed down and made available for collision with another ^{235}U nucleus, such a self-sustaining reaction can be produced (see Fig. 149). In the natural uranium reactor, containing only 0.7% of ^{235}U atoms, the neutrons are slowed down by embedding rods of the metal in a lattice work of graphite. The rods are 'canned' in order to contain the fission products of the reaction and prevent them from con-

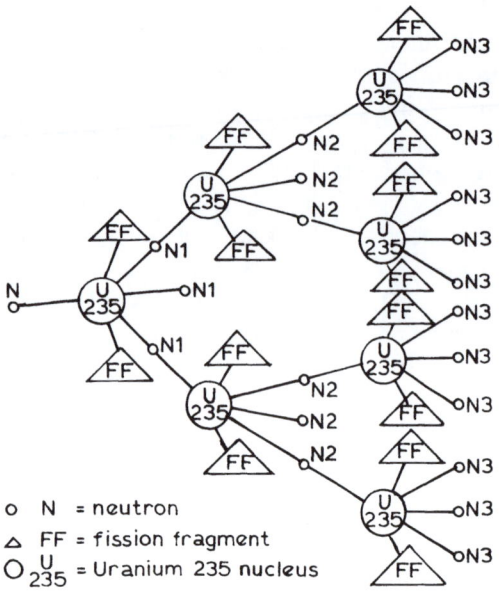

o N = neutron
△ FF = fission fragment
O $^{U}_{235}$ = Uranium 235 nucleus

149 Diagram illustrating the nuclear fission chain reaction in uranium

taminating the reactor as a whole. These cans must be permeable to neutrons without too much loss and, to date, have been made from the light metals such as aluminium, magnesium and zirconium.

The heat produced by this controlled reaction is removed by means of a coolant, which can be gas or liquid and utilized to heat air or convert water into steam in a boiler.

The fission reaction requires initiation by a neutron from within or outside the system which strikes the nucleus of a ^{235}U atom dividing it into two parts, giving off several new neutrons. The fission fragments could be two more metals, isotopes of molybdenum (^{95}Mo) and lanthanum (^{139}La), and the resultant mass after the reaction is slightly less than that before, and the difference represents the mass which has been converted into energy according to Einstein's law. At least two neutrons are available to collide with further ^{235}U nuclei and sustain the reaction. One ounce (28 g) of normal uranium, only 0.7% of which is ^{235}U, is theoretically capable of releasing 640 000 kWh (electrical units) of energy or about three million times the same weight of coal. In fact, the radioactive fission fragments accumulate and poison the reaction so that before all the ^{235}U is converted the uranium rods have to be removed from the reactor and purified.

Many other isotopes are produced in fission reactions and one of these, plutonium, has a half-life long enough for us to determine its mechanical properties, and thus we can say that man has not only been able to produce isotopes of metals unknown to nature but entirely new elements. Plutonium 239 arises from uranium 238 which, as we have seen, is the majority of the natural uranium. This can be removed comparatively easily during the refining process and used like ^{235}U to produce further energy; its half-life is 24 000 years.

While natural uranium has been used as a nuclear fuel, its use has been fraught with difficulties. Structur-

ally, it is like iron with many different allotropes (crystallographic forms) stable at different temperatures, and this has meant that it cannot be used at as high a temperature as one would wish, considering that the efficiency of steam usage increases with temperature. Furthermore, at high temperatures, reactions occur between the uranium and the can which, if allowed to develop to their ultimate conclusions, would lead to the mechanical failure of the can and the escape of radioactive fission products. So, the maximum temperature of the early power producing reactors was limited to 400°C which is below that of normal coal or oil-burning steam raising plant, and therefore the efficiency is lower. This fact has lead to the development of a mechanically more stable uranium fuel, uranium dioxide (UO_2), which is strictly a non-metal. Because its active ^{235}U atoms are now further diluted by oxygen atoms, it has been necessary to artificially increase the number of ^{235}U atoms to obtain a high enough concentration to achieve a self-sustaining reaction. The canning material to withstand the higher temperatures is now stainless steel which, because of its high mechanical strength as compared with magnesium, can be made somewhat thinner and thus allow the neutrons to pass through it sufficiently easily. So, historically, we have already passed through the metallic-fuel using stage to the use of what is virtually a ceramic.

This is a tendency which we see in a number of fields as the demands of modern society increase in the direction of materials which must stand higher and higher temperatures. Just as carbon steel was once hailed as the best tool material following bronze and stone, and then surpassed by Mushet's special steel, now we look to tungsten carbide, aluminium oxide, and other ceramic materials to replace them. It would seem that metallurgy has returned to its origins in a lithic–ceramic environment; but from now on the two, metal and non-metal, will be used together for the good of mankind.

References

1   C. S. SMITH: 'A history of metallography', 1960, Chicago, Chicago University Press
2   J. G. HAWTHORNE and C. S. SMITH (ed. and trans.): 'On divers arts; the treatise of Theophilus', 1963, Chicago, Chicago University Press
3   C. S. SMITH (ed.): 'Sources for the history of steel, 1532–1786', 1968, London, Society for History of Technology and MIT Press
4   C. S. SMITH: op. cit., 89
5   R. A. F. de REAUMUR: 'L'art de convertir le fer forgé en acier et l'art d'adoucir le fer fondu, 1722, Paris
6   A. G. SISCO: 'Réaumur's memoirs on steel and iron', Chicago, 1956, Chicago University Press
7   LU DA: Acta Met. Siniatica, 1966, 9, 1; see also Durrer Festschrift, 1965, 68
8   D. DIDEROT and J. D'ALEMBERT: 'Encyclopédie'
9   T. DESAGULIERS: Phil. Trans. Roy. Soc., 1724, 33, 345
10   P. VAN MUSSCHENBROEK: 'Introductio ad coherentium corporum furnorum', 1729, Leyden
11   These are the small 'assay' pounds; 100 of these are equivalent to 3–22 g
12   C. GUYTON and M. DARCET: Phil. Mag., 1799, 3, 400

13   C. S. SMITH: Four outstanding researches in metal-lurgical history', 35 pp, 1963, **ASTM**

14   R. T. GUNTER: *Nature*, 1939, **143**, 667

15   G. PEARSON: *Phil. Trans. Roy. Soc.*, 1795, **85**, 322

16   M. FARADAY: *Quart. J. of Sci.*, 1819, **7**, 288

17   J. STODART and M. FARADAY: *ibid.*, 1820, **9**, 319

18   J. STODART and M. FARADAY: *Phil. Trans.*, 1822, **112**, 253

19   SIR R. A. HADFIELD: 'Faraday and his metallurgical researches', 1931, London, Chapman and Hall

20   M. BREANT: *Ann. des Mines*, 1824, **9**, 319

21   N. T. BELAIEW: *Rev. Mét.*, 1914, **11**, 221

22   D. K. TSCHERNOFF: *Proc. Inst. Mech. Engrs.*, 1880, 152

23   N. T. BELAIEW: *Mét. et CiviL*, 1950, **1**, 10

24   R. W. BUNSEN and LORD LYON PLAYFAIR: 'The gases evolved from iron furnaces with reference to the theory of the smelting of iron', 1845, Cambridge, British Association; reprinted by the Iron and Steel Institute as a separate report in 1903, 76pp

25   A.G. QUARRELL: 'Metallography' 1963, 1, 1964, London, The Iron and Steel Institute

26   C. S. SMITH: *op. cit.*, 207

27   Anon: *EA News*, 1956, **35**, (408),129

28   SIR R. HADFIELD: *Chem. and Ind*, 1925, **44**, 1

29   M. COHEN and J. M. HARRIS: The Sorby Centennial Symposium On the History of Metallurgy, (ed. C. S. Smith), 209, 1965, London, Gordon and Breach

30   E. OSMOND: *Bull. Soc. d Encouragement pour l'Industrié Nationale*, 1895, **10**, (4), 480

31   W. L. BRAGG: *Phil. Mag.*, 1914, **28**, 355

32   Much of this was due to the efforts of William Hume-Rothery and his colleagues at Oxford

33   H. Y. HUNSICKER and H. C. STUMPF: 'The Sorby Centennial Symposium on the History of Metallurgy', (ed. C. S. Smith), 209,1965, London, Gordon andBreach

34   W. C. ROBERTS-AUSTEN: *Proc. Roy. Soc.*, 1900, 67, 101

35   W. C. ROBERTS-AUSTEN: *Phil. Trans. Roy. Soc.*, 1896, **187**, 383

36   H. C. H. CARPENTER and C. F. ELAM: *Proc. Roy. Soc.*, 1921, 100, 329

37   C. F. ELAM: *J. Inst. Metals*, 1930, **43**, 217

38   C. E. INGLIS: *Trans. Inst. Naval Arch.*, 1913, **55**, 219

39   A. A. GRIFFITH: *PhiL Trans. Roy. Soc.*, 1921, **221A**, 163

40   SIR L. BRAGG and J. F. NYE: *Proc. Roy. Soc.*, 1947, **190A**, 474

41   J. NUTTING: 'Metallography 1963', 154, 1964, London, The Iron and Steel Institute

42   D. A. MELFORD: *ibid.*, 206

43   SIR H. E. ROSCOE: 'John Dalton and the rise of modern chemistry', 1901, London, Cassell

44   LUCRETIUS: 'De Rerum Natura', (ed. C. Bailey), Vol. III, 908,1947, Oxford, Oxford University Press

45   J. NEEDHAM: 'Science and civilization in China', vol. IV, pt. I, Cambridge University Press

Appendixes

APPENDIX 1
Technical Glossary

Alembic The head of a distillation apparatus, which fits over the retort or cucurbit, containing the delivery spout which connects with the condenser.

Alpha Brass Copper alloy containing up to 30% zinc.

Alpha-delta eutectoid A hard constituent normally present in the structure of cast bronze containing more than about 6% tin.

Annealing The process of softening a metal hardened by cold working (e.g. hammering). The lowest temperature at which a metal will soften varies with the degree of old working, greater amounts of work tending to reduce

Argentojarosites Silver-rich clay-like minerals which sometimes appear in the secondary enrichment zone of metal deposits.

Argol The reddish deposit found at the bottom of wine vats; essentially it is a crude sodium and potassium bitartrate. Austenite A non-magnetic form of iron normally existing only at high temperatures (above about 720°C). Carbon can dissolve in it up to about 1.8% at 1 150°C and diffuse readily.

Azurite A blue-green basic carbonate of copper ($2CuCO_3$. $Cu(OH_2)$).

Beaker A type of flat-bottomed clay vessel, with a wide distribution during the Copper Age.

Bloom or bloomery iron Iron that has been produced in a solid condition directly as the result of the reduction (e.g. smelting) of iron ore. Pure iron melts at 1 535°C, but bloomery iron usually has not normally been heated above about 1 250° C. The carbon content is variable but usually low. High carbon bloomery irons have properties similar to modern carbon steels

'Blowing in' The starting up of a blast furnace.

Brazing The joining of two pieces of solid metal by means of a molten alloy of copper and zinc (brass); in modern usage this has been extended to include a wider range of molten metals.

Calamine A mineral containing zinc; in antiquity it was used to signify the carbonate $ZnCO_3$, now called smithsonite.

Carbon[14](^{14}C) isotope *see* Radioactive isotopes.

Cassiterite Tin oxide SnO_2.

Cemented blister bar Iron bar carburized in a cementation furnace, during which process slight evolution of carbon monoxide gives rise to blistering.

Cementing With reference to iron, this is the pick-up of carbon from a carbonaceous medium such as charcoal. In the case of copper (cement copper) it means the precipitatiorl of fine copper on scrap iron from copperbearing solutions.

Cementite A compound or carbide of iron with the formula Fe_3C. Very hard and brittle, forming one of the constituents of pearlite. It also appears as a separate constituent in the grain boundaries of wrought iron containing about 0–0.2% carbon, and in irons containing more than 0.89% carbon. In the latter case it may produce a Widmanstatten structure. It also appears in white cast iron in the pearlite and as a separate constituent. The carbon in cementite is normally referred to as 'combined carbon' to distinguish it from the form of carbon known as graphite.

Cohenite Iron carbide

Collodion plate An early form of photographic plate in which the silver salts were dispersed in a solution of cellulose nitrate in alcohol and ether.

Cope The top part of a two-part mould; in some cases the lower part is a core (v.i.) but it is more common to have a top and a bottom part (drag).

Core Piece of a mould inserted in such a way as to give a hollow in the final casting.

Coring; cored Term used to describe the segregation which occurs during solidification of metallic crystals in an alloy melt. The melt has a uniform composition in the liquid state but segregation (separation) often occurs during cooling, resulting in a clearly visible structure under the microscope.

Cucurbit A retort with a narrow neck which fits into the alembic (see above).

Cupellation The removal of lead by oxidation for the recovery of silver.

Dendrites A fern- or leaf-like growth formed by a solid metal or constituent growing from the liquid. Many pure metals and alloys solidify in this way, as do some constituents of slags, such as wustite and magnetite in fayalite.

Die A matrix usually made of metal for the manufacture of decorative sheet or coins.

Eddy current effect The heating effect of an alternating electric current due to hysteresis.

Electrum A whitish gold–silver alloy, containing more than 40% silver.

Equiaxed Term applying to crystals which are roughly as broad as they are long.

Etching Developing the structure.of a metal by attacking it with acid or other solutions.

Fahlerz An ore from the secondary enrichment zone of a copper deposit comparatively rich in arsenic, antimony, and silver.

Fayalite Iron silicate.

Fenestrated axe A shafthole axe with semicircular openings.

Flan An intermediate product from which coins are struck.

Flapping The stirring of the surface of a bath of molten copper for the removal of impurities by selective oxidation.

Flux Lime or other material added to the smelting charge to render a slag easy-flowing.

Gamma iron *see* Austenite.

Gangue Unwanted mineral.

Gossan That part of a metalliferous deposit from which the wanted metal has been leached and which is rich in iron.

Gunmetal A ternary alloy of copper, zinc, and tin; modern gunmetals usually contain less zinc than tin, but some contain equal proportions together with lead, and have a composition such as 85.5.5.5.

Hardness The hardness of metals is usually measured by indentation tests. The hardness is estimated from the size or depth of an indentation made with a loaded ball or pyramidal diamond. Two popular scales of hardness are the diamond pyramid (HV) and the Brinell (HB). Between 0 and 300 these systems are roughly equivalent. For hardnesses above 300, only the diamond pyramid system has been used in this work.

Hematite Oxide of iron (Fe_2O_3); normally red, occasionally black.

Kish graphite Free graphite flakes which can sometimes be found floating on the surface of cast iron when it leaves the blast furnace or cupola.

Leaching The removal of elements from the soil or a metal deposit by aqueous solutions.

Leaded bronzes Copper-tin alloys containing lead.

Limonite Mixture of hydrated and other oxides of iron, e.g. goethite (FeO.OH) and hematite.

Litharge Lead oxide, PbO.

Martensite A hard product produced by quenching iron containing carbon from temperatures above 720°C. The hardness depends to some extent on the carbon content. In order to produce a structure containing only martensite, the temperature must be above a certain figure, depending on the carbon content, and the rate of quenching must be extremely high. Suitable rates are obtained by quenching into cold water or brine.

Matte A compound of metals and sulphur, often produced in the first state of smelting copper in which case it is a mixture of iron and copper sulphides.

Mild steel Modern equivalent of wrought iron but without the slag which gives the latter its fibrous structure.

Neutron Type of particle present in the atomic nucleus which does not carry a charge.

Niello A mixture of sulphides, usually of copper and silver, which is used as a black decorative inlay on silver and some other metals.

Paktong Corruption of Chinese 'Paitung' used for alloys containing zinc. The other elements are usually copper and nickel; this alloy is synonymous with a nickel-silver or German silver.

Panning A mineral washing process in which the lighter unwanted mineral is removed from the wanted mineral in a shallow vessel (or pan).

Pearlite A structure in the iron-carbon system consisting of alternate laminations of ferrite and cementite.

Placer deposits Deposits released by the weathering of rocks, and concentrated by water action.

Poling Plunging a pole of wood into a bath of molten copper for the purpose of reducing its oxygen content by the evolution of hydrogen and other gases evolved during the distillation of the wood.

Proton Type of particle present in the atomic nucleus carrying a positive charge.

Quench hardening The hardening of steel by plunging it at a red heat into cold liquid such as water, brine, or oil.

Quern stones Stones used for grinding grain.

Radioactive isotopes Elements consist of a number of isotopes which behave in the same way chemically but which have slightly different internal structures. Some of these isotopes are radioactive and disintegrate at a steady and measurable rate, expressed as their half-life which is the length of time in which half the radioactivity has diminished. From the archaeological point of view the most important of these is the isotope carbon ^{14}C. This isotope is taken up by carbon-using material, such as plants and animals, from the carbon dioxide in the earth's atmosphere and, when so fixed in their remains, disintegrates in such a way that the amount remaining is a measure of the date at which the carbon was fixed in the growing plant or animal.

Reamer A tool for cleaning out a hole bored by a drill or auger.

Reverberatory A type of furnace in which the heating flame is reflected on to the charge from the roof.

Retort A vessel in which minerals are heated for the recovery of volatile components.

Riser The hole through which air is vented and the metal rises when it is poured through the runner.

Runner (or Sprue) That part of a mould into which metal is poured. The metal filling these channels is often found amongst the scrap metal.

Rutile Titanium oxide.

Shear steel A mixture of steel and iron designed to give a hard cutting edge without embrittlement.

Smelting Involves a chemical reaction between the ore and the fuel, or between a heated sulphide ore and the atmosphere. Most smelting processes are carried out above the melting point of the metal concerned, the main exception being iron.

Speiss A residue of lead or copper smelting containing a high proportion of metallic arsenides, i.e. compounds of arsenic.

Spiegeleisen Literally mirror iron; cast iron containing a high proportion of manganese which stabilizes the carbon as cementite and which gives it a white fracture.

Stannite Tin sulphide, SnS solid solution with Cu and Fe to give Cu_2FeSnS_4.

Steatite A soft silicate which carves easily; also called soapstone.

Strickle board Wooden board used in the moulding of a circular component; its edge is shaped to the profile of the object and rotated around it.

Tang Projection of a knife or spearhead that goes into the handle.

Tap slag Slag run off from a furnace in liquid condition.

Troilite Iron sulphide with specific reference to meteorites.

Troostite A constituent in the iron-carbon system in which cementite and ferrite have a radial distribution forming spherulites; formed by slow quenching (oil).

Trunnions Pivots which allow a vessel to swing, or a core to be suspended in a mould.

Tutty A white oxide of zinc, ZnO, given off when smelting ores rich in Zn.

Tuyere Tube for blowing air into a furnace.

Twins Faults in crystals which show that the structure has at one time been strained, mostly by hammering or bending.

Vacant lattice sites A point in the regular atomic order of a metal where an atom is missing.

Widmanstatten structure The structure occurring in steels which have been fairly rapidly cooled from high temperatures (about 1 000°C). Precipitation of ferrite or cementite takes place along certain crystal planes forming a mesh-like arrangement. The same type of structure occurs in the octahedrite meteorites.

Work hardening Metals, when hammered at low temperatures, become hardened and stronger. If the temperature of working is increased, a point is reached at which hardening no longer occurs, i.e. the hot-working temperature is reached. The dividing line between hot and cold working for lead is about room temperature; for pure iron it is about 600°C.

APPENDIX 3
Table of elements

Element	Symbol	Specific Gravity	Melting point, °C
Aluminium	Al	2·70	659·7
Antimony	Sb	6·62	630·5
Arsenic	As	5·73	
Beryllium	Be	1·8	1278 ± 5
Bismuth	Bi	9·75	271·3
Boron	B	3·33	
Cadmium	Cd	8·65	320·9
Calcium	Ca	1·54	
Carbon	C	3·52	
Cerium	Ce	6·79	640
Chlorine	Cl		
Chromium	Cr	6·73	1890
Cobalt	Co	8·71	1495
Copper	Cu	8·95	1083
Fluorine	F		
Germanium	Ge	5·46	958·5
Gold	Au	19·3	1063
Hydrogen	H		
Indium	In	7·28	156·1
Iron	Fe	7·88	1535
Lead	Pb	11·34	327·4
Lithium	Li	0·53	186
Magnesium	Mg	1·74	651
Manganese	Mn	7·42	1260
Mercury	Hg	13·60	−38·9
Molybdenum	Mo	9·01	2620 ± 10
Nickel	Ni	8·9	1455
Niobium	Nb	8·4	1950
Nitrogen	N		
Oxygen	O		
Phosphorus	P	1·83	44·1
Platinum	Pt	21·37	1773·5
Potassium	K	0·87	62·3
Selenium	Se	4·8	217
Silicon	Si	2·42	1420
Silver	Ag	10·53	960·8
Sodium	Na	0·97	97·5
Sulphur	S	2·1	112·8
Tellurium	Te	6·25	452
Thallium	Tl	11·86	303·5
Tin	Sn	7·29	231·9
Titanium	Ti	4·5	1800
Tungsten	W	19·1	3370
Vanadium	V	5·69	1710
Zinc	Zn	7·16	419·5
Zirconium	Zr	6·44	1900

APPENDIX 2:
Notes on units of weight, stress, and hardness

In the interest of uniformity and to aid comparison, all original weights have been converted to metric units as accurately as the context justifies. In general, this has meant that avoirdupois tons have become metric tonnes (t) without change. In some cases the 2% difference is significant and has been taken into consideration in the conversion.

Precious metals in the past, and to some extent at present, have been weighed in troy ounces and the amounts of precious metals in ores and in base metals given in troy ounces per avoirdupois ton. In all cases these have been converted to grammes per metric tonne by multiplying by 30.5.

As far as stresses and pressures are concerned SI (Système International) units have been used. For those unfamiliar with such units, 1 ton/in² is equivalent to 154 MN/m², and 1 lb/in² blast pressure is equivalent to 0.069 bar (6.9 kN/m²). Hardness is usually measured by an indentation test in which a specially shaped hard indenter is pressed into the metal surface under controlled conditions of time and load. 'HV' stands for the Vickers diamond pyramid hardness test; the numbers following these letters give the load applied in kg. The hardness figures before the letters are in kg/mm² and are generally comparable, irrespective of load. In some cases the hardness is given as 'HB', which stands for the Brinell system of measurement. The results in the range given are roughly comparable to those in 'HV'.

APPENDIX 4
Approximate date of start of metal ages (BC unless otherwise stated)

Archaeological ages	Chalcolithic		Early Bronze	Middle/Late Bronze	Early Iron	Roman Iron	Late Iron
Metal ages	Native copper†	Smelted copper	'Early Bronze'*	Full Bronze			
Anatolia ⎫ Troad ⎬ Egypt ⎭		5000	3000	2000	1000	30	
Palestine		3500	2900	1900	1000	100	
Mesopotamia	7000	4000	3500	2800	1200	AD 100	
Aegean		3300	2500	2000	1000	150	
Italy		3000	2000	1200	800	250	
Iberia		3000	1500	1000	700	200	
SE Europe	5000	4500	3000	1500	700	AD 100	
NW Europe		2200	1800	1200	500	50	AD 400
E Europe and Russia		2200	1500	700	400	–	
Far East			1500	1300	700	–	
West Africa					500		
East Africa (Nubia, Sudan)					200	AD 1	
South America		AD 700	AD 1000	AD 1400	–	–	
North America	4000						

Note: these dates are very approximate and are only offered as a guide; they are under constant revision; *Includes arsenical copper and low-tin bronzes; †earliest known use of

APPENDIX 5
Chinese chronology

Dynasty	Age	Date
Shang (or Yin) Early (Pre-Anyang)	EBA	1500–1300 BC
Shang Late (Anyang)	LBA	1300–1000
Western Chou	LBA	1027–771
Eastern Chou (Spring and Autumn)	EIA	770–475
Eastern Chou (Warring States)		475–221
Ch'in		221–206
Han		206 BC–AD 220
The Three Kingdoms		AD 221–264
Western Chin-Toba		265–580
Sui		581–618
Tang		618–906
Sung		906–1279
Mongol		1280–1368
Ming		1368–1644
Manchu		1644–1911

APPENDIX 6
Journals consulted and abbreviations

Acta Met. Sin.: Acta Metallurgia Siniatica, Peking

Act. Congr. Int. Sci: Actes du VIIᵉ Congrès International des Sciences Prehistoriques et Protohistoriques; Prague1966

AJA: American Journal of Archaeology, Baltimore, USA

AJS: American Journal of Science, Newhaven, USA

AMNH: American Museum of National History, New York

Ann. des Mines.: Annales des Mines, Paris

Antiquity: Antiquity, Cambridge, UK

Ant. J.: Antiquaries Journal; Journal of the Society of Antiquaries, London

Apulum: Cluj, Romania

Arch.: Archaeologia (Miscellaneous Tracts relating to Antiquity), Society of Antiquaries, London

Archaeology: The Archaeological Institute of America, New York

Arch. Ael.: Archaeologia Aeliana, Society of Antiquaries, Newcastle upon Tyne

Archeom.: Archaeometry, Cambridge, UK

Arch. Aust.: Archaeologica Austriaca, Vienna

Arch. Camb.: ArchaeologiaCambrensis,Cardiff

Arch. Cant.: Archaeologia Cantiana, Kent Archaeological Soc., Maidstone

Arch. Eisenh.: Archiv für das Eisenhüttenwesen, Düsseldorf Verein Deutscher Eisenhüttenleute

Arch. J.: Archaeological Journal, Royal Archaeological Institute of Great Britain and Ireland, London
ASM.: American Society of Metals, Metals Park, Ohio, USA
Atti CISPP: Atti, VI Congresso Internazionale delle Scienze Preistoriche E Protostoriche, Roma
Brit. Assoc.: Reports of the British Association for the Advancement of Science, London
BRGK: Bericht der Römisch-Germanischen Kommission, Frankfurt
BSA: Annual of the British School at Athens, London
Bull. Chem. Soc. Japan: Bulletin of the Chemical Society, Japan
Bull. HMG: Bulletin of the Historical Metallurgy Group, London
Bull. Inst. Metallurgists: Bulletin of the Institution of Metallurgists, London
CAH: Cambridge Ancient History, Cambridge, UK
Chem. Ind.: Chemistry and Industry, London
Copper NA: Copper in Nature, Technics, Art and Economy, Norddeutscher Affinerie, Hamburg, 1966
Copper: Copper Development Association, London
Corn. Arch.: Cornish Archaeology, Camborne
Current Anthrop: Current Anthropology, Chicago, USA
Durrer Festschrift: Vita pro Ferro, Festschrift für Robert Durrer, 1965, Schaffhausen
EA News: Edgar Allen News, Sheffield
Econ. Hist. Rev.: Economic History Revue, London
Eng. and Mining J.: Engineering and Mining Journal, New York
Eng. Hist. Rev.: English Historical Review, London
FHS: Flintshire Historical Society, UK
FTJ: Foundry Trade Journal, London
Gallia: Gallia Préhistoire, Paris
Hesperia: Journal of the American School of Classical Studies at Athens, Cambridge, Mass.
Ind. Arch.: Industrial Archaeology, Newton Abbot, Devon
Inst. Arch.: Annual Report of London University Institute of Archaeology
Iran: Journal of British Institute of Persian Studies, Teheran
Iraq: Iraq, Journal of the British School of Archaeology in Iraq, London
Israel Excav. J.: Israel Excavation Journal, Jerusalem
Israel Explor. J.: Israel Exploration Journal, Jerusalem
J. Amer. Chem. Soc.: Journal of the American Chemical Society, Washington, DC
J. Chem. Educ.: Journal of Chemical Education, Easton, Pennsylvania, USA
J. Chem. Soc.: Journal of the Chemical Society, London
J. Chem. Soc. Japan: Journal of the Chemical Society of Japan
JDAI: Jahrbuch des Deutschen Archäologischen Institutes, Berlin
JEA: Journal of Egyptian Archaeology, London
J. Econ. Hist.: Journal of Economic History, New York
JHMS: Journal of the Historical Metallurgical Society, UK
J. Inst. Metal: Journal of the Institute of Metals, London
J. Iron Steel Inst.: Journal of the Iron and Steel Instittlte, London
JRAI: Journal of the Royal Anthropological Institute, London
J. Roy. Inst. Cornwall: Journal of the Royal Institution of Cornwall, Truro
J. Roy. Soc. Arts: Journal of the Royal Society of Arts, London
JSAP: Journale de Societé des Americanistes de Paris
J. West Mid. Reg. Studies: Journal of the West-Midland Regional Studies, Wolverhampton
Kuml: Aarhus, Denmark
Mainzer Zeit.: Mainzer Zeitschrift, Mainz

Man: Man, Royal Anthropological Institute, London
MAGW: Mitteilungen der Anthropologischen Gesellschaft in Wien
Mariners' Mirror: Journal of the Society of Nautical Research, Cambridge, UK
Med. Arch.: Medieval Archaeology, London
Mét et Civil.: Métaux et Civilisations, Nancy
Met. Ital: Metallurgia Italiana, Milano
Metals and Materials: The Metals Society, London
Metall und Erz: Berlin.
Metaux-Corrosion-Ind.: Metaux Corrosion Industrie, St. Germain en Laye
Metr. Mus.: Metropolitan Museum, New York
Mikrochim. Acta.: Mikrochimica Acta, Wien
MM: Mining Magazine, London
NA: Norfolk Archaeology, Norwich
Nach. Eisen-Biblio.: Nachrichten aus der Eisen-Bibliothek der Georg Fischer AG Schaffausen
Nat. Tech. Mus. Prague: National Technical Museum, Prague
Nature: Weekly Journal of Science, London
North East Ind. Arch. Soc.: North East Industrial Archaeology Society, Teesside
Num. Chron.: Numismatic Chronicle, London
Oxoniensa: Oxford
Palaeohistoria: Groningen
PBA: Proceedings of the British Academy, London
PBNHPS: Proceedings of the Belfast Natural History and Philosophical Society, Belfast
PEQ: Palestine Exploration Quarterly, London
Phil. Mag.: Philosophical Magazine, London
Phil. Trans. Roy. Soc.: Philosophical Transactions of the Royal Society, London
Post Med. Arch.: Post Medieval Archaeology, London
PPS: Proceedings of the Prehistoric Society, Cambridge, UK
Przeglad Arch.: Przeglad Archeologiczny, Poznan
PRIA: Proceedings of the Royal Irish Academy, Dublin
PZ: Praehistorische Zeitschrift, Berlin
Proc. Brit. Acad.: Proceedings of the British Academy, London
Proc. Man. Lit. and Phil. Soc.: Proceedings of the Manchester Literary and Philosophical Society
Proc. Inst. Mech. Engrs.: Proceedings of the Institute of Mechanical Engineers, London
Proc. Roy. Soc.: Proceedings of the Royal Society, London
PSAS: Proceedings of the Society of Antiquaries of Scotland, Edinburgh
P. Thoresby Soc.: Proceedings of the Thoresby Society, Leeds
PUDPS: Proceedings of the University of Durham Philosophical Society
Quart. J. of Sci: Quarterly Journal of Science, London
RA: Revue Archéologique, Paris
Rev. Al.: Revue de l'Aluminium, Paris
RAI: Royal Anthropological Institute, London
Rev. d'Hist. des Mines Mét.: Revue d'Histoire des Mines et Métallurgie, Geneva
Rev. Mét.: Revue de Métallurgie, Paris
RCZM: Römisch-Germanisches Zentralmuseum, Mainz
RHS: Revue d'Histoire de la Sidérurgie, Nancy
Roy. Num. Soc. Mem.: Royal Numismatic Society Memoir, London
Science: American Association for the Advancement of Science, Washington DC
SE: Studi Etruschi, Firenzi, Italy
Sibrium: Centro di Studi Preistorice ed Archeologici Varese, Italy
Slovenska Arch.: Slovenska Archeologice, Bratislava

Soviet Arch.: Soviet Archaeology, Moscow

Stahl und Eisen: Verein Deutscher Eisenhüttenleute, Düsseldorf

TCWAAS: Transactions of the Cumberland and Westmorland Antiquarian & Archaeological Society 12

Tech. et Civil: Techniques et Civilisations, Paris

Tech. Cult.: Technology and Culture, USA 13

THSLC: Transactions of the Historical Society for Lancashire and Cheshire 14

TLSSAS: Transactions of the Lichfield and South Staffs. Archaeological Society 15

TNS: Transactions of the Newcomen Society, London

Trab. Antrop. EtnoL Soc. Portug.: Trabalhos Antropologia e Etnologia Soc. Portugal 16

Trans. AIMME: Transactions of the American Institute of Mining and Metallurgical Engineers, New York 17

Trans. Amer. Phil. Soc.: Transactions of the American Philosophical Society, Philadelphia 18

Trans. Birmingham Arch. Soc.: Transactions of the Birmingham Archaeological Society 19

Trans. Fed. Inst. Min. Engrs.: Transaction of the Federated Institution of Mining Engineers, Newcastle upon Tyne 20

Trans. Inst. Min. Met.: Transactions of the Institution of Mining and Metallurgy

Trans. Inst. Welding: Transactions of the Institute of Welding, London 21

Trans. Roy. Hist. Soc.: Transactions of the Royal Historical Society, London 22

TWS: Transactions of the Woolhope Naturalists' Field Club, Hereford 23

Ulster Arch. J.: Ulster Archaeological Journal, Belfast

Ur-Schweiz: Ur-Schweiz, Mitteilungen zur Ur- und Frühgeschichte der Schweiz, Basel 24

VCH: Victoria County History, London 25

WMF: Welding and Metal Fabrication, London

World Arch.: World Archaeology, London

Zeit. Numismatik.: Zeitschrift Numismatik, Berlin

Zephyrus: Portugal

Z. Otchlani Wiekow: Warsaw

Erzeugung technischen Eisens, Freiberger Forschungshefte, Kultur und Technik, 1954, D.6. Berlin

L. BECK: 'Geschichte des Eisens', (5 vols.) 1884-1903, Braunschweig

J. NEEDHAM: 'Science and civilisation in China', 1976 Vol.5, pt 3, Cambridge University Press

J. R. MARECHAL: 'Prehistoric metallurgy', 1962, Lammersdorf, Otto Junker GMBH

E. R. CALEY: 'Analysis of ancient metals', 1964, Oxford, Pergamon

R. J. FORBES: 'Studies in ancient technology', vols. 7-9,1964, Leiden, Brill

R. F. TYLECOTE: 'Metallurgy in archaeology', 1962, London, Edward Arnold

C. D. DODWELL (ed.): 'Theophilus; De Diversis Artibus', 1961, London, Nelson

J. G. HAWTHORNE and C. S. SMITH: 'On Divers Arts—the treatise of Theophilus', 1963, Chicago, Chicago University Press

C. S. SMITH (ed.): 'The Sorby Centennial Symposium on the History of Metallurgy', 1965, London, Gordon and Breach

G. JARS: 'Voyages Métallurgiques', 3 vols., 1774-81, Lyon

P. A. DUFRENOY et al.: 'Voyage Métallurgique en Angleterre', 2 vols, 1837, Paris

D. DIDEROT and J. d'ALEMBERT: 'Encyclopédie ou Dictionaire raisonné des Sciences, des Arts et des Metiers', 1771-80, Paris

J. D. MUHLY: 'Copper and tin', *Trans. Connecticut Academy of Arts and Sciences:* 1973, **43**, 155

B. ROTHENBERG: 'Timna; valley of the biblical copper mines', 1972, London, Thames and Hudson

Principal Works Consulted

1   H. H. COGHLAN: 'Notes on the prehistoric metallurgy of copper and bronze in the Old World', 1961, Oxford, Pitt Rivers Museum

2   H. H. COGHLAN: 'Notes on prehistoric and Early Iron in the Old World', 1956, Oxford, Pitt Rivers Museum

3   H. R. SCHUBERT: 'History of the British iron and steel industry from c.450 BC to AD 1775', 1957, London, Routledge and Kegan Paul

4   G. AGRICOLA: 'De Re Metallica', (trans. H. C. and L. H. Hoover), 1950, London (1st Edn. Mining Magazine, London, 1912)

5   V. BIRINGUCCIO: 'Pirotechnia' (trans. C. S. Smith and M. T. Gnudi). 1959, New York, Basic Books

6   L. AITCHISON: 'A history of metals', (2 vols.), 1960, London, Macdonald and Evans

7   J. PERCY: 'Metallurgy; fuel; copper; zinc; brass etc.', 1861, London, Murray

8   J. PERCY: 'Metallurgy; iron and steel', 1864, London, Murray

9   J PERCY: 'Metallurgy; lead', 1870, London, Murray

10   J. PERCY: 'Metallurgy; refractory materials and fuel' (revised edn. 1875), London

11   B. NEUMANN: Die ältesten Verfahren der

MAP 1 Aegean and the Near East

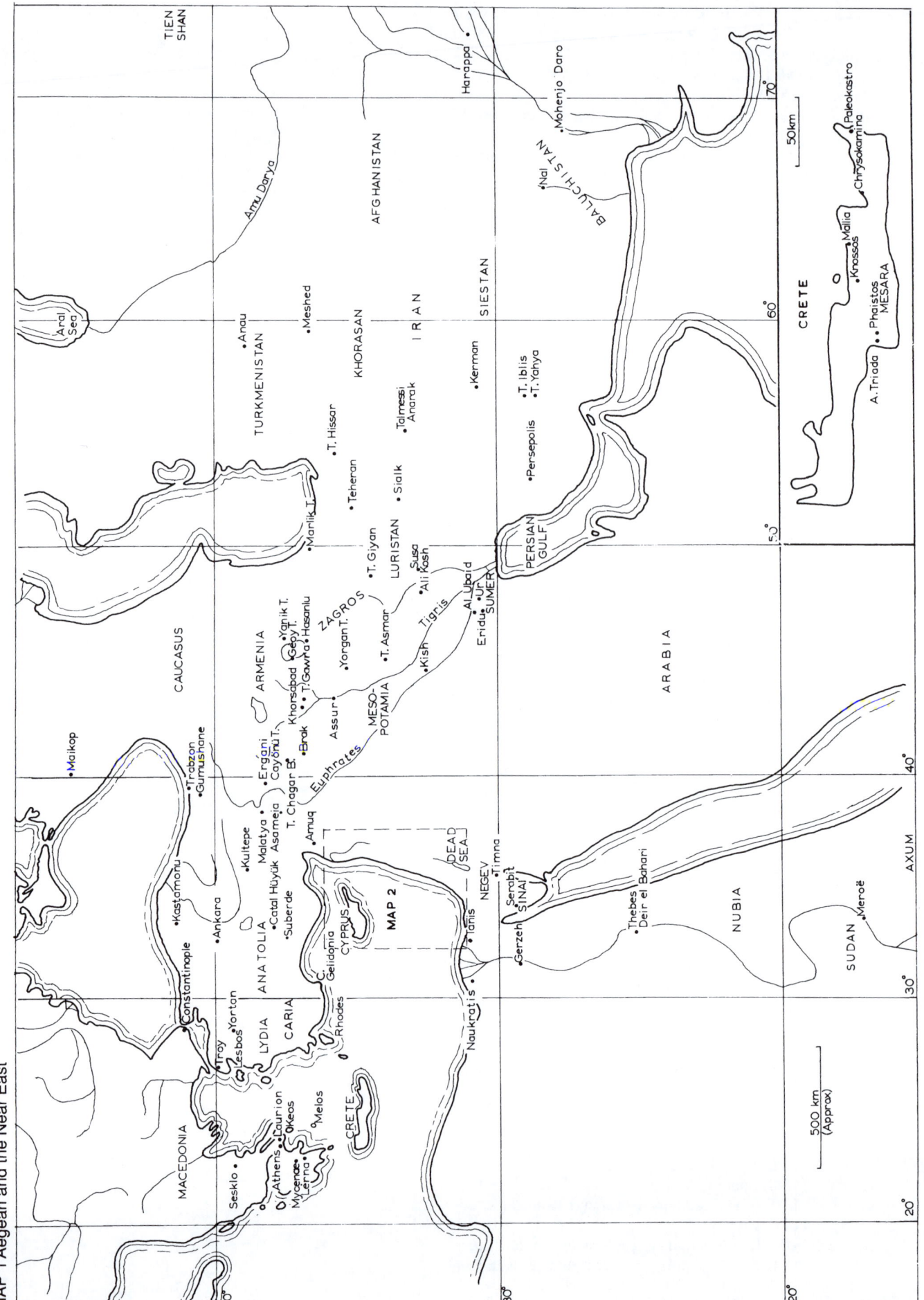

MAP 2 Palestine

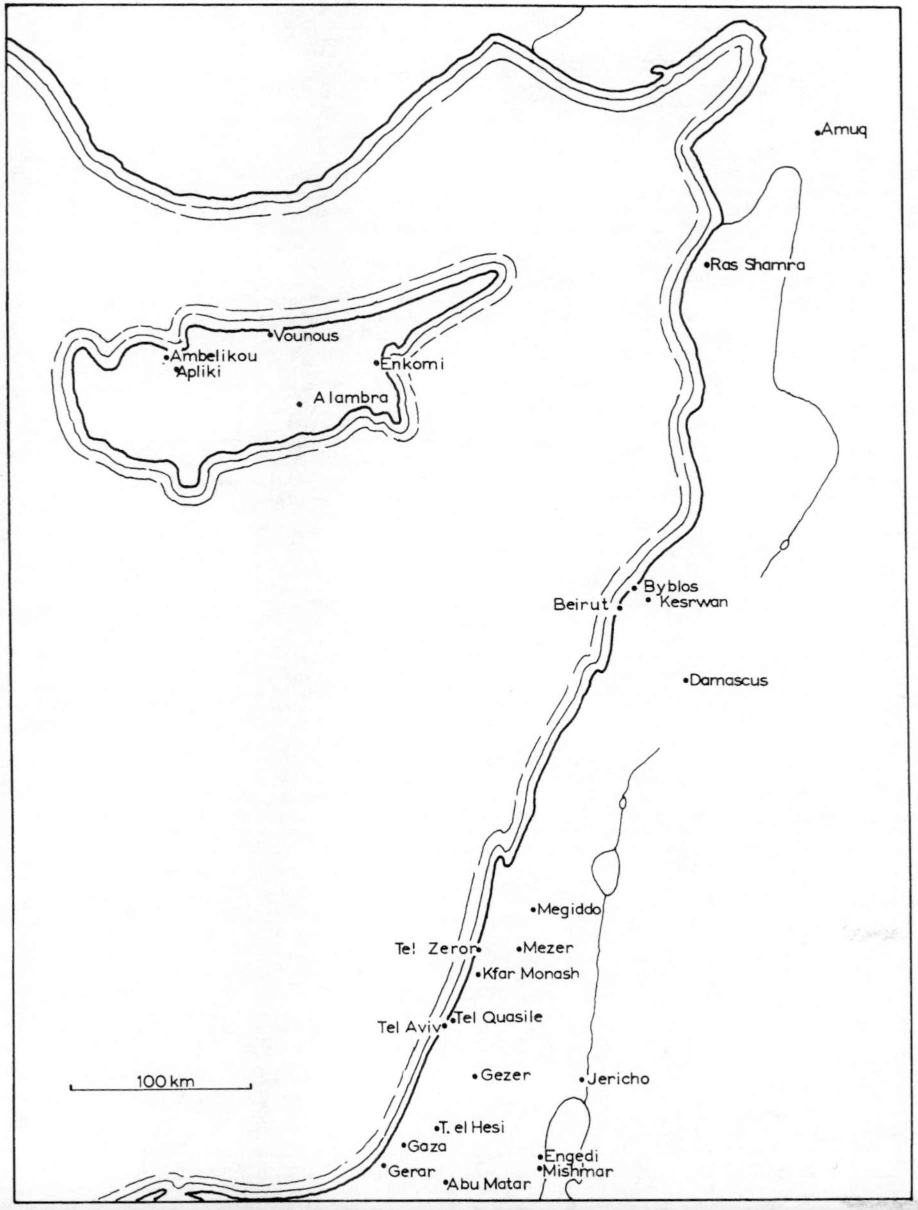

- Amuq
- Ras Shamra
- Byblos
- Beirut
- Kesrwan
- Damascus
- Megiddo
- Mezer
- Tel Zeron
- Kfar Monash
- Tel Aviv
- Tel Quasile
- Gezer
- Jericho
- T. el Hesi
- Engedi
- Mishmar
- Gaza
- Gerar
- Abu Matar
- Enkomi
- Vounous
- A lambra
- Ambelikou
- Apliki

100 km

MAP 3 India and the Far East

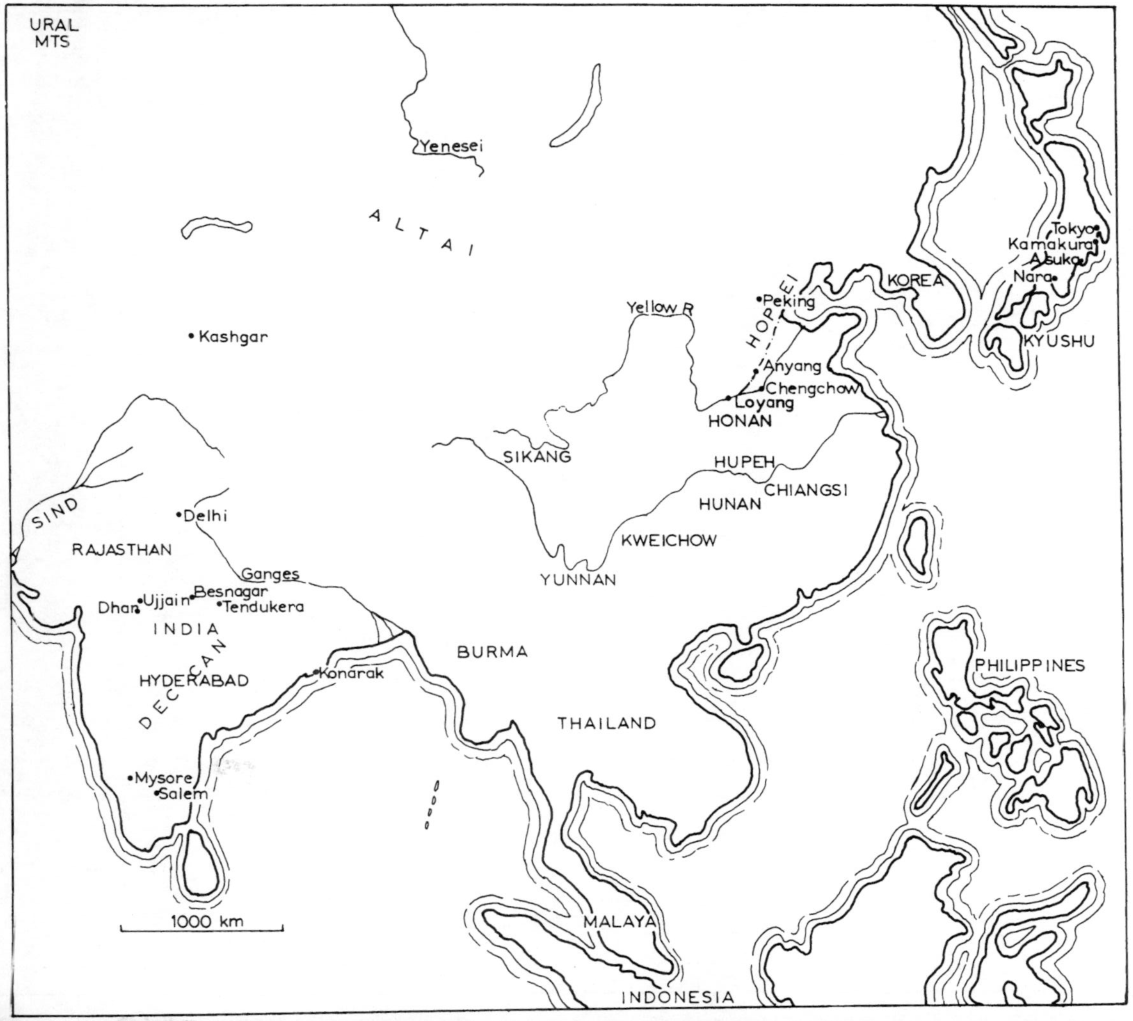

URAL MTS

- Yenesei
- ALTAI
- Yellow R
- Peking
- Anyang
- Chengchow
- Loyang
- HONAN
- HUPEH
- CHIANGSI
- HUNAN
- KWEICHOW
- YUNNAN
- SIKANG
- KOREA
- Tokyo
- Kamakura
- Asuka
- Nara
- KYUSHU
- PHILIPPINES
- THAILAND
- BURMA
- MALAYA
- INDONESIA
- Kashgar
- Delhi
- SIND
- RAJASTHAN
- Dhar
- Ujjain
- Besnagar
- Ganges
- Tendukera
- INDIA
- DECAN
- HYDERABAD
- Konarak
- Mysore
- Salem

1000 km

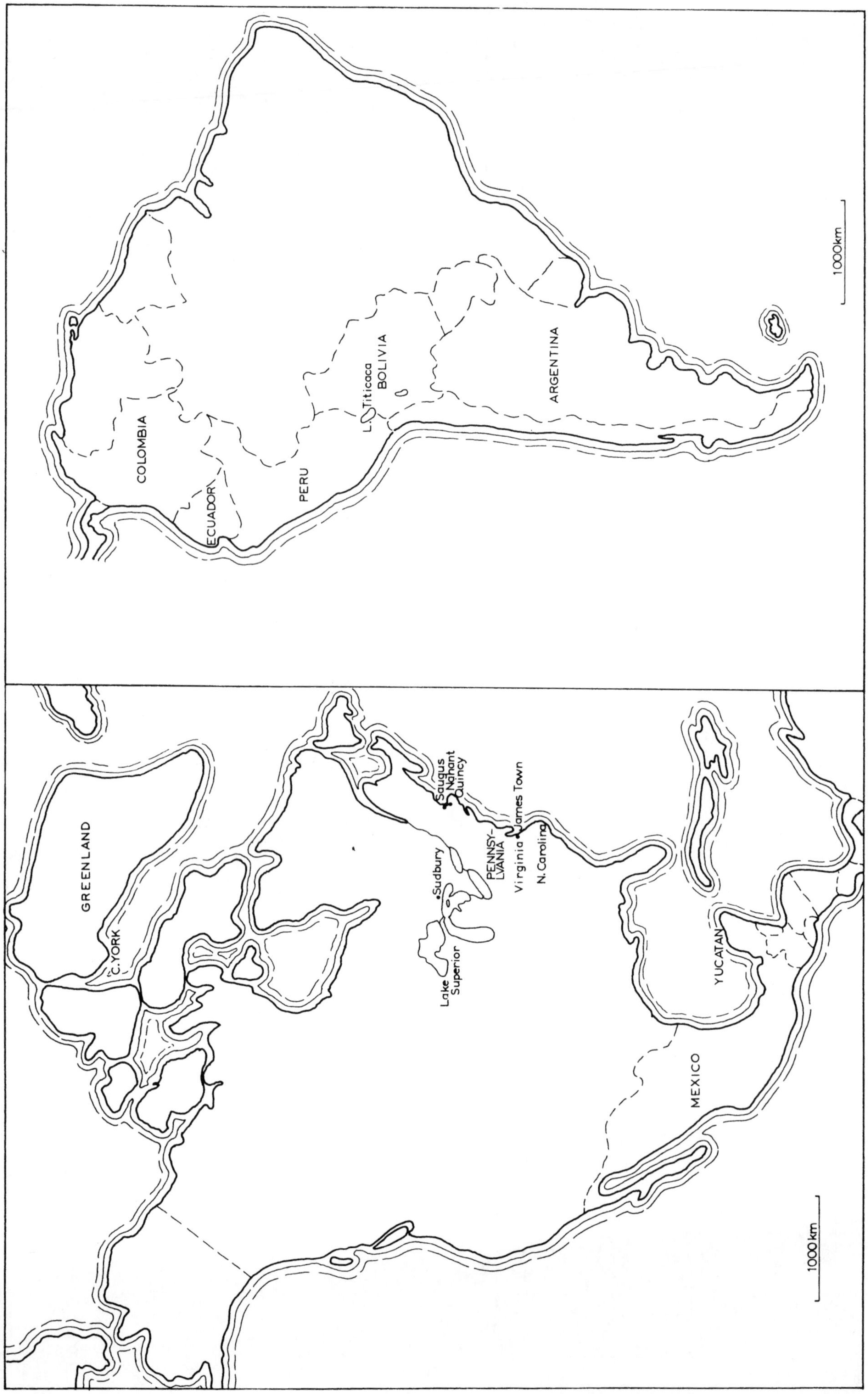

COLOMBIA

ECUADOR

PERU

L.Titicaca

BOLIVIA

ARGENTINA

1000km

GREENLAND

C.YORK

Lake
Superior

•Sudbury

Saugus
Nahant
Quincy

PENNSY-
LVANIA

Virginia James Town

N.Carolina

YUCATAN

MEXICO

1000 km

MAP 4 North and South America

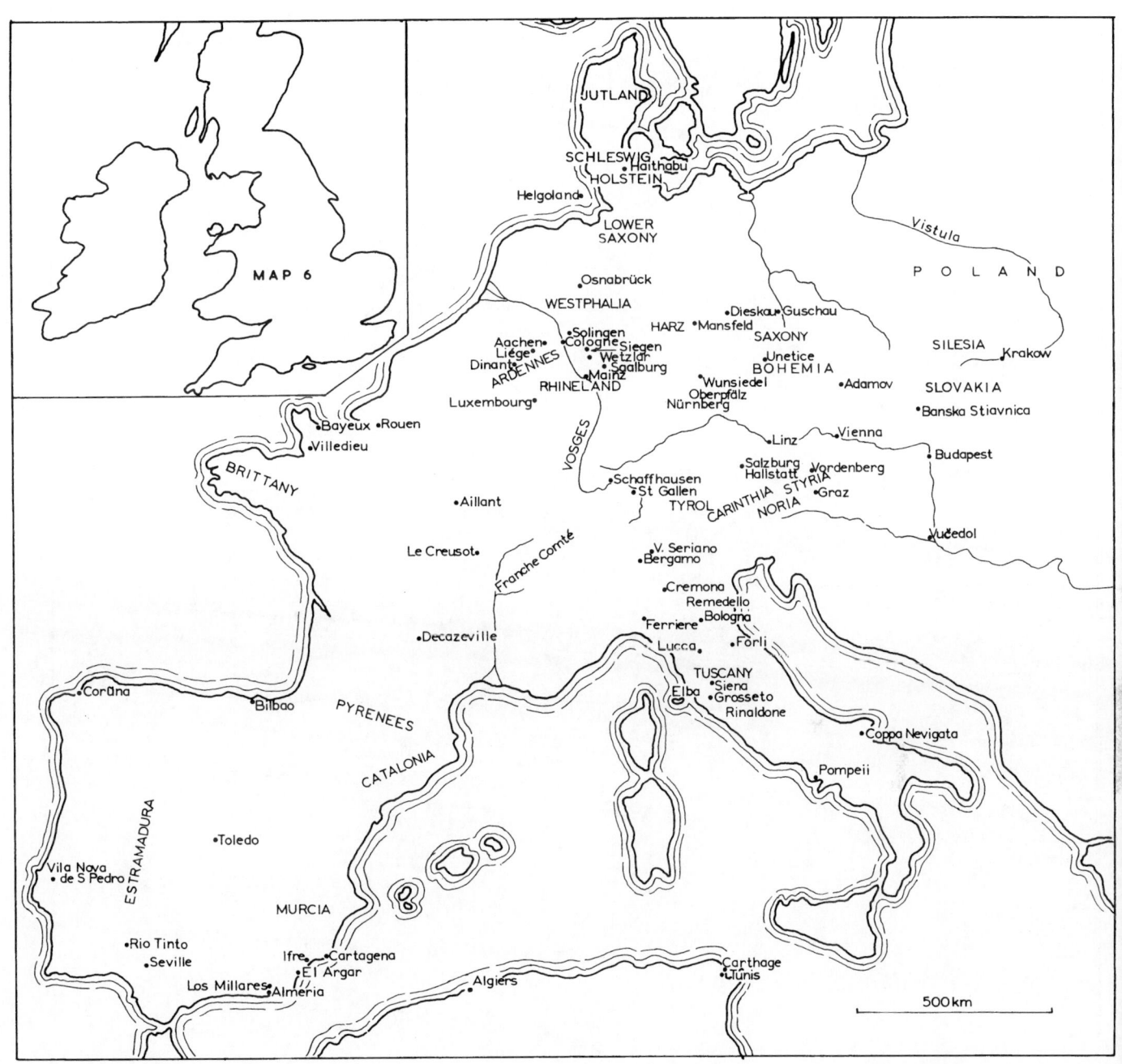

MAP 5 Western Europe

MAP 6 The British Isles

Index